T0299359

A Walk Through
Weak
Hyperstructures
H_v-Structures

A Walk Through
Weak Hyperstructures
H_v-Structures

Bijan Davvaz
Yazd University, Iran

Thomas Vougiouklis
Democritus University of Thrace, Greece

Published by

World Scientific Publishing Co. Pte. Ltd.

5 Toh Tuck Link, Singapore 596224

USA office: 27 Warren Street, Suite 401-402, Hackensack, NJ 07601

UK office: 57 Shelton Street, Covent Garden, London WC2H 9HE

Library of Congress Cataloging-in-Publication Data

Names: Davvaz, Bijan, author. | Vougiouklis, Thomas, author.
Title: A walk through weak hyperstructures : H_v-structures / by Bijan Davvaz
   (Yazd University, Iran), Thomas Vougiouklis (Democritus University of Thrace, Greece).
Description: New Jersey : World Scientific, 2019. | Includes bibliographical
   references and index.
Identifiers: LCCN 2018048126 | ISBN 9789813278868 (hardcover : alk. paper)
Subjects: LCSH: Hypergroups. | Group theory. | Ordered algebraic structures.
Classification: LCC QA174.2 .D3845 2019 | DDC 511.3/3--dc23
LC record available at https://lccn.loc.gov/2018048126

British Library Cataloguing-in-Publication Data
A catalogue record for this book is available from the British Library.

For any available supplementary material, please visit
https://www.worldscientific.com/worldscibooks/10.1142/11229#t=suppl

Printed in Singapore

Preface

Hyperstructures were born as a generalization of an operation by the hyperoperation, from the single-valued operation to the multi-valued one. It was then that the problem of generalizations was transferred into the generalizations of axioms. In 1934 Frederick Marty, who introduced the hyperoperation and gave the definition of the hypergroup, used the 'double' axiom of reproductivity instead of the two axioms: the existence of the unit element and the existence of the inverses. This is a revolutionary generalization since the majority of the hyperstructures do not have unit elements. In mathematics, any generalization of a structure should contain the generalized one as a sub-case. In hyperstructures the problem becomes complicated as in the result we replace the elements by sets, in fact, we replace a set by a power set. Therefore, we need new tools to achieve the connection of the hyperstructures with the classical structures. This new tool is the fundamental relation of each new hyperstructure. It is a fact that any fundamental relation is based on the 'result'. For example, in hypergroups the fundamental relation β^* is the transitive closure of the relation β, where two elements are in β-relation if they belong to a hyperproduct of two elements. In the fundamental relation β^*, introduced by M. Koskas in 1970, a classical group corresponds to any hypergroup. In other words, any hypergroup hides a group. The largest generalization in order to have this correspondence, the existence of the β^* fundamental relation, is the one by using the so called weak axioms. In the weak axioms, defined in all known classical structures as introduced by Vougiouklis in 1990, the 'equality' in any relation is replaced by the 'non-empty intersection' and this leads to the largest class of hyperstructures called H_v-structures. The main way to prove theorems in this topic is the reduction to absurdity. Since the weak generalization is the most general, the number of H_v-structure is dramatically big. Therefore, many problems in life and in other sciences can be expressed by models using H_v-structures. In order to specify the appropriate H_v-structure in models, one can use more restrictions or axioms to reduce the number of possible cases. In generalizations new concepts appear. Moreover, new axioms, new properties and new classes of hyperstructures, are discovered. Consequently, new classifications are needed and very interesting mathematical problems are revealed. The present book consists of seven chapters. Chapter 1 contains a fairly detailed discussion of the basic ideas underlying the fundamentals of algebraic structures such as semigroups, groups, rings, modules and vector spaces. Chapter 2 gives a brief introduction to algebraic hyperstructures to be used in the next chapters. Many readers, already familiar with these theories, may wish to skip them or to regard them as a survey. In Chapter 3, the concept of H_v-semigroups, H_v-groups and some examples are presented. Fundamental relations on H_v-groups are discussed. Reversible H_v-groups, a sequence of finite H_v-groups, fuzzy H_v-groups and H_v-semigroups as noise problem are studied. In Chapter 4, we present the notion of H_v-rings. H_v-rings are the largest class of algebraic hyperstructures that satisfy ring-like axioms. We consider the fundamental relation γ^* defined on H_v-rings and give some properties of this important relation. The fundamental relation on an H_v-ring is the smallest equivalence relation such that the quotient would be the (fundamental) ring. Then, we present several kinds of H_v-rings. In particular, we investigate multiplicative H_v-rings, H_v-fields, P-hyperoperations, ∂-hyperoperations, H_v-ring of fractions, H_v-near rings and fuzzy H_v-ideals. Chapter 5 begins with the definition of H_v-module. Then the concepts of H_v-module of fractions, direct system and direct limit of H_v-modules are provided. It is proved that direct limit always exists in the category of H_v-modules. We study $M[-]$ and $-[M]$ functors and investigate the exactness and some related concepts. Next, we prove Five Short Lemma and Shanuel's lemma in H_v-modules. At the end of this chapter, the concepts of fuzzy and intuitionistic fuzzy H_v-submodules are presented. In Chapter 6, we cover H_v-vector space, hyperalgebra, e-hyperstructures and H_v-matrix representations. Moreover, we study Lie-Santilli theory. In the quiver of hyperstructures Santilli, in early 90'es, tried to find algebraic structures in order to express his pioneer Lie-Santilli's Theory. Santilli's theory on 'isotopies' and 'genotopies', born in 1960's, desperately needs 'units e' on left or right, which are nowhere singular, symmetric, real-valued, positive-defined for n-dimensional matrices based on the so called isofields. These elements can be found in hyperstructure theory, especially in H_v-structure theory introduced. This connection appeared first in 1996 and actually several H_v-fields, the e-hyperfields, can be used as isofields or genofields, in such way that they should cover additional properties and satisfy more restrictions. Several large classes of hyperstructures as the P-hyperfields, can be used in Lie-Santilli's theory when multivalued problems appeared, either in finite or in infinite case. Chapter 7, which is novel in a book of this kind, illustrates the use of weak hyperstructures. We present examples of weak hyperstructures associated with chain reactions and dismutation reactions. For the first time Davvaz and Dehghan-Nezhad provided examples of hyperstructures associated with chain reactions. Also, we investigate the examples of hyperstructures and weak hyperstructures associated with redox reactions and electrochemical cells. Another motivation for the study of hyperstructures comes from biology, more specifically from Mendel, the father of genetics, who took the first steps in defining "contrasting characters, genotypes in F_1 and F_2 ... and setting different laws". The genotypes of F_2 are dependent on the type of its parents genotype and it follows certain rules. Also, inheritance issue based on genetic information is examined carefully via a new hyperalgebraic approach. Several examples are provided from different biology points of view, and we show that the theory of hyperstructures exactly fits the inheritance issue. Moreover, we provide a physical example of hyperstructures associated with the elementary particle physics, the leptons. We consider this important group of the elementary particles and show that this set along with the interactions between its members can be described by the algebraic hyperstructures.

Bijan Davvaz
Department of Mathematics, Yazd University,
Yazd, Iran

Thomas Vougiouklis
School of Science of Education, Democritus University of Thrace,
Alexandroupolis, Greece

Contents

Chapter 1

Fundamentals of algebraic structures

1.1 Semigroups and groups

Let S be a non-empty set and $\zeta : S \times S \to S$ a *binary operation* that maps each ordered pair (x, y) of S to an element $\zeta(x, y)$ of S. The pair (S, ζ) (or just S, if there is no fear of confusion) is called a *groupoid*.

Definition 1.1. A *semigroup* is a pair $(S, \cdot)$ in which S is a non-empty set and $\cdot$ is a binary associative operation on S, i.e., the equation

$$(x \cdot y) \cdot z = x \cdot (y \cdot z)$$

holds for all $x, y, z \in S$.

For an element $x \in S$ we let x^n be the product of x with itself n times. So, $x^1 = x$, $x^2 = x \cdot x$ and $x^{n+1} = x \cdot x^n$ for $n \geq 1$.

A semigroup S is *finite* if it has only a finitely many elements. A semigroup S is *commutative*, if it satisfies

$$x \cdot y = y \cdot x$$

for all $x, y \in S$. If there exists e in S such that for all $x \in S$,

$$e \cdot x = x \cdot e = x$$

we say that S is a *semigroup with identity* or (more usual) a *monoid*. The element e of S is called *identity*.

Proposition 1.1. *A semigroup can have at most one identity.*

Proof. If e and e' are both identities, then $e = e \cdot e' = e'$. ∎

The description of the binary operation in a semigroup $(S, \cdot)$ can be carried out in various ways. The most natural is simply to list all results of

1

the operation for arbitrary pairs of elements. This method of describing the operation can be presented as a *multiplication table*, also called a *Cayley table*.

Example 1.1.

(1) Let $S = \{a,\, b,\, c\}$ be a set of three elements and define the following table.

·	a	b	c
a	a	b	c
b	b	a	c
c	c	b	c

Then, S is a finite semigroup.

(2) Let $\mathbb{N} = \{0,\, 1,\ldots\}$ be the set of all non-negative integers and $\mathbb{N}^* = \{1,\, 2,\ldots\}$ the set of all positive integers. Then, $(\mathbb{N}, \cdot)$ is a semigroup for the usual multiplication of integers. Also, $(\mathbb{N}, +)$ is a semigroup, when $+$ is the ordinary addition of integers. Define $(\mathbb{N}, \star)$ by $n \star m = \max\{n,\, m\}$. Then, $(\mathbb{N}, \star)$ is a semigroup, since

$$n \star (m \star k) = \max\{n,\, \max\{m,\, k\}\} = \max\{n,\, m,\, k\}$$
$$= \max\{\max\{n,\, m\},\, k\} = (n \star m) \star k.$$

(3) The set $M_n(\mathbb{R})$ of $n \times n$ square matrices over real numbers, with matrix multiplication, is a semigroup.

(4) The *direct product* $S \times T$ of two semigroups $(S, \cdot)$ and $(T, \circ)$ is defined by

$$(s_1, t_1) \star (s_2, t_2) = (s_1 \cdot s_2, t_1 \circ t_2) \ \ (s_1, s_2 \in S,\ t_1, t_2 \in T).$$

It is easy to show that the so defined product is associative and hence the direct product is, indeed, a semigroup. The direct product is a convenient way of combining two semigroup operations. The new semigroup $S \times T$ inherits properties of both S and T.

(5) The *bicyclic semigroup* $B = \mathbb{N} \times \mathbb{N}$ with binary operation

$$(a, b) * (c, d) = (a - b + \max\{b, c\}, d - c + \max\{b, c\}).$$

This is a monoid with identity $(0, 0)$.

Proposition 1.2. *Let $(S, \cdot)$ be a finite semigroup. Then, there exists $a \in S$ such that $a^2 = a$.*

Proof. Suppose that x is an arbitrary element of S. Since S is finite, it follows that $x, x^2, x^3, \ldots$ are distinct, where $x^n = x \cdot \ldots \cdot x$ (n times). So, there exist integers m, n with $n < m$ such that $x^m = x^n$. Hence, $x^{n+k} = x^n$, where $k = m - n$. Now, we have

$$x^{2n+k} = x^n \cdot x^{n+k} = x^{2n}.$$

By mathematical induction, we obtain

$$x^{rn+k} = x^{rn}, \text{ for all } r \in \mathbb{N}.$$

Also, we have

$$x^{rn+2k} = x^{rn+k} \cdot x^k = x^{rn} \cdot x^k = x^{rn+k} = x^{rn},$$
$$x^{rn+3k} = x^{rn+2k} \cdot x^k = x^{rn} \cdot x^k = x^{rn+k} = x^{rn},$$

and so on. Again, by mathematical induction, we obtain

$$x^{rn+lk} = x^{rn}, \text{ for all } l \in \mathbb{N}.$$

In particular, we obtain $x^{kn+nk} = x^{kn}$ or $x^{2nk} = x^{kn}$. Now, we set $x^{nk} = a$. ∎

Let us consider the set consisting of all the integers $\{0, \pm 1, \pm 2, \ldots\}$. The sum $m + n$ of any two integers m and n is also an integer, and the following two rules of addition hold for any arbitrary integers m, n and p:

(1) $m + n = n + m$,
(2) $(m + n) + p = m + (n + p)$.

Furthermore, for any two given integers m and n, the equation

$$m + x = n$$

has a unique solution $x = n - m$, which is also an integer.

Similar situations often occur in many different fields of mathematics, and they do not necessarily concern only the integers. Consider, for instance, the set of all nonsingular 2×2 matrices (that is, all 2×2 matrices A such that the determinant of A is not zero). Let A and B be any 2×2 matrices,

$$A = \begin{bmatrix} a_1 & a_2 \\ a_3 & a_4 \end{bmatrix} \text{ and } B = \begin{bmatrix} b_1 & b_2 \\ b_3 & b_4 \end{bmatrix}.$$

Then, the product of A and B is also a 2×2 matrix. This product is defined as

$$AB = \begin{bmatrix} a_1 b_1 + a_2 b_3 & a_1 b_2 + a_2 b_4 \\ a_3 b_1 + a_4 b_3 & a_3 b_2 + a_4 b_4 \end{bmatrix}.$$

With respect to the binary operation of the multiplication of matrices, we note that, in general, AB is not necessarily equal to BA, but the associative law $(AB)C = A(BC)$ is valid for any three 2×2 matrices A, B and C. If the 2×2 matrix A is nonsingular, then the equation $AX = B$ and $YA = B$ have unique solutions, $X = A^{-1}B$ and $Y = BA^{-1}$, respectively, where X and Y are both 2×2 matrices.

A given set of elements together with an operation satisfying the associative law is said to be a group or to form a group if any linear equation has a unique solution which is in the set. Thus, the totality of nonsingular 2×2 matrices together with multiplication is said to be group, as is the set of all the integers with addition.

We will now state the formal definition of a group.

Definition 1.2. Let G be a non-empty set together with a binary operation that assigns to each ordered pair (a, b) of elements of G an element $a \cdot b$ in G. We say G is a *group* under this operation if the following two properties are satisfied:

(1) For any three elements a, b and c of G, the associative law holds: $(a \cdot b) \cdot c = a \cdot (b \cdot c)$.
(2) For two arbitrary elements a and b, there exist x and y of G which satisfy the equations $a \cdot x = b$ and $y \cdot a = b$.

The following properties of a group are important.

Theorem 1.1.

(2′) *There is a unique element e in G such that for all $g \in G$, $g \cdot e = e \cdot g = g$.*
(2″) *For any element $a \in G$, there is a unique element $a' \in G$ such that $a \cdot a' = a' \cdot a = e$, where e is the element of G defined in (2′).*
(3) *The solutions x and y of the equations $a \cdot x = b$ and $y \cdot a = b$ are unique and we have $x = a' \cdot b$ and $y = b \cdot a'$, where a' is the element associated with the element a in (2″).*

Proof.   Since the set G is not empty, we take an element a of G. By the property (2), there are solutions $x = e$ and $y = e'$ of the equations $a \cdot x = a$ and $y \cdot a = a$. Also, if g is an arbitrary element of G, there are elements u and v of G such that $a \cdot u = g = v \cdot a$; so we have $g \cdot e = (v \cdot a) \cdot e = v \cdot (a \cdot e) = v \cdot a = g$. The second equality, $(v \cdot a) \cdot e = v \cdot (a \cdot e)$, follows from the associative law. Similarly, we obtain $e' \cdot g = g$. Since the element g is arbitrary, we may take $g = e$ to obtain $e' \cdot e = e$. On the other hand, the element e satisfies $g \cdot e = g$ for any $g \in G$, so $e' \cdot e = e'$. Therefore, we have $e' = e' \cdot e = e$. This proves that an arbitrary solution of the equation $a \cdot x = a$ is equal to a solution of $y \cdot a = a$. Thus, the uniqueness of the element e is proved, and (2′) holds.

The proof of (2″) is similar. By (2), there are elements a' and a'' of G such that $a \cdot a' = e = a'' \cdot a$. Using (1) and (2′), we have $a'' = a'' \cdot e = a'' \cdot (a \cdot a') = (a'' \cdot a) \cdot a' = e \cdot a' = a'$. Hence, the proof of the uniqueness of a' is similar to that of the uniqueness of e in (2′).

The proof of (3). If $a \cdot x = b$, then the left multiplication of a' of (2″) gives us $a' \cdot b = a' \cdot (a \cdot x) = (a' \cdot a) \cdot x = e \cdot x = x$. Thus, a solution of $a \cdot x = b$ is $x = a' \cdot b$, and it is unique; similarly, the solution of $y \cdot a = b$ is uniquely determined to be $y = b \cdot a'$. ∎

Corollary 1.1. *A non-empty set G with an operation is a group if the conditions (1), (2′) and (2″) are satisfied.*

The element e defined in (2′) is called the *identity* of G, the element a' defined in (2″) is called the *inverse* of a. The inverse of an element a is customary denoted by a^{-1}.

Theorem 1.2. *We have $(a^{-1})^{-1} = a$ and $(a \cdot b)^{-1} = b^{-1} \cdot a^{-1}$, for all $a, b \in G$.*

Proof. The first equality follows from (2″)(the uniqueness of the inverse). The second one is proved by the equality

$$(a{\cdot}b){\cdot}(b^{-1}{\cdot}a^{-1}) = ((a{\cdot}b){\cdot}b^{-1}){\cdot}a^{-1} = (a{\cdot}(b{\cdot}b^{-1})){\cdot}a^{-1} = (a{\cdot}e){\cdot}a^{-1} = a{\cdot}a^{-1} = e$$

and the uniqueness of the inverse. Notice the change in the order of the factors from $a \cdot b$ to $b^{-1} \cdot a^{-1}$. ∎

The product of the elements $a_1, a_2, ..., a_n$ ($n \geq 3$) of G is defined inductively by $a_1 \cdot ... \cdot a_n = (a_1 \cdot ... \cdot a_{n-1}) \cdot a_n$. The *general associative law* holds in any group. That is, if $x_1, x_2, ..., x_n$ are n arbitrary elements of a group, then the product of $x_1, ..., x_n$ is uniquely determined irrespective of the ways the product is taken, provided that the order of factors is unchanged. For example, $(x \cdot y) \cdot ((z \cdot (u \cdot v)) \cdot w) = x \cdot (((y \cdot (z \cdot u)) \cdot v) \cdot w)$.

If $a_1 = a_2 = ... = a_n$, then we use the power notation (for $a = a_1$), $a_1 \cdot a_2 \cdot ... \cdot a_n = a^n$. If $n = -m$ is a negative integer, then we define $a^n = (a^{-1})^m$; also, we define $a^0 = e$. The formulas $a^m \cdot a^n = a^{m+n}$ and $(a^m)^n = a^{mn}$ hold for any element a of G and any pair of integers m and n.

We say that two elements a and b of a group G are commutative or commute if $a \cdot b = b \cdot a$. A group is said to be *abelian* or *commutative*, if any two elements commute.

The number of elements in a group G is called the *order* of G and is denoted by $|G|$. If $|G|$ is finite, then G is said to be a finite group; otherwise G is an infinite group.

Before going on to work out some properties of groups, we pause to examine some examples. Motivated by these examples we shall define various special types of groups which are important.

Example 1.2.

(1) The set of integers $\mathbb{Z}$, the set of rational numbers $\mathbb{Q}$ and the set of real numbers $\mathbb{R}$ are all groups under ordinary addition.

(2) The set $\mathbb{Z}_n = \{0, 1, ..., n - 1\}$ for $n \geq 1$ is a group under addition modulo n. For any i in $\mathbb{Z}_n$, the inverse of i is $n - i$. This group usually referred to as the *group of integers modulo n*.

(3) For a positive integer n, consider the set $C_n = \{a^0, a^1, ..., a^{n-1}\}$. On C_n define a binary operation as follows:

$$a^l a^m = \begin{cases} a^{l+m} & \text{if } l + m < n \\ a^{(l+m)-n} & \text{if } l + m \geq n. \end{cases}$$

For every positive integer n, C_n is an abelian group. The group C_n is called the *cyclic group* of order n.

(4) In mathematics, a dihedral group is the group of symmetries of a regular polygon, including both rotations and reflections. Dihedral groups are among the simplest examples of finite groups, and they play an important role in group theory, geometry, and chemistry.

(5) The *quaternion group* is a non-abelian group of order 8. It is often denoted by Q or Q_8 and written in multiplicative form, with the following 8 elements

$$Q = \{1, -1, i, -i, j, -j, k, -k\}.$$

Here 1 is the identity element, $(-1)^2 = 1$ and $(-1) \cdot a = a \cdot (-1) = -a$ for all a in Q. The remaining multiplication rules can be obtained from the following relation:

$$i^2 = j^2 = k^2 = i \cdot j \cdot k = -1.$$

The concept of subgroups is one of the most basic ideas in group theory.

Definition 1.3. A non-empty subset H of a group G is said to be a *subgroup* of G if the following conditions are satisfied:

(1) $a, b \in H$ implies $a \cdot b \in H$;

(2) $a \in H$ implies $a^{-1} \in H$.

If H is a subgroup of G, then H is a group in its own right.

Lemma 1.1. *Let G be a group and H be a non-empty subset of G. Then, H is a subgroup of G if H is closed under division, i.e., if $a \cdot b^{-1}$ is in H, whenever a, b are in H.*

Proof. It is straightforward. ∎

Lemma 1.2. *Let H be a non-empty finite subset of a group G. Then, H is a subgroup of G if H is closed under the operation of G.*

Proof. It is straightforward. ∎

Example 1.3.

(1) Let G be the group of all real numbers under addition, and let H be the set of all integers. Then, H is a subgroup of G.

(2) Let G be the group of all nonzero real numbers under multiplication, and let H be the set of positive rational numbers. Then, H is a subgroup of G.

(3) Let G be the group of all non-zero complex numbers under multiplication, and let $H = \{a + bi \mid a^2 + b^2 = 1\}$. Then, H is a subgroup of G.

(4) Let G be an abelian group. Then, $H = \{x \in G \mid x^2 = e\}$ is a subgroup of G.

(5) The center $Z(G)$ of a group G is the subset of elements in G that commute with every element of G. In symbols, $Z(G) = \{a \in G \mid a \cdot x = x \cdot a, \text{ for all } x \in G\}$. The center of a group G is a subgroup of G.

(6) If G is a group and $a \in G$, then the *cyclic subgroup generated by a*, denoted by $\langle a \rangle$, is the set of all the powers of a.

Among the subgroups of G, the subgroups G and $\{e\}$ are said to be *trivial*. A subgroup H is said to be a *proper subgroup* of G if $H \neq G$. If M is a proper subgroup of G and if $M \subseteq H \subseteq G$ for a subgroup H of G

implies that $G = H$ or $H = M$, then M is said to be a *maximal subgroup* of G.

Proposition 1.3. *Let H and K be two subgroups of a group G. The intersection $H \cap K$ of H and K is a subgroup of G. In general if $\{H_i\}_{i \in I}$ is a family of subgroups of G, then $\bigcap_{i \in I} H_i$ is a subgroup of G.*

Proof.   It is straightforward.   ∎

Definition 1.4. If X is a subset of a group G, then the smallest subgroup of G containing X, denoted by $\langle X \rangle$, is called the *subgroup generated by X.* If X consists of a single element a, then $\langle X \rangle = \langle a \rangle$, the cyclic subgroup generated by a.

Theorem 1.3. *If X is a non-empty subset of a group G, then the subgroup $\langle X \rangle$ is the set of all finite products of the form $u_1 \cdot u_2 \cdot ... \cdot u_n$, where for each i, either $u_i \in X$ or $u_i^{-1} \in X$.*

Proof.   Let H be the set of all finite products of the form $u_1 \cdot u_2 \cdot ... \cdot u_n$, where u_i or $u_i^{-1} \in X$ and n any positive integer. Consider $x = a_1 \cdot a_2 \cdot ... \cdot a_n$ and $y = b_1 \cdot b_2 \cdot ... \cdot b_m$ in H. Then, $x \cdot y = a_1 \cdot a_2 \cdot ... \cdot a_n \cdot b_1 \cdot b_2 \cdot ... \cdot b_m$ is a product of finite number of elements a_i, b_j such that either the factor or its inverse is in X, consequently $x \cdot y \in H$. Further, $x^{-1} = a_n^{-1} \cdot ... \cdot a_2^{-1} \cdot a_1^{-1}$. Since a_i or a_i^{-1} is in X, and $a_i = (a_i^{-1})^{-1}$, we see that either a_i^{-1} or $(a_i^{-1})^{-1}$ is in X, and so $x^{-1} \in H$. This proves that H is a subgroup of G. Clearly, $X \subseteq H$. Consider any subgroup K of G containing X. Then, for each $u \in X$, $u \in K$ and hence $u^{-1} \in K$. Thus, if $x = u_1 \cdot u_2 \cdot ... \cdot u_n$, where $u_i \in X$ or $u_i^{-1} \in X$, is any element of H, then $x \in K$, since $u_i \in K$ for all i. Hence, $H \subseteq K$. This proves that H is the subgroup of G generated by X.   ∎

Definition 1.5. Let G be a group and H be a subgroup of G. For $a, b \in G$ we say a *is congruent to b* mod H, written as $a \equiv b \ mod \ H$ if $a \cdot b^{-1} \in H$.

Lemma 1.3. *The relation $a \equiv b \ mod \ H$ is an equivalence relation.*

Proof.   It is straightforward.   ∎

Definition 1.6. If H is a subgroup of G and $a \in G$, then $H \cdot a = \{h \cdot a \mid h \in H\}$. Then, $H \cdot a$ is called a *right coset* of H in G. A *left coset* $a \cdot H$ is defined similarly. The number of distinct right cosets of H is called the *index* of H in G and denoted by $[G : H]$.

Lemma 1.4. *For all* $a \in G$, *we have* $H \cdot a = \{x \in G \mid a \equiv x \bmod H\}$.

Proof. It is straightforward. ∎

The following corollary contains the basic properties of right cosets and it is useful in many applications.

Corollary 1.2. *Let* H *be a subgroup of* G.

(1) *Every element* a *of* G *contained in exactly one coset of* H. *This coset is* $H \cdot a$.
(2) *Two distinct cosets of* H *have no common element.*
(3) *The group* G *is partitioned into a disjoint union of cosets of* H.
(4) *There is a one to one correspondence between any two right cosets of* H *in* G.
(5) *There is a one to one correspondence between the set of left cosets of* H *in* G *and the set of right cosets of* H *in* G.

Theorem 1.4. (Lagrange's theorem) *If* G *is a finite group and* H *is a subgroup of* G, *then* $|H|$ *is a divisor of* $|G|$.

The above theorem of Lagrange is one of the basic results in finite group theory.

Definition 1.7. If G is a group and $a \in G$, then the *order* of a is the least positive integer n such that $a^n = e$. If no such integer exists we say that a is of *infinite order*.

There are many useful corollaries of Lagrange theorem.

Corollary 1.3. *A finite cyclic group of prime order contains no non-trivial subgroup.*

Corollary 1.4. *The order of an element of a finite group* G *divides the order* $|G|$.

Definition 1.8. Let A and B be two subsets of a group G. The set $A \cdot B = \{a \cdot b \mid a \in A,\ b \in B\}$ consisting of the products of elements $a \in A$ and $b \in B$ is said to be the *product* of A and B.

The associative law of multiplication gives us $(A \cdot B) \cdot C = A \cdot (B \cdot C)$ for any three subsets A, B and C.

The product of two subgroups is not necessarily a subgroup. We have the following theorem.

Theorem 1.5. *Let A and B be subgroups of a group G. Then, the following two conditions are equivalent.*

(1) *The product $A \cdot B$ is a subgroup of G;*
(2) *We have $A \cdot B = B \cdot A$.*

Proof. Suppose that $A \cdot B$ is a subgroup of G. Then, for any $a \in A$, $b \in B$, we have $a^{-1} \cdot b^{-1} \in A \cdot B$ and so $b \cdot a = (a^{-1} \cdot b^{-1})^{-1} \in A \cdot B$. Thus, $B \cdot A \subseteq A \cdot B$. Now, if x is any element of $A \cdot B$, then $x^{-1} = a \cdot b \in A \cdot B$ and so $x = (x^{-1})^{-1} = (a \cdot b)^{-1} = b^{-1} \cdot a^{-1} \in B \cdot A$, so $A \cdot B \subseteq B \cdot A$. Thus, $A \cdot B = B \cdot A$.

On the other hand, suppose that $A \cdot B = B \cdot A$, i.e., if $a \in A$ and $b \in B$, then $a \cdot b = b_1 \cdot a_1$ for some $a_1 \in A$, $b_1 \in B$. In order to prove that $A \cdot B$ is a subgroup we must verify that it is closed and every element in $A \cdot B$ has its inverse in $A \cdot B$. Suppose that $x = a \cdot b \in A \cdot B$ and $y = a' \cdot b' \in A \cdot B$. Then, $x \cdot y = a \cdot b \cdot a' \cdot b'$, but since $b \cdot a' \in B \cdot A = A \cdot B$, $b \cdot a' = a_2 \cdot b_2$ with $a_2 \in A$, $b_2 \in B$. Hence, $x \cdot y = a \cdot (a_2 \cdot b_2) \cdot b' = (a \cdot a_2) \cdot (b_2 \cdot b') \in A \cdot B$. Clearly, $x^{-1} = b^{-1} \cdot a^{-1} \in B \cdot A = A \cdot B$. Thus, $A \cdot B$ is a subgroup of G. ∎

There is one kind of subgroup that is especially interesting. If G is a group and H is a subgroup of G, it is not always true that $a \cdot H = H \cdot a$ for all $a \in G$. There are certain situations where this does hold, however, and these cases turn out to be of critical importance in the theory of groups. It was Galois, who first recognized that such subgroups were worthy of special attention.

Definition 1.9. A subgroup N of a group G is called a *normal subgroup* of G if $aN = Na$ for all $a \in G$.

A group G is said to be *simple* if $G \neq \{e\}$ and G contains no non-trivial normal subgroup.

The only simple abelian groups are $\mathbb{Z}_p$ with p prime.

There are several equivalent formulations of the definition of normality.

Lemma 1.5. *Let G be a group and N be a subgroup of G. Then,*

(1) *N is normal in G if and only if $a^{-1}na \in N$ for all $a \in G$ and $n \in N$.*
(2) *N is normal in G if and only if the product of two right cosets of N in G is again a right coset of N in G.*

Proof. It is straightforward. ∎

Example 1.4.

(1) The center $Z(G)$ of a group is always normal. Indeed, any subgroup of $Z(G)$ is normal in G.

(2) If H has only two left cosets in G, then H is normal in G.

(3) Let G be the set of all real matrices $\begin{bmatrix} a & b \\ 0 & d \end{bmatrix}$ where $ad \neq 0$, under matrix multiplication. Then, $N = \left\{ \begin{bmatrix} 1 & b \\ 0 & 1 \end{bmatrix} \mid b \in \mathbb{R} \right\}$ is a normal subgroup of G.

(4) Let $\mathbb{Q}$ be the set of all rational numbers and $G = \{(a, b) \mid a, b \in \mathbb{Q}, a \neq 0\}$. Define $\star$ on G as follows: $(a, b) \star (c, d) = (ac, ad + b)$. Then, $(G, \star)$ is a non-abelian group. If we consider $N = \{(1, b) \mid b \in \mathbb{Q}\}$, then N is a normal subgroup of G.

Theorem 1.6. *Let N be a normal subgroup of a group G, and let $\overline{G}$ denote the set of all cosets of N. For any two elements X and Y of $\overline{G}$, we define their product $X \cdot Y$ as the subset of G obtained by taking the product of the two subsets X and Y of G. Then, $X \cdot Y$ is a coset of N. With respect to this multiplication on $\overline{G}$, the set $\overline{G}$ forms a group.*

Proof. Let X and Y be two elements of $\overline{G}$. Then, there are elements x and y of G such that $X = N \cdot x$ and $Y = N \cdot y$. By assumption, N is normal so that $N \cdot x = x \cdot N$ for any $x \in G$. Whence

$$X \cdot Y = (N \cdot x) \cdot (N \cdot y) = N \cdot (x \cdot N) \cdot y = N \cdot (N \cdot x) \cdot y = N \cdot x \cdot y.$$

This proves that $X \cdot Y$ is a coset of N. If $Z \in \overline{G}$, then by the associative law for the product of subsets, we have $(X \cdot Y) \cdot Z = X \cdot (Y \cdot Z)$. Thus, the multiplication defined on $\overline{G}$ satisfies the associative law. By definition, we obtain $(N \cdot x) \cdot (N \cdot y) = N \cdot x \cdot y$; so the coset which contains the identity e of G, namely N, is the identity of $\overline{G}$, and the inverse of $N \cdot x$ is the coset $N \cdot x^{-1}$. Therefore, $\overline{G}$ forms a group. ∎

The group $\overline{G}$ which was defined in the above theorem is called the *quotient group* of G by N and is written $\overline{G} = G/N$. The mapping $x \to Nx$ from G into $\overline{G}$ is called the *canonical map*. The order of the quotient group G/N is equal to the index of the normal subgroup N, i.e., $|G/N| = [G : N]$.

Example 1.5.

(1) Let $G = \mathbb{Z}_{18}$ and $N = <6>$. Then, $G/N = \{0 + N, 1 + N, 2 + N, 3 + N, 4 + N, 5 + N\}$.

(2) Let G be a group such that $(ab)^p = a^p b^p$ for all $a, b \in G$, where p is a prime number. Let $N = \{x \in G \mid x^{p^m} = e$ for some m depending on $x\}$. Then, N is a normal subgroup of G. If $\overline{G} = G/N$ and if $\overline{x} \in \overline{G}$ is such that $\overline{x}^p = \overline{e}$, then $\overline{x} = \overline{e}$.

We close this section with the following correspondence theorem.

Theorem 1.7. *Let N be a normal subgroup of a group G, and let $\overline{G} = G/N$. For any subgroup $\overline{V}$ of $\overline{G}$, there corresponds a subgroup V of G such that*

$$N \subseteq V \text{ and } \overline{V} = V/N.$$

The subgroup V consists of those elements of G which are contained in some elements of $\overline{V}$ and is uniquely determined by $\overline{V}$. Thus, between the set $\overline{\mathcal{G}}$ of subgroups of $\overline{G}$ and the set $\mathcal{G}$ of subgroups of G which contain N, there exists a one to one correspondence, $\overline{V} \leftrightarrow V$.

Proof. It is straightforward. ∎

Let G be a finite group with n elements $a_1, a_2, ..., a_n$. A multiplication table for G is the $n \times n$ matrix with i, j entry $a_i * a_j$:

G	a_1	a_2	...	a_n
a_1	$a_1 * a_1$	$a_1 * a_2$	...	$a_1 * a_n$
a_2	$a_2 * a_1$	$a_2 * a_2$	...	$a_2 * a_n$
...	...	...	...	...
a_n	$a_n * a_1$	$a_n * a_2$	...	$a_n * a_n$

Informally, we say that we "know" a finite group G if we can write a multiplication table for it.

In this section, we consider one of the most fundamental notions of group theory-"homomorphism". The homomorphism term comes from the Greek words "homo", which means like and "morphe", which means form. In our presentation about groups we see that one way to discover information about a group is to examine its interaction with other groups using homomorphisms. A group homomorphism preserves the group operation.

Let us consider two almost trivial examples of groups. Let G be the group whose elements are the numbers 1 and -1 with operation multiplication and let $\hat{G}$ be the additive group $\mathbb{Z}_2$. Compare multiplication tables of these two groups:

G	1	-1
1	1	-1
-1	-1	1

$\hat{G}$	0	1
0	0	1
1	1	0

It is quite clear that G and $\hat{G}$ are distinct groups; on the other hand, it is equally clear that there is no significant difference between them. Let us make this idea precise.

Definition 1.10. A function f defined on a group G to a group $\hat{G}$ (not necessarily distinct from G) is said to be a (*group*) *homomorphism* from G into $\hat{G}$ if $f(xy) = f(x)f(y)$ for all $x, y \in G$. If f is a surjective homomorphism, i.e., $f(G) = \hat{G}$, then $\hat{G}$ is said to be *homomorphic* to G. If f is a surjective and one to one homomorphism, then f is called an *isomorphism* from G onto $\hat{G}$. If there is an isomorphism from G onto $\hat{G}$, we say that G and $\hat{G}$ are *isomorphic* and write $G \cong \hat{G}$.

Let f be a homomorphism from G into $\hat{G}$. The subset

$$H = \{x \in G \mid f(x) \text{ is the identity of } \hat{G}\}$$

is called the *kernel* of f and is denoted by $Kerf$.

Example 1.6.

(1) Every canonical mapping is a homomorphism.
(2) Let G be the group of all positive real numbers under the multiplication of the real numbers and let $\hat{G}$ be the group of all real numbers under addition. Let $f : G \to \hat{G}$ be defined by $f(x) = log_{10}x$ for all $x \in G$. Since $log_{10}(xy) = log_{10}x + log_{10}y$, we have $f(xy) = f(x) + f(y)$, so f is a homomorphism. Also, it happens to be onto and one to one.
(3) Let $GL(n)$ be the multiplicative group of all nonsingular $n \times n$ matrices over the real numbers. Let $\mathbb{R}^*$ be the multiplicative group of all nonzero real numbers. We define $f : G \to \mathbb{R}^*$ by $f(A) = detA$ for all $A \in GL(n)$. Since for any two $n \times n$ matrices A, B, $det(AB) = detA \cdot detB$, we obtain $f(AB) = f(A)f(B)$. Hence, f is a homomorphism of $GL(n)$ into $\mathbb{R}^*$. Also, f is onto.
(4) Let D_{2n} be the dihedral group defined as the set of all formal symbols $a^i b^j$, $i = 0, 1$, $j = 0, 1, ..., n-1$, where $a^2 = e$, $b^n = e$ and $ab = b^{-1}a$. Then, the subgroup $N = \{e, b, b^2, ..., b^{n-1}\}$ is normal in G and $D_{2n}/N \cong H$, where $H = \{1, -1\}$ is the group under the multiplication of the real numbers.

Proposition 1.4. *Let f be a homomorphism from a group G into a group $\hat{G}$. The following propositions hold.*

(1) $f(e) = e'$, *the identity element of $\hat{G}$.*

(2) $f(x^{-1}) = f(x)^{-1}$ for all $x \in G$.

(3) *The kernel of f is a normal subgroup of G.*

(4) *Let H be a subgroup of G. The image $f(H) = \{f(x) \mid x \in H\}$ is a subgroup of $\hat{G}$. For a subgroup $\hat{H}$ of $\hat{G}$, the inverse image*

$$f^{-1}(\hat{H}) = \{x \in G \mid f(x) \in \hat{H}\}$$

is a subgroup of G.

(5) *For two elements x and y of G, $f(x) = f(y)$ if and only if x and y lie in the same coset of the kernel f. In particular, if f is surjective, then f is an isomorphism if and only if the kernel of f is $\{e\}$.*

(6) *If H is a normal subgroup of G, then $f(H)$ is a normal subgroup of $f(G)$.*

Proof.   It is straightforward.   ∎

We are in a position to establish an important connection between homomorphisms and quotient groups. Many authors prefer to call the next theorem the Fundamental theorem of group homomorphism.

Theorem 1.8. *Let f be a homomorphism from a group G onto a group $\hat{G}$. Then, there exists an isomorphism g from $G/\ker f$ onto $\hat{G}$ such that $f = g\varphi$, where φ is the canonical homomorphism from G onto $G/\ker f$. In this case, we say the following diagram is commutative.*

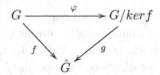

Proof.   Suppose that $K = \ker f$. Define a function g from G/K to $\hat{G}$ by

$$g(K \cdot x) = f(x).$$

The function g is well defined and does not depends on the choice of a representative from $K \cdot x$. Since f is a homomorphism, we have

$$g(K \cdot x \cdot K \cdot y) = g(K \cdot x \cdot y) = f(x \cdot y) = f(x) \cdot f(y) = g(K \cdot x) \cdot g(K \cdot y).$$

Hence, g is a homomorphism from G/K onto $\hat{G}$. If the coset Kx lies in the kernel of g, then $e = g(K \cdot x) = f(x)$, so that $x \in K$. Thus, g is an isomorphism. It is clear that $f = g\varphi$ holds; so the diagram is commutative.   ∎

Corollary 1.5. *Let N be a normal subgroup of a group G, and let φ be the canonical homomorphism from G onto G/N. Let f be a homomorphism from G into a group $\hat{G}$. Then, there exists a homomorphism g from G/N into $\hat{G}$ such that $f = g\varphi$ if and only if $N \subseteq \ker f$. In this case, we have $f(G) \cong (G/N)/(\ker f/N)$.*

Proof. If there is a homomorphism g satisfying $f = g\varphi$, we have $f(N) = e$; so $N \subseteq \ker f$. Conversely, suppose that $N \subseteq \ker f$. Set $K = \ker f$ and $\overline{G} = G/N$. We define a function g from $\overline{G}$ into $\hat{G}$ by

$$g(N \cdot x) = f(x).$$

The function g is uniquely determined independent of the choice of a representative x from $N \cdot x$, and g is a homomorphism from $\overline{G}$ into $\hat{G}$. Clearly, $f = g\varphi$ by definition, and we have $\ker g = \overline{K}$. Now, we obtain

$$G/K = f(G) \cong g(\overline{G}) \cong \overline{G}/\overline{K} = (G/N)/(K/N). \qquad \blacksquare$$

Corollary 1.6. *Let f be a homomorphism from a group G onto a group $\hat{G}$. Let $\hat{N}$ be a normal subgroup of $\hat{G}$ and set $N = f^{-1}(\hat{N})$. Then, N is a normal subgroup of G and $G/N \cong \hat{G}/\hat{N}$.*

Theorem 1.9. *Let H be a normal subgroup of a group G, and let K be any subgroup of G. Then, the following hold.*

    (1) *HK is a subgroup of G;*
    (2) *$H \cap K$ is a normal subgroup of K;*
    (3) *$H \cdot K/H \cong K/H \cap K$.*

Proof. Since H is a normal subgroup of G, $H \cdot K = K \cdot H$ and so according to Theorem 1.5, $H \cdot K$ is a subgroup of G.

Now, consider the canonical mapping from G onto the factor group G/H. Let f be the restriction of the canonical mapping on K. Then, f is a homomorphism from K into G/H. By definition, f is a homomorphism from K onto $H \cdot K/H$. Thus, we have

$$H \cdot K/H \cong K/\ker f.$$

But, the kernel of the canonical homomorphism is K, whence we get $\ker f = H \cap K$. This completes the proof. $\qquad \blacksquare$

Example 1.7.

    (1) Any finite cyclic group of order n is isomorphic to $\mathbb{Z}_n$. Any infinite cyclic group is isomorphic to $\mathbb{Z}$.

(2) Let G be the group of all real valued functions on the unit interval $[0,1]$, where for any $f, g \in G$, we define addition $f + g$ by

$$(f + g)(x) = f(x) + g(x) \text{ for all } x \in [0,1].$$

If $N = \{f \in G \mid f(\frac{1}{4}) = 0\}$, then $G/N \cong \mathbb{R}$ under $+$.

(3) The quotient group $\mathbb{R}/\mathbb{Z}$ is isomorphic to the group S^1 of complex numbers of absolute value 1 (with multiplication). An isomorphism is given by $f(x + \mathbb{Z}) = e^{2\pi x i}$ for all $x \in \mathbb{R}$.

(4) The quotient groups and subgroups of a cyclic groups are cyclic.

Definition 1.11. By an *automorphism* of a group G we shall mean an isomorphism of G onto itself. If g is an element of G, then the function $i_g : x \mapsto g^{-1}xg$ is an automorphism of G, which is called the *inner automorphism* by g.

Let $Aut(G)$ denote the set of all automorphisms of G. For the product of elements of $Aut(G)$ we can use the composition of mappings.

Lemma 1.6. *If G is a group, then $Aut(G)$ is also a group.*

$Aut(G)$ is called the *group of automorphism* of G. The set of all inner automorphisms is a subgroup of $Aut(G)$, which is written $Inn(G)$, and is called the *group of inner automorphism* of G. If G is abelian, then $Inn(G) = \{e\}$.

Theorem 1.10. *The group of inner automorphisms of G is isomorphic to the quotient group $G/Z(G)$, where $Z(G)$ is the center of G. Furthermore, $Inn(G)$ is a normal subgroup of $Aut(G)$.*

Proof. The function $g \mapsto i_g$ is a homomorphism from G onto $Inn(G)$. Thus, $G/K \cong Inn(G)$, where K is the kernel of the above homomorphism. An element g of G lies in the kernel K if and only if $i_g = e$, i.e., $g^{-1} \cdot x \cdot g = x$, for all $x \in G$. Therefore, we have $K = Z(G)$.

The last assertion follows immediately from the formula

$$\sigma^{-1} i_g \sigma = i_{\sigma(g)},$$

which holds for any $g \in G$ and $\sigma \in Aut(G)$.   ∎

Example 1.8.

(1) $Aut(\mathbb{Z}) \cong \mathbb{Z}_2$;
(2) $Aut(G) = 1$ if and only if $|G| \leq 2$.

1.2 Rings

In this section, the definition of a ring and numerous examples are given. A ring is a two-operational system and these operations are usually called *addition* and *multiplication*.

Definition 1.12. A nonempty set R is said to be a *ring* if in R there are defined two binary operations, denoted by $+$ and $\cdot$ respectively, such that for all a, b, c in R:

(1) $a + b = b + a$,
(2) $(a + b) + c = a + (b + c)$,
(3) there is an element 0 in R such that $a + 0 = a$,
(4) there exists an element $-a$ in R such that $a + (-a) = 0$,
(5) $(a \cdot b) \cdot c = a \cdot (b \cdot c)$,
(6) $\cdot$ is distributive with respect to $+$, i.e., $x \cdot (y + z) = x \cdot y + x \cdot z$ and $(x + y) \cdot z = x \cdot z + y \cdot z$.

Axioms (1) through (4) merely state that R is an abelian group under the operation $+$. The additive identity of a ring is called the *zero element*. If $a \in R$ and $n \in \mathbb{Z}$, then na has its usual meaning for additive groups.

If in addition:

(7) $a \cdot b = b \cdot a$, for all a, b in R,

then R is said to be a *commutative ring*. If R contains an element 1 such that

(8) $1 \cdot a = a \cdot 1 = a$ for all a in R,

then R is said to be a *ring with unit element*.

If R is a system with unit satisfying all the axioms of a ring expect possibly $a + b = b + a$ for all $a, b \in R$, then one can show that R is a ring.

For any two elements a, b of a ring R, we shall denote $a + (-b)$ by $a - b$ and for convenience sake we shall usually write ab instead of $a \cdot b$.

Before going on to work out some properties of rings, we pause to examine some examples. Motivated by these examples we shall define various special types of rings which are of importance.

Example 1.9. (Some examples of commutative rings).

(1) Each of the number sets $\mathbb{Z}$, $\mathbb{Q}$, $\mathbb{R}$ and $\mathbb{C}$ forms a ring with respect to ordinary addition and multiplication.

(2) For every $m \in \mathbb{Z}$, $\{ma \mid a \in \mathbb{Z}\}$ forms a ring with respect to ordinary addition and multiplication.

(3) The set $\mathbb{Z}_n$ is a ring with respect to addition and multiplication modulo n.

(4) We say that a ring R is a *Boolean ring* (after the English Mathematician George Boole) if $x^2 = x$ for all $x \in R$. A Boolean ring is commutative. Let X be a set and A, B be subsets of X. The symmetric difference between two subsets A and B, denoted by $A \triangle B$, is the set of all x such that either $x \in A$ or $x \in B$ but not both. The set $\mathcal{P}(X)$ of all subsets of a set X is a ring. The addition is the symmetric difference $\triangle$ and the multiplication is the set operation intersection $\cap$. Its zero element is the empty set, and its unit element is the set X. This is an example of a Boolean ring.

(5) Let $\mathbb{Z}[i]$ denote the set of all complex numbers of the form $a + bi$ where a and b are integers. Under the usual addition and multiplication of complex numbers, $\mathbb{Z}[i]$ forms a ring called the *ring of Gaussian integers*.

(6) The set of all continuous real-valued functions defined on the interval $[a, b]$ forms a ring, the operations are addition and multiplication of functions.

(7) A *polynomial* is a formal expression of the form

$$p(x) = a_0 + a_1 x + ... + a_{n-1} x^{n-1} + a_n x^n,$$

where $a_0, ..., a_n \in \mathbb{R}$ and x is a variable. Polynomials can be added and multiplied as usual. With these operations the set $\mathbb{R}[x]$ of all polynomials is a ring. In fact, given any commutative ring R, one can construct the ring $R[x]$ of polynomials over R in a similar way. We define now the ring of polynomials in the n-variables $x_1, ..., x_n$ over R, $R[x_1, ..., x_n]$, as follows: let $R_1 = [x_1]$, $R_2 = R_1[x_2],...,$ $R_n = R_{n-1}[x_n]$. R_n is called the *ring of polynomials in $x_1, ..., x_n$ over R*.

(8) Let R be a commutative ring with unit element and denoted by $R[[x]]$ the set of all formal power series over the ring R. Then $R[[x]]$ is a ring with addition and multiplication defined by

$$\sum_{i=0}^{\infty} a_i x^i + \sum_{i=0}^{\infty} b_i x^i = \sum_{i=0}^{\infty} (a_i + b_i) x^i,$$

$$\sum_{i=0}^{\infty} a_i x^i \cdot \sum_{i=0}^{\infty} b_i x^i = \sum_{i=0}^{\infty} c_i x^i,$$

where $c_n = \sum\limits_{i=0}^{n} a_i b_{n-i}$. The ring $R[[x]]$ is called the *ring of power series*.

(9) Let R be a commutative ring with unit. A nonempty subset S of R is called a *multiplicative subset* if $0 \notin S$ and $s_1, s_2 \in S$ implies $s_1 s_2 \in S$. Let $R \times S$ be the set of all ordered pairs (r, s) where $r \in R$ and $s \in S$. In $R \times S$ we define now a relation as follows: $(r_1, s_1) \sim (r_2, s_2)$ if and only if there exists $s \in S$ such that $s(r_1 s_2 - s_1 r_2) = 0$. The relation $\sim$ is an equivalence relation on $R \times S$. Let $[r, s]$ be the equivalence class of (r, s) in $R \times S$, and let $S^{-1}R$ be the set of all such equivalence classes $[r, s]$ where $r \in R$ and $s \in S$. The quotient set $S^{-1}R$ is a commutative ring with unit under addition and multiplication defined by

$$[r_1, s_1] + [r_2, s_2] = [r_1 s_2 + r_2 s_1, s_1 s_2],$$
$$[r_1, s_1] \cdot [r_2, s_2] = [r_1 r_2, s_1 s_2],$$

for all $r_1, r_2 \in R$ and $s_1, s_2 \in S$. $S^{-1}R$ is usually called the *ring of fractions* of R. In the special case in which R is the ring of integers, the $S^{-1}R$ so constructed is, of course, the ring of rational numbers.

Example 1.10. (Some examples of noncommutative rings).

(1) One of the smallest noncommutative rings is the Klein 4-ring $(R, +, \cdot)$, where $(R, +)$ is the Klein 4-group $\{0, a, b, c\}$ with 0 the neutral element and the binary operation $\cdot$ given by the following table:

$\cdot$	0	a	b	c
0	0	0	0	0
a	0	a	0	a
b	0	b	0	b
c	0	c	0	c

(2) The set $M_n(\mathbb{R})$ of all $n \times n$ matrices with entries from $\mathbb{R}$ forms a ring with respect to the usual addition and multiplication of matrices. In fact, given an arbitrary ring R, one can consider the ring $M_n(R)$ of $n \times n$ matrices with entries from R.

(3) If G is an abelian group, then $End(G)$, the set of endomorphisms of G, forms a ring, the operations in this ring are the addition and composition of endomorphisms.

(4) Let Ω consist of all complex valued functions f of real variable x such that

$$\int_{-\infty}^{\infty} |f(x)|dx < \infty.$$

Ω is an additive abelian group with respect to the ordinary addition. We consider the binary operation $*$ called *convolution*,

$$h = f * g,$$

where $(f * g)(x)$ is defined by the equation

$$(f * g)(x) = \int_{-\infty}^{\infty} f(x - t)g(t)dt.$$

It can be shown that if f and g are in Ω, then h is also in Ω (it follows from Fubini's theorem in analysis). The remaining axioms are easy to verify and we conclude that Ω is a ring with respect to the ordinary addition $+$ and convolution $*$. This ring lacks a unit element.

(5) Let G be a group and R a ring. Firstly, we define the set $R[G]$ to be one of the following:

• The set of all formal R-linear combinations of elements of G.

• The set of all functions $f : G \to R$ with $f(g) = 0$ for all but finitely many g in G.

No matter which definition is used, we can write the elements of $R[G]$ in the form $\sum_{g \in G} a_g g$, with all but finitely many of the a_g being 0, and the addition on $R[G]$ is the addition of formal linear combinations or addition of functions, respectively. The multiplication of elements of $R[G]$ is defined by setting

$$\left(\sum_{g \in G} a_g g\right)\left(\sum_{h \in G} b_h h\right) = \sum_{g,h \in G} (a_g b_h)gh.$$

If R has a unit element, this is the unique bilinear multiplication for which $(1g)(1h) = (1gh)$. In this case, G is commonly identified with the set of elements $1g$ of $R[G]$. The identity element of G then serves as the 1 in $R[G]$. It is not difficult to verify that $R[G]$ is a ring. This ring is called the *group ring* of G over R.

Note that: If R and G are both commutative (i.e., R is commutative and G is an abelian group), then $R[G]$ is commutative.

(6) This last example is often called the *ring of real quaternions*. This ring was firstly described by the Irish mathematician Hamilton. Initially it was extensively used in the study of mechanics; today its primary interest is that of an important example, although still it plays key roles in geometry and number theory.

Let Q be the set of all symbols $a_0 + a_1 i + a_2 j + a_3 k$, where all the numbers a_0, a_1, a_2 and a_3 are real numbers. Define the equality between two elements of Q as follows: $a_0 + a_1 i + a_2 j + a_3 k = b_0 + b_1 i + b_2 j + b_3 k$ if and only if $a_0 = b_0$, $a_1 = b_1$, $a_2 = b_2$ and $a_3 = b_3$. We define the addition and multiplication on Q by

$$(a_0 + a_1 i + a_2 j + a_3 k) + (b_0 + b_1 i + b_2 j + b_3 k)$$
$$= (a_0 + b_0) + (a_1 + b_1)i + (a_2 + b_2)j + (a_3 + b_3)k,$$
$$(a_0 + a_1 i + a_2 j + a_3 k) \cdot (b_0 + b_1 i + b_2 j + b_3 k)$$
$$= (a_0 b_0 - a_1 b_1 - a_2 b_2 - a_3 b_3) + (a_0 b_1 + a_1 b_0 + a_2 b_3 - a_3 b_2)i$$
$$+ (a_0 b_2 + a_2 b_0 + a_3 b_1 - a_1 b_3)j + (a_0 b_3 + a_3 b_0 + a_1 b_2 - a_2 b_1)k.$$

It is easy to see that Q is a noncommutative ring in which $0 = 0 + 0i + 0j + 0k$ and $1 = 1 + 0i + 0j + 0k$ are the zero and unit elements respectively. Note that the set $\{1, -1, i, -i, j, -j, k, -k\}$ forms a non-abelian group of order 8 under this product.

(7) (Differential operator rings). Consider the homogeneous linear differential equation $a_n(x)D^n y + ... + a_1(x)Dy + a_0(x)y = 0$, where the solution $y(x)$ is a polynomial with complex coefficients, and also the terms $a_i(x)$ belong to $\mathbb{C}[x]$. The equation can be written in compact form as $L(y) = 0$, where L is the differential operator $a_n(x)D^n + ... + a_1(x)D + a_0(x)$, with $D = \frac{d}{dx}$. Thus the differential operator can be thought as a polynomial in the two indeterminates x and D, but in this case the indeterminates do not commute, since $D(xy(x)) = y(x) + xD(y(x))$, yielding the identity $Dx = 1 + xD$. The repeated use of this identity makes possible to write the composition of two differential operators in the standard form $a_0(x) + a_1(x)D + ... + a_n(x)D^n$, and we denote the resulting ring by $\mathbb{C}[x][D]$.

We wish to be able to compute in rings in the same manner in which we compute with real numbers, keeping in mind always that there are different. It may happen that $ab \neq ba$, or a does not divide b. To this end we mention some preliminary results, which assert that certain something we should like to be true in rings are indeed true.

Lemma 1.7. *Let R be a ring. Then*

(1) *Since a ring is an abelian group under $+$, there are certain things we know from the group theory, for instance, $-(-a) = a$ and $-(a + b) = -a - b$, for all a, b in R and so on,*

(2) $0a = a0 = 0$, *for all a in R,*

(3) $(-a)b = a(-b) = -(ab)$, *for all a, b in R,*

(4) $(-a)(-b) = ab$, *for all a, b in R,*

(5) $(na)b = a(nb) = n(ab)$, *for all $n \in \mathbb{Z}$ and a, b in R,*

(6) $\left(\sum\limits_{i=1}^{n} a_i \right) \left(\sum\limits_{j=1}^{m} b_j \right) = \sum\limits_{i=1}^{n} \sum\limits_{j=1}^{m} a_i b_j$, *for all a_i, b_j in R.*

Moreover, if R has a unit element 1, then

(7) $(-1)a = -a$, *for all $a \in R$,*

(8) $(-1)(-1) = 1$.

Proof.   It is straightforward.   ∎

In dealing with an arbitrary ring R there may exist non-zero elements a and b in R, such that their product is zero. Such elements are called *zero-divisors*.

Definition 1.13. A non-zero element a is called a *zero-divisor* if there exists a non-zero element $b \in R$ such that either $ab = 0$ or $ba = 0$.

Example 1.11. As examples of such rings, we have

(1) In the ring $\mathbb{Z}_6$ we have $2 \cdot 3 = 0$ and so 2 and 3 are zero-divisors. More generally, if n is not prime then $\mathbb{Z}_n$ contains zero-divisors.

(2) Consider the ring R of all order pairs of real numbers (a, b). If (a, b) and (c, d) are two elements in R, we define the addition and multiplication in R by the equalities: $(a, b) + (c, d) = (a + c, b + d)$ and $(a, b) \cdot (c, d) = (ac, bd)$. Then R is a ring. The zero element is $(0, 0)$ and the ring has zero-divisors.

Definition 1.14. A commutative ring is an *integral domain* if it has no zero-divisors.

The ring of integers, is an example of an integral domain. It is easy to verify that a ring R has no zero-divisors if and only if the right and left cancellation laws hold in R.

Definition 1.15. If the non-zero elements of a ring R form a multiplicative group, i.e., R has unit element and every element except the zero element has an inverse, then we shall call the ring a skew field or a *division ring*.

Definition 1.16. A *field* is a commutative division ring.

The inverse of an element a under multiplication will be denoted by a^{-1}.

Example 1.12.

(1) If p is prime, then $\mathbb{Z}_p$ is a field.

(2) $\mathbb{Q}$, $\mathbb{R}$ and $\mathbb{C}$ are examples of fields whereas $\mathbb{Z}$ is not.

(3) In Example 1.9(9), let R be an integral domain and $S = R \backslash \{0\}$. Then $S^{-1}R = F$ is a field. F is usually called the *field of fractions*.

(4) Consider the set $\{a + bx \mid a, b \in \mathbb{Z}_2\}$ with x a "indeterminate". We use the arithmetic addition modulo 2 and multiplication using the "rule" $x^2 = x + 1$. Then we obtain a field with 4 elements: $\{0, 1, x, 1 + x\}$.

(5) Consider the set $\{a + bx + cx^2 \mid a, b, c \in \mathbb{Z}_2\}$, where we now use the rule $x^3 = 1 + x$. This gives a field with 8 elements:

$$\{0, 1, x, 1 + x, x^2, 1 + x^2, x + x^2, 1 + x + x^2\}.$$

(6) Consider the set $\{a + bx \mid a, b \in \mathbb{Z}_3\}$ with arithmetic modulo 3 and the "rule" $x^2 = -1$ (so it is similar as the multiplication in $\mathbb{C}$). Then we obtain a field with 9 elements:

$$\{0, 1, 2, x, 1 + x, 2 + x, 2x, 1 + 2x, 2 + 2x\}.$$

More generally, using "tricks" like the above ones, we can construct a finite field with p^k elements for any prime p and positive integer k. This is denoted by $GF(p^k)$ and it is called the *Galois Field* named after the French mathematician Evariste Galois.

The ring of quaternions is a division ring which is not a field. Many other examples of non-commutative rings exist, for instance see the following example.

Example 1.13. Consider the set $M = \left\{ \begin{bmatrix} a & b \\ -\bar{b} & \bar{a} \end{bmatrix} \mid a, b \in \mathbb{C} \right\}$, where $\bar{a}, \bar{b}$ are conjugates of a, b. M is a ring with unit under matrix addition and multiplication. If $A = \begin{bmatrix} x + iy & u + iv \\ -u + iv & x - iy \end{bmatrix}$ is a nonzero matrix in M, then

$$A^{-1} = \begin{bmatrix} \dfrac{x - iy}{x^2 + y^2 + u^2 + v^2} & -\dfrac{u + iv}{x^2 + y^2 + u^2 + v^2} \\ \dfrac{u - iv}{x^2 + y^2 + u^2 + v^2} & \dfrac{x + iy}{x^2 + y^2 + u^2 + v^2} \end{bmatrix}.$$

Hence M is a division ring, but is not commutative, since

$$\begin{bmatrix} 0 & 1 \\ -1 & 0 \end{bmatrix} \begin{bmatrix} i & 0 \\ 0 & -i \end{bmatrix} \neq \begin{bmatrix} i & 0 \\ 0 & -i \end{bmatrix} \begin{bmatrix} 0 & 1 \\ -1 & 0 \end{bmatrix}.$$

Clearly, every field is an integral domain, but, in general, an integral domain is not a field. For example, the ring of integers is an integral domain, but not all nonzero elements have under multiplication inverse. However, for finite domains of integrity, we have the following theorem.

Theorem 1.11. *Any finite ring without zero-divisors is a division ring.*

Proof. Let $R = \{x_1, x_2, ..., x_n\}$ be a finite ring without zero-divisors and suppose that $a(\neq 0) \in R$. Then $ax_1, ax_2, ..., ax_n$ are all n distinct elements lying in R, as cancellation laws hold in R. Since $a \in R$, there exists $x_i \in R$ such that $a = ax_i$. Then we have $a(x_i a - a) = a^2 - a^2 = 0$, and so $x_i a = a$. Now, for every $b \in R$ we have $ab = (ax_i)b = a(x_i b)$, hence $b = x_i b$ and further $ba = b(x_i a) = (bx_i)a$ which implies that $b = bx_i$. Hence x_i is the unit element for R and we denote it by 1. Now, $1 \in R$, so there exists $c \in R$ such that $1 = ac$. Also $a(ca - 1) = (ac)a - a = a - a = 0$, and so $ca = 1$. Consequently, R is a division ring. ∎

Corollary 1.7. *A finite integral domain is a field.*

By a famous theorem of Wedderburn, "every finite division ring is a field". Therefore, we can say that "any finite ring without zero-divisors is a field.

Definition 1.17. Let R be a ring. Then D is said to be of *finite characteristic* if there exists a positive integer n such that $na = 0$ for all $a \in R$. If no such of n exists, R is said to be of *characteristic* 0. If R is of finite characteristic, then we define the *characteristic* of R as the smallest positive integer n such that $na = 0$ for all $a \in R$.

The characteristic of $\mathbb{Z}_n$ is equal to n, whereas $\mathbb{Z}$, $\mathbb{Q}$, $\mathbb{R}$ and $\mathbb{C}$ are of characteristic 0.

Proposition 1.5.

(1) *Any finite field is of finite characteristic. However, an integral domain may be infinite and with a finite characteristic.*

(2) *The characteristic of an integral domain with unit element is either zero or a prime number.*

(3) *If D is an integral domain and if $na = 0$ for some $a \neq 0$ in D and some integer $n \neq 0$, then D is of finite characteristic. Note that, it is not true for an arbitrary ring; it is enough to consider the ring $\mathbb{Z}_2 \times \mathbb{Z}$.*

(4) *Let R be a ring with unit element. Then the characteristic of R is equal to n if and only if n is the least positive integer such that $n \cdot 1 = 1$.*

Proof. It is straightforward. ∎

In the study of groups, subgroups play a crucial role. Subrings, the analogous notion in ring theory, play a much less important role than their counterparts in group theory. Nevertheless, subrings are important.

Definition 1.18. Let R be a ring and S be a nonempty subset of R, which is closed under the addition and multiplication in R. If S is itself a ring under these operations then S is called a *subring* of R; more formally, S is a subring of R if the following conditions hold.

$$a, b \in S \implies a - b \in S \text{ and } a \cdot b \in S.$$

Example 1.14.

(1) For each positive integer n, the set $n\mathbb{Z} = \{0, \pm n, \pm 2n, \pm 3n, ...\}$ is a subring of $\mathbb{Z}$.

(2) $\mathbb{Z}$ is a subring of the ring of real numbers and also a subring of the ring of polynomials $\mathbb{Z}[X]$.

(3) The ring of Gaussian integers is a subring of the complex numbers.

(4) The set A of all 2×2 matrices of the type $\begin{bmatrix} a & 0 \\ b & c \end{bmatrix}$, where a, b and c are integers, is a subring of the ring $M_2(\mathbb{Z})$.

(5) The polynomial ring $R[x]$ is a subring of $R[[x]]$.

(6) If R is any ring, then the *center* of R is the set $Z(R) = \{x \in R \mid xy = yx, \forall y \in R\}$. Clearly, the center of R is a subring of R.

Let R be a ring and S be a proper subring of it. Then there exists the following five cases:

- R and S have a common unit element.
- R has a unit element but S does not.
- R and S both have their own nonzero unities but these are distinct.

- R has no unit element but S has a unit element.
- Neither R nor S have unit element.

Example 1.15.

(1) The ring $\mathbb{Q}$ and its subring $\mathbb{Z}$ have the common unit element 1.
(2) The subring S of even integers of the ring $\mathbb{Z}$ has no unit element. Actually, the only subring with unit of $\mathbb{Z}$ is $\mathbb{Z}$.
(3) Let S be the subring of all pairs $(a, 0)$ of the ring $\mathbb{Z} \times \mathbb{Z}$ for which the operations $+$ and $\cdot$ are defined component by component. Then S and $\mathbb{Z} \times \mathbb{Z}$ have the unities $(1, 0)$ and $(1, 1)$, respectively.
(4) Let S be the subring of all pairs $(a, 0)$ of the ring $R = \{(a, 2b) \mid a, b \in \mathbb{Z}\}$ (operations are defined component by component). Now S has the unit element $(1, 0)$ but R has no unit element.
(5) Neither the ring $\{(2a, 2b) \mid a, b \in \mathbb{Z}\}$ (operations are defined component by component) nor its subring consisting of the pairs $(2a, 0)$ have unit element.

In group theory, normal subgroups play a special role, they permit us to construct quotient groups. Now, we introduce the analogous concept for rings.

Definition 1.19. A non-empty subset I of a ring R is said to be an *ideal* of R if

(1) I is a subgroup of R under addition,
(2) for every $a \in I$ and $r \in R$, both ar and ra are in I.

Clearly, each ideal is a subring. For any ring R, $\{0\}$ and R are ideals of R. The ideal $\{0\}$ is called the *trivial ideal*. An ideal I of R such that $I \neq 0$ and $I \neq R$ is called a *proper ideal*. Observe that if R has a unit element and I is an ideal of R, then $I = R$ if and only if $1 \in I$. Consequently, a nonzero ideal I of R is proper if and only if I contains no invertible elements of R. It is easy to see that the intersection of any family of ideals of R is also an ideal.

Example 1.16.

(1) For any positive integer n, the set $n\mathbb{Z}$ is an ideal of $\mathbb{Z}$. In fact, every ideal of $\mathbb{Z}$ has this form, for suitable n.
(2) Let I be the set of all polynomials over $\mathbb{R}$ with zero constant term. Then I is an ideal of $\mathbb{R}[x]$.

(3) Let R be the ring of all real-valued functions of a real variable. The subset S of all differentiable functions is a subring of R but not an ideal of R.

(4) Let $f \in \mathbb{Q}[x]$. Then the set $\{fg \mid g \in \mathbb{Q}[x]\}$ is an ideal of $\mathbb{Q}[x]$. In fact, every ideal, though not every subring, of $\mathbb{Q}[x]$ has this form.

(5) Let

$$R = \left\{ \begin{bmatrix} a & b & c \\ d & e & f \\ 0 & 0 & g \end{bmatrix} \mid a, b, c, d, e, f, g \in \mathbb{Z} \right\}$$

then R is a ring under matrix addition and multiplication. The set

$$I = \left\{ \begin{bmatrix} 0 & 0 & x \\ 0 & 0 & y \\ 0 & 0 & 0 \end{bmatrix} \mid x, y \in \mathbb{Z} \right\}$$

is an ideal of R.

(6) Let R be a ring and let $M_n(R)$ be the ring of matrices over R. If I is an ideal of R then the set $M_n(I)$ of all matrices with entries in I is an ideal of $M_n(R)$. Conversely, every ideal of $M_n(R)$ is of this type.

(7) Let m be a positive integer such that m is not a square in $\mathbb{Z}$. If $R = \{a + \sqrt{m}b \mid a, b \in \mathbb{Z}\}$, then R is a ring under the operations of sum and product of real numbers. If p is an odd prime number, consider the set $I_p = \{a + \sqrt{m}b \mid p|a \text{ and } p|b\}$, where $a + \sqrt{m}b \in R$. Then I_p is an ideal of R.

(8) For ideals I_1, I_2 of a ring R define $I_1 + I_2$ to be the set $\{a+b \mid a \in I_1, b \in I_2\}$ and $I_1 I_2$ to be the set $\left\{ \sum_{i=1}^{n} a_i b_i \mid n \in \mathbb{Z}^+, a_i \in I_1, b_i \in I_2 \right\}$. Then $I_1 + I_2$ and $I_1 I_2$ are ideals of R.

(9) Let R be an arbitrary ring and let $a_1, a_2, ... a_m \in R$. Then the set of all elements of the form

$$\sum_{i=1}^{m} z_i a_i + \sum_{i=1}^{m} s_i a_i + \sum_{i=1}^{m} a_i t_i + \sum_{i=1}^{m} \left(\sum_{i=1}^{n_i} u_{i,k} a_i v_{i,k} \right),$$

where $m, z_i, n_i \in \mathbb{Z}$, $s_i, t_i, u_{i,k}, v_{i,k} \in R$, is an ideal. In fact it is the smallest ideal of R which contains $a_1, a_2, ... a_m$. Hence it is called the *ideal generated* by $a_1, a_2, ... a_m$.

If R is commutative and has a unit element, the above set reduces to the set $\{a_1 r_1 + a_2 r_2 + ... + a_m r_m \mid r_i \in R\}$. We denote this ideal briefly by $\langle a_1, a_2, ... a_m \rangle$. If $m = 1$ the ideal $\langle a_1 \rangle$ is called the *principal ideal* generated by a_1. In particular, $\langle 1 \rangle = R$.

(10) The subset E of $\mathbb{Z}[x]$ composed by all polynomials with even constant term is an ideal of $\mathbb{Z}[x]$. In fact $E = \langle x, 2 \rangle$ and it is not principal.

(11) Let X be a nonempty set and $\mathcal{P}(X)$ denotes the ring of power set of X. Then a nonempty subset I of $\mathcal{P}(X)$ is an ideal of $\mathcal{P}(X)$ if and only if $\mathcal{P}(A \cup B) \subseteq I$ for all $A, B \in I$.

(12) Let R be a commutative ring and let A be an arbitrary subset of R. Then the *annihilator* of A, $Ann(A) = \{r \in R \mid ra = 0 \text{ for all } a \in A\}$ is an ideal.

Lemma 1.8. *Let R be a commutative ring with unit element whose only ideals are the trivial ideal and R. Then R is a field.*

Proof. In order to prove this lemma, for any nonzero element $a \in R$ we must find an element $b \in R$ such that $ab = 1$. The set $Ra = \{xa \mid x \in R\}$ is an ideal of R. By our assumptions on R, $Ra = \{0\}$ or $Ra = R$. Since $0 \neq a = 1 \cdot a \in Ra$, $Ra \neq \{0\}$, and so $Ra = R$. Since $1 \in R$, there exists $b \in R$ such that $1 = ba$. ∎

Definition 1.20. (Quotient ring) Let R be a ring and let I be an ideal of R. In order to define the quotient ring, we consider firstly an equivalence relation on R. We say that the elements $a, b \in R$ are equivalent, and we write $a \sim b$, if and only if $a - b \in I$. If a is an element of R, we denote the corresponding equivalence class by $[a]$. The *quotient ring* of modulo I is the set $R/I = \{[a] \mid a \in R\}$, with a ring structure defined as follows. If $[a], [b]$ are equivalence classes in R/I, then

$$[a] + [b] = [a + b] \quad \text{and} \quad [a] \cdot [b] = [ab].$$

Since I is closed under addition and multiplication, it follows that the ring structure in R/I is well defined. Clearly, $a + I = [a]$.

Example 1.17. Let us present some quotient rings.

(1) $\mathbb{Z}/6\mathbb{Z} = \{6\mathbb{Z}, 1 + 6\mathbb{Z}, 2 + 6\mathbb{Z}, 3 + 6\mathbb{Z}, 4 + 6\mathbb{Z}, 5 + 6\mathbb{Z}\}$.

(2) We consider the ring of polynomials $\mathbb{R}[x]$ with real coefficients and $\langle x^2 + 1 \rangle$ generated by $x^2 + 1$. Then

$$\mathbb{R}[x]/\langle x^2 + 1 \rangle = \{ax + b + \langle x^2 + 1 \rangle \mid a, b \in \mathbb{R}\}.$$

(3) If $R = \mathbb{Z}[x, y]$ and $I = \langle x^2, y^2 + 1 \rangle$, then every element of R/I has the form $a + bx + cy + dxy + I$, where $a, b, c, d \in \mathbb{Z}$.

Now, as a group homomorphism preserves the group operation, a ring homomorphism preserves the ring operations.

Definition 1.21. A mapping φ from the ring R into the ring R' is said to be a *ring homomorphism* if

(1) $\varphi(a + b) = \varphi(a) + \varphi(b)$,
(2) $\varphi(ab) = \varphi(a)\varphi(b)$,

for all $a, b \in R$.

If φ is a ring homomorphism from R to R', then $\varphi(0) = 0$ and $\varphi(-a) = -\varphi(a)$ for every $a \in R$.

A ring homomorphism $\varphi : R \to R'$ is called an *epimorphism* if φ is onto. It is called a *monomorphism* if it is one to one, and an *isomorphism* if it is both one to one and onto. A homomorphism φ of a ring R into itself is called an *endomorphism*. An endomorphism is called an *automorphism* if it is an isomorphism. The rings R and R' are said to be *isomorphic* if there exists an isomorphism between them, in this case, we write $R \cong R'$.

Before going on we examine these concepts for certain examples.

Example 1.18.

(1) For any positive integer n, the mapping $k \to k$ mod n is a ring homomorphism from $\mathbb{Z}$ onto $\mathbb{Z}_n$.
(2) Let I be an ideal of a ring R. We define $\varphi : R \to R/I$ by $\varphi(a) = a + I$ for all $a \in R$. Then φ is an epimorphism. This map is called a *natural homomorphism*.
(3) Let $\mathbb{Z}(\sqrt{2})$ be the set of real numbers of the form $m + n\sqrt{2}$ where m, n are integers; $\mathbb{Z}(\sqrt{2})$ forms a ring under the usual addition and multiplication of real numbers. We define $\varphi : \mathbb{Z}(\sqrt{2}) \to \mathbb{Z}(\sqrt{2})$ by $\varphi(m + n\sqrt{2}) = m - n\sqrt{2}$. Then φ is an automorphism.

Lemma 1.9. *Let φ be a homomorphism from the ring R to the ring R'. Let S be a subring of R, I an ideal of R and J an ideal of R'.*

(1) *$\varphi(S) = \{\varphi(a) \mid a \in S\}$ is a subring of R'.*
(2) *If φ is onto, then $\varphi(I)$ is an ideal of R'.*
(3) *$\varphi^{-1}(J) = \{r \in R \mid \varphi(r) \in J\}$ is an ideal of R.*
(4) *If R is commutative then $\varphi(R)$ is commutative.*
(5) *If R has a unit element 1 and φ is onto, then $\varphi(1)$ is the unit element of R'.*

(6) *If φ is an isomorphism from R to R', then φ^{-1} is an isomorphism from R' to R.*

Proof. It is straightforward. ∎

Now, we introduce an important ideal that is intimately related to the image of a homomorphism.

Definition 1.22. If φ is a ring homomorphism of R into R', then the *kernel* of φ is defined by $\{x \in R \mid \varphi(x) = 0\}$.

Corollary 1.8. *If φ is a ring homomorphism from R to R', then $ker\varphi$ is an ideal of R.*

Theorem 1.12. *A ring homomorphism φ from R to R' is one to one if and only if $ker\varphi = \{0\}$.*

Proof. It is straightforward. ∎

We are in a position to establish an important connection between homomorphisms and quotient rings. Many authors prefer to call the next theorem the fundamental theorem of ring isomorphism.

Theorem 1.13. *(First isomorphism theorem). Let $\varphi : R \to R'$ be a homomorphism from R to R'. Then $R/ker\varphi \cong \varphi(R)$; in fact, the mapping $\psi : R/ker\varphi \to \varphi(R)$ defined by $\psi(a + ker\varphi) = \varphi(a)$ defines an isomorphism from $R/ker\varphi$ onto $\varphi(R)$. Moreover there is a one to one correspondence between the set of ideals of R' and the set of ideals of R which contain $ker\varphi$. This correspondence can be achieved by associating with an ideal J in R', the ideal I in R defined by $I = \{x \in R \mid \varphi(x) \in J\}$. With I so defined, R/I is isomorphic to R'/J.*

We go on to the next isomorphism theorem.

Theorem 1.14. *(Second isomorphism theorem). Let I and J be two ideals of a ring R. Then $(I + J)/I \cong J/(I \cap J)$.*

Finally, we come to the last of the isomorphism theorem that we wish to state.

Theorem 1.15. *(Third isomorphism theorem). Let I and J be two ideals of a ring R such that $J \subseteq I$. Then $R/I \cong (R/J)/(I/J)$.*

Example 1.19.

(1) $\mathbb{Z}/n\mathbb{Z} \cong \mathbb{Z}_n$.

(2) Let $R = \left\{ \begin{bmatrix} a & b \\ -b & a \end{bmatrix} \mid a, b \in \mathbb{R} \right\}$. We define $\psi : R \to \mathbb{C}$ by

$$\psi\left(\begin{bmatrix} a & b \\ -b & a \end{bmatrix} \right) = a + bi.$$ Then ψ is an isomorphism and so R is isomorphic to the field of complex numbers.

(3) Let R be the ring of all real valued continuous functions defined on the closed unit interval. Then $I = \{f \in R \mid f(\frac{1}{2}) = 0\}$ is an ideal of R. One can shows that R/I is isomorphic to the real field.

Lemma 1.10. *Let R be a ring with unit element 1. The mapping $\varphi : \mathbb{Z} \to R$ given by $\varphi(n) = n1$ is a ring homomorphism.*

Proposition 1.6. *If R is a ring with unit element and the characteristic of R is $n > 0$, then R contains a subring isomorphic to $\mathbb{Z}_n$. If the characteristic of R is 0, then R contains a subring isomorphic to $\mathbb{Z}$.*

Proof. The set $S = \{n1 \mid n \in \mathbb{Z}\}$ is a subring of R. Lemma 1.10 shows that the mapping φ from $\mathbb{Z}$ onto S given by $\varphi(n) = n1$ is a homomorphism, and by the first isomorphism theorem, we have $\mathbb{Z}/ker\varphi \cong S$. But, clearly $ker\varphi = n\mathbb{Z}$. So $S \cong \mathbb{Z}_n$ if $n > 0$, whereas $S \cong \mathbb{Z}/\langle 0 \rangle \cong \mathbb{Z}$ if $n = 0$. ∎

Proposition 1.7. *If F is a field of characteristic p, then F contains a subfield isomorphic to $\mathbb{Z}_p$. If F is a field of characteristic 0, then F contains a subfield isomorphic to $\mathbb{Q}$.*

Proof. By Proposition 1.6, F contains a subring isomorphic to $\mathbb{Z}_p$ if F has characteristic p and F has a subring S isomorphic to $\mathbb{Z}$ if F has characteristic 0. In the latter case, let $K = \{ab^{-1} \mid a, b \in S, b \neq 0\}$. Then K is isomorphic to $\mathbb{Q}$. ∎

Now, we define some special ideals of a ring and we give some important results about them. Firstly, we begin with the definition of maximal ideal of a ring.

Definition 1.23. A proper ideal M of R is said to be a *maximal ideal* of R if whenever U is an ideal of R and $M \subseteq U \subseteq R$ then $U = M$ or $U = R$.

Example 1.20. Examples of maximal ideals.

(1) In a division ring, $< 0 >$ is a maximal ideal.

(2) In the ring of even integers, $< 4 >$ is a maximal ideal.
(3) In the ring of integers, an ideal $n\mathbb{Z}$ is maximal if and only if n is a prime number.
(4) The ideal $\langle x^2 + 1 \rangle$ is maximal in $\mathbb{R}[x]$.
(5) Let R be the ring of continuous functions from $\mathbb{R}$ to $\mathbb{R}$. The set $M = \{f \in R \mid f(0) = 0\}$ is a maximal ideal of R.

Zorn's lemma is a form of the axiom of choice which is technically very useful for proving existence theorems. For instance, from Zorn's lemma it follows directly that every ring has a maximal ideal.

Theorem 1.16. *If R is a commutative ring with unit element and M is an ideal of R, then M is a maximal ideal of R if and only if R/M is a field.*

Proof. Suppose that M is a maximal ideal and let $a \in R$ but $a \notin M$. It suffices to show that $a + M$ has a multiplicative inverse. Consider

$$U = \{ar + b \mid r \in R, b \in M\}.$$

This is an ideal of R that contains M properly. Since M is maximal, we have $U = R$. Thus $1 \in U$, so there exist $c \in R$ and $d \in M$ such that $1 = ac + d$. Then $1 + M = ac + d + M = ac + M = (a + M)(c + M)$.

Now, suppose that R/M is a field and U is an ideal of R that contains M properly. Let $a \in U$ but $a \notin M$. Then $a + M$ is a nonzero element of R/M and so there exists an element $b + M$ such that $(a + M)(b + M) = 1 + M$. Since $a \in U$, we have $ab \in U$. Also, we have $1 - ab \in M \subseteq U$. So $1 = (1 - ab) + ab \in U$ which implies that $U = R$. ∎

The motivation for the definition of a prime ideal comes from the integers.

Definition 1.24. An ideal P in a ring R is said to be *prime* if $P \neq R$ and for any ideals A, B in R,

$$AB \subseteq P \Rightarrow A \subseteq P \text{ or } B \subseteq P.$$

The definition of a prime ideal excludes the ideal R for both historical and technical reasons. The following corollary is a very useful characterization of prime ideals.

Corollary 1.9. *Let R be a commutative ring. An ideal P of R is prime if $P \neq R$ and for any $a, b \in R$,*

$$ab \in P \Rightarrow a \in P \text{ or } b \in P.$$

Example 1.21. Examples of prime ideals.

(1) A positive integer n is a prime number if and only if the ideal $n\mathbb{Z}$ is a prime ideal in $\mathbb{Z}$.

(2) In the ring $\mathbb{Z}[x]$ of all polynomials with integer coefficients, the ideal generated by 2 and x is a prime ideal.

(3) The prime ideals of $\mathbb{Z} \times \mathbb{Z}$ are $\{0\} \times \mathbb{Z}$, $\mathbb{Z} \times \{0\}$, $p\mathbb{Z} \times \mathbb{Z}$, $\mathbb{Z} \times q\mathbb{Z}$, where p and q are primes.

(4) If R denotes the ring $\mathbb{C}[x,y]$ of polynomials in two variables with complex coefficients, then the ideal generated by the polynomial $y^2 - x^3 - x - 1$ is a prime ideal. Also the ideals $\langle 0 \rangle \subseteq \langle y - x - 1 \rangle \subseteq \langle x - 2, y - 3 \rangle$ are all prime.

(5) In $\mathbb{Z}[x,y,z]$, the ideals $\langle x \rangle \subseteq \langle x, y \rangle \subseteq \langle x, y, z \rangle$ are all prime, but none is maximal.

Theorem 1.17. *If R is a commutative ring with unit element and P is an ideal of R, then P is a prime ideal of R if and only if R/P is an integral domain.*

Proof. Suppose that R/P is an integral domain and $ab \in P$. Then, $(a + P)(b + P) = ab + P = P$. So either $a + P = P$ or $b + P = P$; that is either $a \in P$ or $b \in P$. Hence P is prime.

Now, suppose that P is prime and $(a + P)(b + P) = 0 + P = P$. Then $ab \in P$ and therefore $a \in P$ or $b \in P$. Thus one of $a + P$ or $b + P$ is zero. ∎

Theorem 1.18. *Let R be a commutative ring with unit element. Each maximal ideal of R is a prime ideal.*

Proof. Suppose that M is maximal in R but not prime, so there exist $a, b \in R$ such that $a \notin M$, $b \notin M$ but $ab \in M$. Then each of the ideals $M + \langle a \rangle$ and $M + \langle b \rangle$ contains M properly. By maximality we obtain $M + \langle a \rangle = R = M + \langle b \rangle$. Therefore,

$$R^2 = (M + \langle a \rangle)(M + \langle b \rangle)$$
$$\subseteq M^2 + \langle a \rangle M + M \langle b \rangle + \langle a \rangle \langle b \rangle$$
$$\subseteq M \subseteq R.$$

This is a contradiction. ∎

Definition 1.25. The *radical* of an ideal I in a commutative ring R, denoted by $Rad(I)$ is defined as

$$Rad(I) = \{r \in R \mid r^n \in I \text{ for some positive integer } n\}.$$

Intuitively, one can think that the radical of I is obtained by taking all the possible roots of elements of I. $Rad(I)$ turns out to be an ideal itself, containing I. The above definition is equivalent to: The radical of an ideal I in a commutative ring R is

$$Rad(I) = \bigcap_{\substack{P \in Spec(R) \\ I \subseteq P}} P,$$

where $Spec(R)$ is the set of all prime ideals of R.

Lemma 1.11. *If $J, I_1, ..., I_n$ are ideals in a commutative ring R, then*

(1) $Rad(Rad(J)) = Rad(J)$,

(2) $Rad(I_1...I_n) = Rad\left(\bigcap_{i=1}^{n} I_i \right) = \bigcap_{i=1}^{n} Rad(I_i)$.

Example 1.22. In the ring of integers

(1) $Rad(12\mathbb{Z}) = 2\mathbb{Z} \cap 3\mathbb{Z} = 6\mathbb{Z}$,

(2) let $n = p_1^{k_1}...p_r^{k_r}$, where p_i's are distinct prime numbers. Then we have $Rad(n\mathbb{Z}) = \langle p_1, ..., p_r \rangle$.

The concept of a maximal ideal in a commutative ring leads immediately to the very important notion of a Jacobson radical of that ring.

Definition 1.26. Let R be a commutative ring. We define the Jacobson radical of R, denoted by $Jac(R)$, as the intersection of all the maximal ideals of R.

We can provide a characterization for the Jacobson radical of a commutative ring.

Lemma 1.12. *(Nakayama's lemma) Let R be a commutative ring, and let $r \in R$. Then $r \in Jac(R)$ if and only if for every $a \in R$, the element $1 - ra$ is an invertible element of R.*

1.3 Modules

Modules over a ring are generalization of abelian groups (which are modules over $\mathbb{Z}$).

Definition 1.27. Let R be a ring. A non-empty subset M is said to be an R-module (or, a a module over R) if M is an abelian group under an operation $+$ such that for every $r \in R$ and $m \in M$ there exists an element rm in M subject to

(1) $r(a + b) = ra + rb$;
(2) $r(sa) = (rs)a$;
(3) $(r + s)a = ra + sa$;

for all $a, b \in M$ and $r, s \in R$.

If R has a unit element 1, and if $1m = m$ for every element m in M, then M is called a *unitary R-module*.

Example 1.23.

(1) Every abelian group G is a module over the ring of integers. Addition is carried out according to the group structure of G; the key point is that we can multiply $x \in G$ by the integer n. If $n > 0$, then $nx = x + x + \ldots + x$ (n times); if $n < 0$, then $nx = -x - x - \ldots - x$ ($|n|$ times).

(2) Let R be a ring and M be left ideal of R. For $r \in R$, $m \in M$, let rm be the product of these elements as elements in R. The definition of left ideal implies that $rm \in M$, while the axioms defining a ring insure us that M is an R-module.

(3) The special case in which $M + R$, any ring R is an R-module over itself.

(4) If S is a ring and R is a subring, then S is an R-module with rm ($r \in R$, $m \in S$) being multiplication in S. In particular, $R[x_1, \ldots, x_n]$ and $R[[x]]$ are R-modules.

(5) Let R be a ring and I be a left ideal of R. Let M consists of all cosets, $a + I$, where $a \in R$. In M define $(a+I)+(b+I) = (a+b)+I$ and $r(a + I) = ra + I$. Then, M is an R-module. M is usually written as R/I and is called the *quotient module* of R by I.

(6) Let A be an abelian group and $End(A)$ its endomorphism ring. Then A is a $End(A)$-module, with fa defined to be $f(a)$, for $a \in A$, $f \in End(A)$.

(7) Let $M = M_{mn}(R)$ be the set of all $m \times n$ matrices with entries in R. Then, M is an R-module, where addition is ordinary matrix addition, and multiplication of the scalar c by the matrix A means multiplication of each entry of A by c.

$M = \{0\}$ is called *zero module* and is often written simply as 0.

Let M be an R-module. The results given for rings in Lemma 1.7 can be applied to establish the following results, which hold for any $x \in M$ and $r \in R$. We distinguish the vector 0_M from the zero scalar 0_R.

(1) $r0_M = 0_M$;
(2) $0_R x = 0_M$;
(3) $(-r)x = r(-x) = -(rx)$;
(4) If R is a field, or more generally a division ring, then $rx = 0_M$ implies that either $r = 0_R$ or $x = 0_M$.

Definition 1.28. An additive subgroup A of the R-module M is called a *submodule* of M if whenever $r \in R$ and $a \in A$, then $ra \in A$.

Example 1.24.

(1) Let I be a left ideal of the ring R, M be an R-module and S a non-empty subset of M. Then,

$$IS = \Big\{ \sum_{i=1}^{n} r_r a_i \mid r_i \in I, a_i \in S, n \in \mathbb{N} \Big\}$$

is a submodule of M. Similarly, if $a \in M$, then $Ia = \{ra \mid r \in I\}$ is a submodule of M.
(2) If $\{N_i \mid i \in I\}$ is a family of submodules of an R-module M, then $\bigcap_{i \in I} N_i$ is a submodule of M.

Definition 1.29. If X is a subset of an R-module M, then the intersection of all submodules of M containing X is called the *submodule generated by* X.

If X consists of a single element, $X = \{a\}$, then the submodule generated by X is called the *cyclic submodule* generated by a. An R-module M is *cyclic* if it is generated by a single element a. Finally, if $\{N_i \mid i \in I\}$ is a family of submodules of an R-module M, then the submodule generated by $X = \bigcup_{i \in I}$ is called the *sum* of the modules N_i. If the index set I is finite, the sum of $N_1, ..., N_n$ is denoted by $N_1 + ... + N_n$.

Theorem 1.19. *Let R be a ring, M be an R-module, X be a subset of M, $\{N_i \mid i \in I\}$ is a family of submodules M, and $a \in M$.*

(1) *Let $Ra = \{ra \mid r \in R\}$. Then, Ra is a submodule of M and the map $R \to Ra$ given by $r \mapsto ra$ is an R-module epimorphism.*

(2) *The cyclic submodule A generated by a is $\{ra + na \mid r \in R, n \in \mathbb{Z}\}$.*
If R has an identity and A is unitary, then $A = Ra$.
(3) *The submodule N generated by X is*

$$\Big\{ \sum_{i=1}^{s} r_i a_i + \sum_{j=1}^{t} n_j b_j \mid s,t \in \mathbb{N}, a_i, b_j \in X, r_i \in R, n_j \in \mathbb{Z} \Big\}.$$

If R has an identity and M is unitary, then

$$N = RX = \Big\{ \sum_{i=1}^{s} r_i a_i \mid s \in \mathbb{N}, a_i \in X, r_i \in R \Big\}.$$

(4) *The sum of the family $\{N_i \mid i \in I\}$ consists of all finite sums*
$a_{i_1} + ... + a_{i_n}$ with $a_{i_k} \in N_{i_k}$.

Proof. It is straightforward. Note that if R has a unit 1_R and M is unitary, then $n1_R \in R$, for all $n \in \mathbb{Z}$ and $na = (n1_R)a$, for all $a \in M$. ∎

Given an R-module M and a submodule A, we could constract the quotient module M/A in a manner similar to the way we constructed quotient group and quotient ring. One could also talk about homomorhisms of one R-module into another, and prove the appropriate homomorphism theorems.

Theorem 1.20. *Let A be a submodule of an R-module M. Then, the quotient group M/A is an R-module with the following external operation*

$$r(m + A) = rm + A, \text{ for all } r \in R \text{ and } m \in M.$$

Proof. Since M is an additive abelian group, A is a normal subgroup, and M/A is well defined abelian group. If $m+A = m'+A$, then $m-m' \in A$. Since A is a submodule, it follows that $rm - rm' = r(m - m') \in A$, for all $r \in R$. Thus, $rm + A = rm' + A$ and the external operation of R on M/A is well defined. The remainder of the proof is now easy. ∎

Definition 1.30. Let M and M' be two R-modules. A function $f : M \to M'$ is an *R-module homomorphism* provided that for all $a, b \in M$ and $r \in R$,

$$f(a + b) = f(a) + f(b) \text{ and } f(ra) = rf(a).$$

When the context is clear R-module homomorphisms are called simply *homomorphism*. An R-module homomorphism $f : M \to M'$ is called an *epimorphism* in case it is onto (surjective). It is called a *monomorphism* in case it is one to one (injective).

Observe that an R-module homomorphism $f : M \to M'$ is necessary a homomorphism of additive abelian groups. Consequently, the same terminology is used. f is an R-module isomorphism if it is one to one and onto too. The *kernel* of f is its kernel as a homomorphism of abelian groups, namely $ker f = \{m \in M \mid f(a) = 0\}$. Similarly, the *image* of f is the set $Im f = \{y \in M' \mid y = f(x) \text{ for some } x \in M\}$. Finally, we conclude that

(1) f is one to one if and only if $ker f = \{0\}$.
(2) $f : M \to M'$ is an R-module isomorphism if and only if there is an R-module homomorphism $g : M' \to M$ such that $gf = 1_M$ and $fg = 1_{M'}$.

Example 1.25. For any modules the zero map $0 : M \to M'$ given by $a \mapsto 0$ (for $a \in M$) is a module homomorphism. Every homomorphism of abelian groups is a $\mathbb{Z}$-module homomorphism. If R is a ring, the map $R[x] \to R[x]$ given by $f \mapsto xf$ is an R-module homomorphism, but not a ring homomorphism.

A homomorphism $f : M \to N$ that is the composite of homomorphisms

$$f = gh$$

is said to *factor through* g and h. The following result essentially says that a homomorphism f factors uniquely through every epimorphism whole kernel is contained in that of f and through every monomorphism whose image contains the image of f.

Theorem 1.21. *Let M, M', N and N' be R-modules and let $f : M \to N$ be an R-module homomorphism.*

(1) *If $g : M \to M'$ is an epimorphism with $ker g \subseteq ker f$, then there exists a unique homomorphism $h : M' \to N$ such that*

$$f = hg.$$

Moreover, $ker h = g(Ker f)$ and $Im h = Im f$, so that h is monomorphism if and only if $ker g = ker f$ and h is epimorphism if and only if f is epimorphism.

(2) *If $g : N' \to N$ is a monomorphism with $Im f \subseteq g$, then there exists a unique homomorphism $h : M \to N'$ such that*

$$f = gh.$$

Moreover, $ker h = ker f$ and $Im h = g^{-1}(Im f)$, so that h is monomorphism if and only if f is monomorphism and h is epimorphism if and only if $Im g = Im f$.

Proof. (1) Since $g : M \to M'$ is epimorphism, for each $m' \in M'$ there is at least one $m \in M$ with $g(m) = m'$. Also, if $l \in M$ with $g(l) = m'$, then clearly $m - l \in \ker g$. But since $\ker g \subseteq \ker f$, we have that $f(m) = f(l)$. Thus, there is a well defined function $h : M' \to N$ such that $f = hg$. To see that h is actually an R-module homomorphism, let $x', y' \in M'$ and $x, y \in M$ with $g(x) = x'$, $g(y) = y'$. Then, for each $r \in R$, $g(rx + y) = rx' + y'$, so that

$$\begin{aligned} h(rx' + y') &= f(rx + y) \\ &= rf(x) + f(y) \\ &= rh(x') + h(y'). \end{aligned}$$

The uniqueness of h with these properties is assured, since g is epimorphism. The final assertion is trivial.

(2) For each $m \in M$, $f(m) \in Imf \subseteq Img$. So since g is monomorphism, there is a unique $n' \in N'$ such that $g(n') = f(m)$. Therefore, there is a function $h : M \to N'$ with $m \mapsto n'$ such that $f = gh$. The rest of the proof is also easy. ∎

In view of the preceding results it is not surprising that the various isomorphism theorems for groups are valid for modules. One need only check at each stage of the proof to see that every subgroup or homomorphism is in fact a submodule or module homomorphism. For convenience we list these results here.

Theorem 1.22. *Let R be a ring, M and M' be two R-modules. If $f : M \to M'$ is an R-module homomorphism and N is a submodule of $\ker f$, then there is a unique R-module homomorphism $\phi : M/N \to M'$ such that $\phi(a + N) = f(a)$ for all $a \in M$; $Im\phi = Imf$ and $\ker\phi = \ker f/N$. Moreover, ϕ is an R-module isomorphism if and only if f is an onto R-module homomorphism and $N = \ker f$. In particular, $M/\ker f \cong Imf$.*

Theorem 1.23. *Let R be a ring, M and M' be two R-modules. If N is a submodule of M, N' is a submodule of M' and $f : M \to M'$ is an R-module homomorphism such that $f(N) \subseteq N'$, then f induces an R-module isomorphism $\phi : M/N \to M'/N'$ given by $a + N \mapsto f(a) + N'$. Moreover, ϕ is an R-module isomorphism if and only if $Imf + N' = M'$ and $f^{-1}(N') \subseteq N$. In particular, if f is an onto R-module homomorphism such that $f(N) = N'$ and $\ker f \subseteq N$, then ϕ is an R-module isomorphism.*

Theorem 1.24. *Let A and B be two submodules of an R-module M.*

(1) *There is an R-module isomorphism $A/(A \cap B) \cong (A + B)/B$.*

(2) *If $B \subseteq A$, then A/B is a submodule of M/B, and there is an R-module isomorphism $(M/B)/(A/B) \cong M/A$.*

Proof.   (1) Define a map $f : A \to M/B$ by $f(x) = x + B$. Then, f is an R-module homomorphism whose kernel is $A \cap B$ and whose image is $\{x + B \mid x \in A\} = (A + B)/B$. The first isomorphism theorem for modules gives the desired result.

(2) Define $f : M/B \to M/A$ by $f(x + B) = x + A$. Then, f is an R-module homomorphism whose kernel is $\{x + B \mid x \in A\} = A/B$, and the image is $\{x + A \mid x \in M\} = M/A$. The result follows from the first isomorphism theorem for modules.   ∎

A pair of R-module homomorphisms

$$M' \xrightarrow{f} M \xrightarrow{g} M''$$

is said to be *exact* at M in case $Imf = Kerg$. Also, a sequence (finite or infinite) of R-module homomorphisms

$$\cdots \xrightarrow{f_{n-1}} M_{n-1} \xrightarrow{f_n} M_n \xrightarrow{f_{n+1}} M_{n+1} \longrightarrow \cdots$$

is *exact* in case it is exact at each M_n, i.e., in case for each successive pair f_n and f_{n+1},

$$Imf_n = kerf_{n+1}.$$

Immediate from the definition is the following set of special cases.

Proposition 1.8. *Given modules M and N and an R-module homomorphism $f : M \to N$, the sequence*

(1) *$0 \to M \xrightarrow{f} N$ is exact if and only if f is a monomorphism;*
(2) *$M \xrightarrow{f} N \to 0$ is exact if and only if f is an epimorphism;*
(3) *$0 \to M \xrightarrow{f} N \to 0$ is exact if and only if f is an isomorphism.*

Proof.   It is straightforward.   ∎

1.4   Vector space

Note that if the ring R is a field, a unital R-module is nothing more than a vector space over R. Indeed:

Definition 1.31. A non-empty set V is said to be a *vector space* over a field F if V is an abelian group under an operation which we denote by $+$, and if for every $\alpha \in F$, $v \in V$ there is defined an element, written αv, in V subject to

(1) $\alpha(v + w) = \alpha v + \alpha w$;
(2) $(\alpha + \beta)v = \alpha v + \beta v$;
(3) $\alpha(\beta v) = (\alpha \beta)v$;
(4) $1v = v$;

for all $\alpha, \beta \in F$, $v, w \in V$ (where 1 represents the unit element of F under multiplication).

Example 1.26.

(1) Let F be a field and K be a field which contains F as a subfield. We consider K as a vector space over F, using as the $+$ of the vector space the addition of elements of K, and by defining, for $\alpha \in F$, $v \in V$, αv to be the products of α and v as elements in the field K. Axioms (1), (2), (3) for a vector space are then consequence of the right distributive law, left distributive law, and associative law, respectively, which hold for K as a field.

(2) Let F be a field and let V be the totality of all ordered n-tuples $(\alpha_1, ..., \alpha_n)$ where $\alpha_i \in F$. Two elements $(\alpha_1, ..., \alpha_n)$ and $(\beta_1, ..., \beta_n)$ are declared to be equal if and only if $\alpha_i = \beta_i$ for each $i = 1, ..., n$. We now introduce the requisite operations in V to make of it a vector space by defining:

$$(\alpha_1, ..., \alpha_n) + (\beta_1, ..., \beta_n) = (\alpha_1 + \beta_1, ..., \alpha_n + \beta_n),$$
$$\gamma(\alpha_1, ..., \alpha_n) = (\gamma\alpha_1, ..., \gamma\alpha_n), \text{ for } \gamma \in F.$$

It is easy to verify that with these operations, V is a vector space over F. Since it will keep reappearing, we assign a symbol to it, namely F^n.

If V is a vector space over F and if $W \subseteq V$, then W is a *subspace* of V if under the operations of V, W, itself, forms a vector space over F. Equivalently, W is a subspace of V whenever $\alpha \in F$, $v, w \in W$ implies that $\alpha v + w \in W$. As in our previous models, a homomorphism is a mapping preserving all the algebraic structure of our system.

Definition 1.32. If V is a vector space over F and if $v_1, ..., v_n \in V$ then any element of the form $\alpha_1 v_1 + \alpha_2 v_2 + ... + \alpha_n v_n$, where $\alpha_i \in F$, is a *linear combination* over F of $v_1, ..., v_n$.

Definition 1.33. The vector space V is said to be *finite dimensional* over F, if there is a finite subset X in V such that $V = \langle X \rangle$.

Note that F^n is finite dimensional over F, for if X consists of the n vectors $(1, 0, ..., 0)$, $(0, 1, 0, ..., 0)$, ..., $(0, 0, ..., 1)$, then $V = \langle X \rangle$.

Definition 1.34. If V is a vector space and if $v_1, ..., v_n$ are in V, we say that they are *linearly dependent* over F if there exist elements $c_1, ..., c_n$ in F, not all of them 0, such that $c_1 v_1 + c_2 v_2 + ... + c_n v_n = 0$.

If the vectors $v_1, ..., v_n$ are not linearly dependent over F, they are said to be *linearly independent* over F. Note that if $v_1, ..., v_n$ are linearly independent then none of them can be 0, for if $v_1 = 0$, say, then $\alpha v_1 + 0 v_2 + ... + 0 v_n = 0$ for any $\alpha \neq 0$ in F.

Lemma 1.13. *If $v_1, ..., v_n \in V$ are linearly independent, then every element in $\langle v_1, ..., v_n \rangle$ has a unique representation in the form $c_1 v_1 + ... + c_n v_n$ with $c_i \in F$.*

Proof. It is straightforward.   ■

Lemma 1.14. *If $v_1, ..., v_n \in V$ are in V, then either they are linearly independent or some v_k is a linear combination of the preceding ones, $v_1, ..., v_{k-1}$.*

Proof. It is straightforward.   ■

Definition 1.35. A subset X of a vector space V is called a *basis* of V if S consists of linearly independent elements (that is, any finite number of elements in X is linearly independent) and $V = \langle X \rangle$.

Chapter 2

Algebraic hyperstructures

2.1 Semihypergroup

The concept of a semihypergroup is a generalization of the concept of a semigroup. Many authors studied different aspects of semihypergroups.

A *hypergroupoid* $(H, \star)$ is a non-empty set H together with a map $\star :$ $H \times H \to \mathcal{P}^*(H)$ called *(binary) hyperoperation*, where $\mathcal{P}^*(H)$ denotes the set of all non-empty subsets of H. The image of the pair (x, y) is denoted by $x \star y$.

If A, B are non-empty subsets of H and $x \in H$, then by $A \star B$, $A \star x$ and $x \star B$ we mean

$$A \star B = \bigcup_{\substack{a \in A \\ b \in B}} a \star b, \ A \star x = A \star \{x\} \text{ and } x \star B = \{x\} \star B.$$

Definition 2.1. A hypergroupoid $(H, \star)$ is called a *semihypergroup* if

$$(x \star y) \star z = x \star (y \star z),$$

for all $x, y, z \in H$. This means that

$$\bigcup_{u \in x \star y} u \star z = \bigcup_{v \in y \star z} x \star v.$$

A semihypergroup H is *finite* if it has only a finitely many elements. A semihypergroup H is *commutative* if it satisfies

$$x \star y = y \star x,$$

for all $x, y \in H$.

Remark 2.1. Every semigroup is a semihypergroup.

Remark 2.2. The associativity for semihypergroups can be applied for subsets, i.e., if $(H, \star)$ is a semihypergroup, then for all non-empty subsets A, B, C of H, we have $(A \star B) \star C = A \star (B \star C)$.

The element $a \in H$ is called *scalar* if

$$|a \star x| = |x \star a| = 1,$$

for all $x \in H$. An element e in a semihypergroup $(H, \star)$ is called *scalar identity* if

$$x \star e = e \star x = \{x\},$$

for all $x \in H$. An element e in a semihypergroup $(H, \star)$ is called *identity* if

$$x \in e \star x \cap x \star e,$$

for all $x \in H$. An element $a' \in H$ is called an *inverse* of $a \in H$ if there exists an identity $e \in H$ such that

$$e \in a \star a' \cap a' \star a.$$

An element 0 in a semihypergroup $(H, \star)$ is called *zero element* if $x \star 0 = 0 \star x = \{0\}$, for all $x \in H$.

Let $(H, \star)$ be a hypergroupoid. The element $a \in H$ is called *right simplifiable* element (respectively, *left*) if for all $x, y \in H$ the following is valid.

$$x \star a = y \star a \Rightarrow x = y \text{ (respectively, } a \star x = a \star y \Rightarrow x = y).$$

Moreover, if $x \in x \star y$ (respectively, $x \in y \star x$) for all $y \in H$, then x is called *left absorbing-like* element (respectively, *right absorbing-like element*).

Similar to semigroups, we can describe the hyperoperation on a semi-hypergroup by Cayley table.

Example 2.1.

(1) Let $H = \{a,\ b,\ c,\ d\}$. Define the hyperoperation $\star$ on H by the following table.

$\star$	a	b	c	d
a	a	$\{a,b\}$	$\{a,c\}$	$\{a,d\}$
b	a	$\{a,b\}$	$\{a,c\}$	$\{a,d\}$
c	a	b	c	d
d	a	b	c	d

Then, $(H, \star)$ is a semihypergroup.

(2) Let $H = \{a,\ b,\ c,\ d,\ e\}$. Define the hyperoperation $\star$ on H by the following table.

$\star$	a	b	c	d	e
a	a	$\{a,b,d\}$	a	$\{a,b,d\}$	$\{a,b,d\}$
b	a	b	a	$\{a,b,d\}$	$\{a,b,d\}$
c	a	$\{a,b,d\}$	$\{a,c\}$	$\{a,b,d\}$	$\{a,b,c,d,e\}$
d	a	$\{a,b,d\}$	a	$\{a,b,d\}$	$\{a,b,d\}$
e	a	$\{a,b,d\}$	$\{a,c\}$	$\{a,b,d\}$	$\{a,b,c,d,e\}$

Then, $(H, \star)$ is a semihypergroup.

(3) Let H be the unit interval $[0,1]$. For every $x, y \in H$, we define

$$x \star y = \left[0,\ \frac{xy}{2}\right].$$

Then, $(H, \star)$ is a semihypergroup.

(4) Let $\mathbb{N}$ be the set of non-negative integers. We define the following hyperoperation on $\mathbb{N}$,

$$x \star y = \{z \in \mathbb{N} \mid z \geq \max\{x, y\}\},$$

for all $x, y \in \mathbb{N}$. Then, $(\mathbb{N}, \star)$ is a semihypergroup.

(5) Let $(S, \cdot)$ be a semigroup and K be any subsemigroup of S. Then, the set $S/K = \{x \cdot K \mid x \in S\}$ becomes a semihypergroup, where the hyperoperation is defined in a usual manner $\overline{x} \star \overline{y} = \{\overline{z} \mid z \in \overline{x} \cdot \overline{y}\}$ with $\overline{x} = x \cdot K$.

(6) The set of real numbers $\mathbb{R}$ with the following hyperoperation

$$a \star b = \begin{cases} (a,b) & \text{if } a < b \\ (b,a) & \text{if } b < a \\ \{a\} & \text{if } a = b, \end{cases}$$

for all $a, b \in \mathbb{R}$ is a semihypergroup, where (a, b) is the open interval $\{x \mid a < x < b\}$.

(7) Let $(S, \cdot)$ be a semigroup and P a non-empty subset of S. We define the following hyperoperation on S,

$$x \star_P y = x \cdot P \cdot y,$$

for all $x, y \in S$. Then, $(S, \star_P)$ is a semihypergroup. The hyperoperation $\star_P$ is called *P-hyperoperation*. This hyperoperation is introduced in [112].

(8) Let $(S, \cdot)$ be a semigroup and for all $x, y \in S$, $\langle x, y \rangle$ denotes the subsemigroup generated by x and y. We define $x \star y = \langle x, y \rangle$. Then, $(S, \star)$ is a semihypergroup.

(9) Let $(H, \star)$ and $(H', \star)$ be two semihypergroups. Then, the Cartesian product of these two semihypergroups is a semihypergroup with the following hyperoperation

$$(x, y) \otimes (x', y') = \{(a, b) \mid a \in x \star x', \; b \in y \star y'\},$$

for all $(x, y), (x', y') \in H \times H'$.

2.2   Hypergroups

A hypergroupoid $(H, \star)$ is called a *quasihypergroup* if for all a of H we have $a \star H = H \star a = H$. This condition is also called the *reproduction axiom*.

Definition 2.2. A semihypergroup $(H, \star)$ is called a *hypergroup* if

$$a \star H = H \star a = H,$$

for all $a \in H$.

A hypergroup is called *regular* if it has at least one identity and each element has at least one inverse.

A hypergroup for which the hyperproduct of any two elements has exactly one element is a group. Indeed, let $(H, \star)$ be a hypergroup, such that for all x, y of H, we have $|x \star y| = 1$. Then, $(H, \star)$ is a semigroup, such that for all a, b in H, there exist x and y for which we have $a = b \star x$ and $a = y \star b$. It follows that $(H, \star)$ is a group.

Now, we look at some examples of hypergroups.

Example 2.2.

(1) If H is a non-empty set and for all x, y of H, we define $x \star y = H$, then $(H, \star)$ is a hypergroup, called the *total hypergroup*.

(2) Let $(S, \cdot)$ be a semigroup and P be a non-empty subset of S. For all x, y of S, we define $x \star y = xPy$. Then, $(S, \star)$ is a semihypergroup. If $(S, \cdot)$ is a group, then $(S, \star)$ is a hypergroup, called a *P-hypergroup*.

(3) If G is a group and for all x, y of G, $< x, y >$ denotes the subgroup generated by x and y, then we define $x \star y = \langle x, y \rangle$. We obtain that $(G, \star)$ is a hypergroup.

(4) If $(G, \cdot)$ is a group, H is a normal subgroup of G and for all x, y of G, we define $x \star y = xyH$, then $(G, \star)$ is a hypergroup.

(5) Let $(G, \cdot)$ be a group and let H be a non-normal subgroup of it. If we denote $G/H = \{xH \mid x \in G\}$, then $(G/H, \star)$ is a hypergroup, where for all xH, yH of G/H, we have $xH \star yH = \{zH \mid z \in xHy\}$.

(6) If $(G, +)$ is an abelian group, ρ is an equivalence relation in G, which has classes $\overline{x} = \{x, -x\}$, then for all $\overline{x}, \overline{y}$ of G/ρ, we define $\overline{x} \star \overline{y} = \{\overline{x+y}, \overline{x-y}\}$. We obtain that $(G/\rho, \star)$ is a hypergroup.

(7) Let D be an integral domain and let F be its field of fractions. If we denote by U the group of the invertible elements of D, then we define the following hyperoperation on F/U: for all $\overline{x}, \overline{y}$ of F/U, we have $\overline{x} \star \overline{y} = \{\overline{z} \mid \exists(u, v) \in U^2 \text{ such that } z = ux + vy\}$. We obtain that $(F/U, \star)$ is a hypergroup.

(8) Let $(L, \wedge, \vee)$ be a lattice with a minimum element 0. If for all $a \in L$, $F(a)$ denotes the principal filter generated from a, then we obtain a hypergroup $(L, \star)$, where for all a, b of L, we have $a \star b = F(a \wedge b)$.

(9) Let $(L, \wedge, \vee)$ be a modular lattice. If for all x, y of L, we define $x \star y = \{z \in L \mid z \vee x = x \vee y = y \vee z\}$, then $(L, \star)$ is a hypergroup.

(10) Let $(L, \wedge, \vee)$ be a distributive lattice. If for all x, y of L, we define $x \star y = \{z \in L \mid x \wedge y \le z \le x \vee y\}$, then $(L, \star)$ is a hypergroup.

(11) Let H be a non-empty set and $\mu : H \to [0, 1]$ be a function. If for all x, y of H we define $x \star y = \{z \in L \mid \mu(x) \wedge \mu(y) \le \mu(z) \le \mu(x) \vee \mu(y)\}$, then $(H, \star)$ is a hypergroup.

(12) Let H be a non-empty set and μ, λ be two functions from H to $[0, 1]$. For all x, y of H we define

$$x \star y = \{u \in H \mid \mu(x) \wedge \lambda(x) \wedge \mu(y) \wedge \lambda(y) \le \mu(u) \wedge \lambda(u)$$
$$\text{and } \mu(u) \vee \lambda(u) \le \mu(x) \vee \lambda(x) \vee \mu(y) \vee \lambda(y)\}.$$

Then, the hyperstructure $(H, \star)$ is a commutative hypergroup.

(13) Define the following hyperoperation on the real set $\mathbb{R}$: for all $x \in \mathbb{R}$, $x \star x = x$ and for all different real elements x, y, $x \star y$ is the open interval between x and y. Then, $(\mathbb{R}, \star)$ is a hypergroup.

A non-empty subset K of a semihypergroup $(H, \star)$ is called a *subsemihypergroup* if it is a semihypergroup. In other words, a non-empty subset K of a semihypergroup $(H, \star)$ is a subsemihypergroup if $K \star K \subseteq K$.

Definition 2.3. A non-empty subset K of a hypergroup $(H, \star)$ is called a *subhypergroup* if it is a hypergroup.

Hence, a non-empty subset K of a hypergroup $(H, \star)$ is a subhypergroup if for all a of K we have $a \star K = K \star a = K$. There are several kinds of subhypergroups. In what follows, we introduce closed, invertible, ultraclosed and conjugable subhypergroups and some connections among them.

Definition 2.4. Let $(H, \star)$ be a hypergroup and $(K, \star)$ be a subhypergroup of it. We say that K is:

- *closed on the left* (*on the right*) if for all k_1, k_2 of K and x of H, from $k_1 \in x \star k_2$ ($k_1 \in k_2 \star x$, respectively), it follows that $x \in K$;
- *invertible on the left* (*on the right*) if for all x, y of H, from $x \in K \star y$ ($x \in y \star K$), it follows that $y \in K \star x$ ($y \in x \star K$, respectively);
- *ultraclosed on the left* (*on the right*) if for all x of H, we have $K \star x \cap (H \backslash K) \star x = \emptyset$ ($x \star K \cap x \star (H \backslash K) = \emptyset$);
- *conjugable on the right* if it is closed on the right and for all $x \in H$, there exists $x' \in H$ such that $x' \star x \subseteq K$. Similarly, we can define the notion of *conjugable on the left*.

We say that K is *closed* (*invertible, ultraclosed, conjugable*) if it is closed (invertible, ultraclosed, conjugable, respectively) on the left and on the right.

Example 2.3.

(1) Let $(A, \star)$ be a hypergroup, $H = A \cup T$, where T is a set with at least three elements and $A \cap T = \emptyset$. We define the hyperoperation $\otimes$ on H, as follows:

> if $(x, y) \in A^2$, then $x \otimes y = x \star y$;
> if $(x, t) \in A \times T$, then $x \otimes t = t \otimes x = t$;
> if $(t_1, t_2) \in T \times T$, then $t_1 \otimes t_2 = t_2 \otimes t_1 = A \cup (T \setminus \{t_1, t_2\})$.

Then, $(H, \otimes)$ is a hypergroup and $(A, \otimes)$ is an ultraclosed, non-conjugable subhypergroup of H.

(2) Let $(A, \star)$ be a total hypergroup, with at least two elements and let $T = \{t_i\}_{i \in \mathbb{N}}$ such that $A \cap T = \emptyset$ and $t_i \neq t_j$ for $i \neq j$. We define the hyperoperation $\otimes$ on $H = A \cup T$ as follows:

> if $(x, y) \in A^2$, then $x \otimes y = A$;
> if $(x, t) \in A \times T$, then $x \otimes t = t \otimes x = (A \setminus \{x\}) \cup T$;
> if $(t_i, t_j) \in T \times T$, then $t_i \otimes t_j = t_j \otimes t_i = A \cup \{t_{i+j}\}$.

Then, $(H, \otimes)$ is a hypergroup and $(A, \otimes)$ is a non-closed subhypergroup of H.

(3) Let us consider the group $(\mathbb{Z}, +)$ and the subgroups $S_i = 2^i \mathbb{Z}$, where i is a non-negative integer. For any $x \in \mathbb{Z} \setminus \{0\}$, there exists a unique integer $n(x)$, such that $x \in S_{n(x)} \setminus S_{n(x)+1}$. Define the following commutative hyperoperation on $\mathbb{Z} \setminus \{0\}$:

> if $n(x) < n(y)$, then $x \star y = x + S_{n(y)}$;
> if $n(x) = n(y)$, then $x \star y = S_{n(x)} \setminus \{0\}$;

if $n(x) > n(y)$, then $x \star y = y + S_{n(x)}$.

Notice that if $n(x) < n(y)$, then $n(x+y) = n(x)$. Then, $(\mathbb{Z} \setminus \{0\}, \star)$ is a hypergroup and for all $i \in \mathbb{N}$, $(S_i \setminus \{0\}, \star)$ is an invertible subhypergroup of $\mathbb{Z} \setminus \{0\}$.

Lemma 2.1. *A subhypergroup K is invertible on the right if and only if $\{x \star K\}_{x \in H}$ is a partition of H.*

Proof. If K is invertible on the right and $z \in x \star K \cap y \star K$, then $x, y \in z \star K$, whence $x \star K \subseteq z \star K$ and $y \star K \subseteq z \star K$. It follows that $x \star K = z \star K = y \star K$. Conversely, if $\{x \star K\}_{x \in H}$ is a partition of H and $x \in y \star K$, then $x \star K \subseteq y \star K$, whence $x \star K = y \star K$ and so we have $x \in y \star K = x \star K$. Hence, for all x of H we have $x \in x \star K$. From here, we obtain that $y \in y \star K = x \star K$. ∎

Similar to Lemma 2.1, we can give a necessary and sufficient condition for invertible subhypergroups on the left.

The following theorems present some connections among the above types of subhypergroups.

If A and B are subsets of H such that we have $H = A \cup B$ and $A \cap B = \emptyset$, then we denote $H = A \oplus B$.

Theorem 2.1. *If a subhypergroup K of a hypergroup $(H, \star)$ is ultraclosed, then it is closed and invertible.*

Proof. First we check that K is closed. For $x \in K$, we have $K \cap x \star (H \setminus K) = \emptyset$ and from $H = x \star K \cup x \star (H \setminus K)$, we obtain $x \star (H \setminus K) = H \setminus K$, which means that $K \star (H \setminus K) = H \setminus K$. Similarly, we obtain $(H \setminus K) \star K = H \setminus K$, hence K is closed. Now, we show that $\{x \star K\}_{x \in H}$ is a partition of H. Let $y \in x \star K \cap z \star K$. It follows that $y \star K \subseteq x \star K$ and $y \star (H \setminus K) \subseteq x \star K \star (H \setminus K) = x \star (H \setminus K)$. From $H = x \star K \oplus x \star (H \setminus K) = y \star K \oplus y \star (H \setminus K)$, we obtain $x \star K = y \star K$. Similarly, we have $z \star K = y \star K$. Hence, $\{x \star K\}_{x \in H}$ is a partition of H, and according to the above lemma, it follows that K is invertible on the right. Similarly, we can show that K is invertible on the left. ∎

Theorem 2.2. *If a subhypergroup K of a hypergroup $(H, \star)$ is invertible, then it is closed.*

Proof. Let $k_1, k_2 \in K$. If $k_1 \in x \star k_2 \subseteq x \star K$, then $x \in k_1 \star K \subseteq K$. Similarly, from $k_1 \in k_2 \star x$, we obtain $x \in K$. ∎

We denote the set $\{e \in H \mid \exists x \in H$, such that $x \in x \star e \cup e \star x\}$ by I_p and we call it the *set of partial identities* of H.

Theorem 2.3. *A subhypergroup K of a hypergroup $(H, \star)$ is ultraclosed if and only if K is closed and $I_p \subseteq K$.*

Proof. Suppose that K is closed and $I_p \subseteq K$. First, we show that K is invertible on the left. Suppose there are x, y of H such that $x \in K \star y$ and $y \notin K \star x$. Hence, $y \in (H \setminus K) \star x$, whence $x \in K \star (H \setminus K) \star x \subseteq (H \setminus K) \star x$, since K is closed. We obtain that $I_p \cap (H \setminus K) \neq \emptyset$, which is a contradiction. Hence, K is invertible on the left. Now, we check that K is ultraclosed on the left. Suppose that there are a and x in H such that $a \in K \star x \cap (H \setminus K) \star x$. It follows that $x \in K \star a$, since K is invertible on the left. We obtain $a \in (H \setminus K) \star x \subseteq (H \setminus K) \star K \star a \subseteq (H \setminus K) \star a$, since K is closed. This means that $I_p \cap (H \setminus K) \neq \emptyset$, which is a contradiction. Therefore, K is ultraclosed on the left and similarly it is ultraclosed on the right.

For the converse, suppose that K is ultraclosed. According to Theorem 2.3.5, K is closed. Now, suppose that $I_p \cap (H \setminus K) \neq \emptyset$, which means that there is $e \in H \setminus K$ and there is $x \in H$, such that $x \in e \star x$, for instance. We obtain $x \in (H \setminus K) \star x$, whence $K \star x \subseteq (H \setminus K) \star x$, which contradicts that K is ultraclosed. Hence, $I_p \subseteq K$. ∎

Theorem 2.4. *If a subypergroup K of a hypergroup $(H, \star)$ is conjugable, then it is ultraclosed.*

Proof.   Let $x \in H$. Denote $B = x \star K \cap x \star (H \setminus K)$. Since K is conjugable it follows that K is closed and there exists $x' \in H$, such that $x' \star x \subseteq K$. We obtain
$$\begin{aligned} x' \star B &= x' \star (x \star K \cap x \star (H \setminus K)) \\ &\subseteq K \cap x' \star x \star (H \setminus K) \\ &\subseteq K \cap K \star (H \setminus K) \\ &\subseteq K \cap (H \setminus K) = \emptyset. \end{aligned}$$
Hence, $B = \emptyset$, which means that K is ultraclosed on the right. Similarly, we check that K is ultraclosed on the left. ∎

Definition 2.5. Let $(H_1, \circ)$ and $(H_2, \star)$ be two hypergroups. A map $f : H_1 \to H_2$ is called a *homomorphism* or a *good homomorphism* if
$$f(x \circ y) = f(x) \star f(y), \text{ for all } x, y \in H_1.$$
f is called an *inclusion homomorphism* if
$$f(x \circ y) \subseteq f(x) \star f(y), \text{ for all } x, y \in H_1.$$

By using a certain type of equivalence relations, we can connect semi-hypergroups to semigroups and hypergroups to groups. These equivalence relations are called strong regular relations. More exactly, by a given (semi)hypergroup and by using a strong regular relation on it, we can construct a (semi)group structure on the quotient set. A natural question arises: Do they also exist regular relations? The answer is positive, regular relations provide us new (semi)hypergroup structures on the quotient sets. Let us define these notions. First, we do some notations.

Let $(H, \star)$ be a semihypergroup and R be an equivalence relation on H. If A and B are non-empty subsets of H, then

$$A\overline{R}B \text{ means that } \forall a \in A, \exists b \in B \text{ such that } aRb \text{ and}$$
$$\forall b' \in B, \exists a' \in A \text{ such that } a'Rb';$$
$$A\overline{\overline{R}}B \text{ means that } \forall a \in A, \forall b \in B, \text{ we have } aRb.$$

Definition 2.6. The equivalence relation R is called

(1) *regular on the right* (*on the left*) if for all x of H, from aRb, it follows that $(a \star x)\overline{R}(b \star x)$ $((x \star a)\overline{R}(x \star b)$ respectively);

(2) *strongly regular on the right* (*on the left*) if for all x of H, from aRb, it follows that $(a \star x)\overline{\overline{R}}(b \star x)$ $((x \star a)\overline{\overline{R}}(x \star b)$ respectively);

(3) R is called *regular* (*strongly regular*) if it is regular (strongly regular) on the right and on the left.

Theorem 2.5. *Let $(H, \star)$ be a semihypergroup and R be an equivalence relation on H.*

(1) *If R is regular, then H/R is a semihypergroup, with respect to the following hyperoperation: $\overline{x} \otimes \overline{y} = \{\overline{z} \mid z \in x \star y\}$;*

(2) *If the above hyperoperation is well defined on H/R, then R is regular.*

Proof. (1) First, we check that the hyperoperation $\otimes$ is well defined on H/R. Consider $\overline{x} = \overline{x_1}$ and $\overline{y} = \overline{y_1}$. We check that $\overline{x} \otimes \overline{y} = \overline{x_1} \otimes \overline{y_1}$. We have xRx_1 and yRy_1. Since R is regular, it follows that $(x \star y)\overline{R}(x_1 \star y)$, $(x_1 \star y)\overline{R}(x_1 \star y_1)$ whence $(x \star y)\overline{R}(x_1 \star y_1)$. Hence, for all $z \in x \star y$, there exists $z_1 \in x_1 \star y_1$ such that zRz_1, which means that $\overline{z} = \overline{z_1}$. It follows that $\overline{x} \otimes \overline{y} \subseteq \overline{x_1} \otimes \overline{y_1}$ and similarly we obtain the converse inclusion. Now, we check the associativity of $\otimes$. Let $\overline{x}, \overline{y}, \overline{z}$ be arbitrary elements in H/R and $\overline{u} \in (\overline{x} \otimes \overline{y}) \otimes \overline{z}$. This means that there exists $\overline{v} \in \overline{x} \otimes \overline{y}$ such that $\overline{u} \in \overline{v} \otimes \overline{z}$. In other words, there exist $v_1 \in x \star y$ and $u_1 \in v \star z$, such that vRv_1 and uRu_1. Since R is regular, it follows that there exists $u_2 \in v_1 \star z \subseteq x \star (y \star z)$

such that $u_1 R u_2$. From here, we obtain that there exists $u_3 \in y \star z$ such that $u_2 \in x \star u_3$. We have $\overline{u} = \overline{u_1} = \overline{u_2} \in \overline{x} \otimes \overline{u_3} \subseteq \overline{x} \otimes (\overline{y} \otimes \overline{z})$. It follows that $(\overline{x} \otimes \overline{y}) \otimes \overline{z} \subseteq \overline{x} \otimes (\overline{y} \otimes \overline{z})$. Similarly, we obtain the converse inclusion.

(2) Let aRb and x be an arbitrary element of H. If $u \in a \star x$, then $\overline{u} \in \overline{a} \otimes \overline{x} = \overline{b} \otimes \overline{x} = \{\overline{v} \mid v \in b \star x\}$. Hence, there exists $v \in b \star x$ such that uRv, whence $(a \star x)\overline{R}(b \star x)$. Similarly we obtain that R is regular on the left. ∎

Corollary 2.1. *If $(H, \star)$ is a hypergroup and R is an equivalence relation on H, then R is regular if and only if $(H/R, \otimes)$ is a hypergroup.*

Proof. If H is a hypergroup, then for all x of H we have $H \star x = x \star H = H$, whence we obtain $H/R \otimes \overline{x} = \overline{x} \otimes H/R = H/R$. According to the above theorem, it follows that $(H/R, \otimes)$ is a hypergroup. ∎

Notice that if R is regular on a (semi)hypergroup H, then the canonical projection $\pi : H \to H/R$ is a good epimorphism. Indeed, for all x, y of H and $\overline{z} \in \pi(x \star y)$, there exists $z' \in x \star y$ such that $\overline{z} = \overline{z'}$. We have $\overline{z} = \overline{z'} \in \overline{x} \otimes \overline{y} = \pi(x) \otimes \pi(y)$. Conversely, if $\overline{z} \in \pi(x) \otimes \pi(y) = \overline{x} \otimes \overline{y}$, then there exists $z_1 \in x \star y$ such that $\overline{z} = \overline{z_1} \in \pi(x \star y)$.

Theorem 2.6. *If $(H, \star)$ and $(K, *)$ are semihypergroups and $f : H \to K$ is a good homomorphism, then the equivalence ρ^f associated with f, that is $x\rho^f y \Leftrightarrow f(x) = f(y)$, is regular and $\varphi : f(H) \to H/\rho^f$, defined by $\varphi(f(x)) = \overline{x}$, is an isomorphism.*

Proof. Let $h_1 \rho^f h_2$ and a be an arbitrary element of H. If $u \in h_1 \star a$, then

$$f(u) \in f(h_1 \star a) = f(h_1) * f(a) = f(h_2) * f(a) = f(h_2 \star a).$$

Then, there exists $v \in h_2 \star a$ such that $f(u) = f(v)$, which means that $u\rho^f v$. Hence, ρ^f is regular on the right. Similarly, it can be shown that ρ^f is regular on the left. On the other hand, for all $f(x), f(y)$ of $f(H)$, we have

$$\varphi(f(x) * f(y)) = \varphi(f(x \star y)) = \{\overline{z} \mid z \in x \star y\} = \overline{x} \otimes \overline{y} = \varphi(f(x)) \otimes \varphi(f(y)).$$

Moreover, if $\varphi(f(x)) = \varphi(f(y))$, then $x\rho^f y$, so φ is injective and clearly, it is also surjective. Finally, for all $\overline{x}, \overline{y}$ of H/ρ^f we have

$$\begin{aligned} \varphi^{-1}(\overline{x} \otimes \overline{y}) &= \varphi^{-1}(\{\overline{z} \mid z \in x \star y\}) = \{f(z) \mid z \in x \star y\} \\ &= f(x \star y) = f(x) * f(y) = \varphi^{-1}(\overline{x}) * \varphi^{-1}(\overline{y}). \end{aligned}$$

Therefore, φ is an isomorphism. ∎

Theorem 2.7. *Let $(H,\star)$ be a semihypergroup and R be an equivalence relation on H.*

(1) *If R is strongly regular, then H/R is a semigroup, with respect to the following operation: $\overline{x} \otimes \overline{y} = \overline{z}$, for all $z \in x \star y$;*

(2) *If the above operation is well defined on H/R, then R is strongly regular.*

Proof. (1) For all x, y of H, we have $(x \star y)\overline{\overline{R}}(x \star y)$. Hence, $\overline{x} \otimes \overline{y} = \{\overline{z} \mid z \in x \star y\} = \{\overline{z}\}$, which means that $\overline{x} \otimes \overline{y}$ has exactly one element. Therefore, $(H/R, \otimes)$ is a semigroup.

(2) If aRb and x is an arbitrary element of H, we check that $(a \star x)\overline{\overline{R}}(b \star x)$. Indeed, for all $u \in a \star x$ and all $v \in b \star x$ we have $\overline{u} = \overline{a} \otimes \overline{x} = \overline{b} \otimes \overline{x} = \overline{v}$, which means that uRv. Hence, R is strongly regular on the right and similarly, it can be shown that it is strongly regular on the left. ∎

Corollary 2.2. *If $(H,\star)$ is a hypergroup and R is an equivalence relation on H, then R is strongly regular if and only if $(H/R, \otimes)$ is a group.*

Theorem 2.8. *If $(H,\star)$ is a semihypergroup, $(S,*)$ is a semigroup and $f : H \to S$ is a homomorphism, then the equivalence ρ^f associated with f is strongly regular.*

Proof. Let $a\rho^f b$, $x \in H$ and $u \in a \star x$. It follows that

$$f(u) = f(a) * f(x) = f(b) * f(x) = f(b \star x).$$

Hence, for all $v \in b \star x$, we have $f(u) = f(v)$, which means that $u\rho^f v$. Hence, ρ^f is strongly regular on the right and similarly, it is strongly regular on the left. ∎

The fundamental relation has an important role in the study of semihypergroups and especially of hypergroups.

Definition 2.7. For all $n > 1$, we define the relation β_n on a semihypergroup H, as follows:

$$a \,\beta_n\, b \Leftrightarrow \exists(x_1, \ldots, x_n) \in H^n \;:\; \{a, b\} \subseteq \prod_{i=1}^{n} x_i,$$

and $\beta = \bigcup_{n \geq 1} \beta_n$, where $\beta_1 = \{(x, x) \mid x \in H\}$ is the diagonal relation on H. Clearly, the relation β is reflexive and symmetric. Denote by β^* the transitive closure of β.

Theorem 2.9. *β^* is the smallest strongly regular relation on H.*

Proof.   We show that

(1) β^* is a strongly regular relation on H;

(2) If R is a strongly regular relation on H, then $\beta^* \subseteq R$.

(1) Let $a \, \beta \,^* b$ and x be an arbitrary element of H. It follows that there exist $x_0 = a, x_1, \ldots, x_n = b$ such that for all $i \in \{0, 1, \ldots, n-1\}$ we have $x_i \, \beta \, x_{i+1}$. Let $u_1 \in a \star x$ and $u_2 \in b \star x$. We check that $u_1 \, \beta^* \, u_2$. From $x_i \, \beta \, x_{i+1}$ it follows that there exists a hyperproduct P_i, such that $\{x_i, x_{i+1}\} \subseteq P_i$ and so $x_i \star x \subseteq P_i \star x$ and $x_{i+1} \star x \subseteq P_i \star x$, which means that $x_i \star x \overline{\overline{\beta}} x_{i+1} \star x$. Hence, for all $i \in \{0, 1, \ldots, n-1\}$ and for all $s_i \in x_i \star x$ we have $s_i \, \beta \, s_{i+1}$. If we consider $s_0 = u_1$ and $s_n = u_2$, then we obtain $u_1 \, \beta^* \, u_2$. Then, β^* is strongly regular on the right and similarly, it is strongly regular on the left.

(2) We have $\beta_1 = \{(x, x) \mid x \in H\} \subseteq R$, since R is reflexive. Suppose that $\beta_{n-1} \subseteq R$ and show that $\beta_n \subseteq R$. If $a \beta_n b$, then there exist $x_1, \ldots, x_n$ in H, such that $\{a, b\} \subseteq \prod_{i=1}^{n} x_i$. Hence, there exists u, v in $\prod_{i=1}^{n-1} x_i$, such that $a \in u \star x_n$ and $b \in v \star x_n$. We have $u \beta_{n-1} v$ and according to the hypothesis, we obtain uRv. Since R is strongly regular, it follows that aRb. Hence, $\beta_n \subseteq R$. By induction, it follows that $\beta \subseteq R$, whence $\beta^* \subseteq R$.   ■

Hence, the relation β^* is the smallest equivalence relation on H, such that the quotient H/β^* is a group. β^* is called the *fundamental equivalence relation* on H and H/β^* is called the *fundamental group*. If H is a hypergroup, then $\beta = \beta^*$. Consider the canonical projection $\varphi_H : H \to H/\beta^*$. The *heart* of H is the set $\omega_H = \{x \in H \mid \varphi_H(x) = 1\}$, where 1 is the identity of the group H/β^*. This relation was introduced by Koskas and studied mainly by Corsini, Davvaz, Freni, Leoreanu-Fotea, Vougiouklis and many others.

2.3   Hyperrings

The more general structure that satisfies the ring-like axioms is the hyperring in the general sense: $(R, +, \cdot)$ is a hyperring if $+$ and $\cdot$ are two hyperoperations such that $(R, +)$ is a hypergroup and $\cdot$ is an associative hyperoperation, which is distributive with respect to $+$. There are different types of hyperrings. If only the addition $+$ is a hyperoperation and the multiplication $\cdot$ is a usual operation, then we say that R is an additive hyperring. A special case of this type is the hyperring introduced by Krasner.

Definition 2.8. A *Krasner hyperring* is an algebraic structure $(R, +, \cdot)$ which satisfies the following axioms:

(1) $(R, +)$ is a *canonical hypergroup*, i.e.,

    (a) for every $x, y, z \in R$, $x + (y + z) = (x + y) + z$,
    (b) for every $x, y \in R$, $x + y = y + x$,
    (c) there exists $0 \in R$ such that $0 + x = \{x\}$ for every $x \in R$,
    (d) for every $x \in R$ there exists a unique element $x' \in R$ such that $0 \in x + x'$;
       (We shall write $-x$ for x' and we call it the *opposite* of x.)
    (e) $z \in x + y$ implies $y \in -x + z$ and $x \in z - y$.

(2) $(R, \cdot)$ is a semigroup having zero as a bilateral absorbing element, i.e., $x \cdot 0 = 0 \cdot x = 0$.

(3) The multiplication is distributive with respect to the hyperoperation $+$.

The following elementary facts follow easily from the axioms: $-(-x) = x$ and $-(x + y) = -x - y$, where $-A = \{-a \mid a \in A\}$. Also, for all $a, b, c, d \in R$ we have $(a + b) \cdot (c + d) \subseteq a \cdot c + b \cdot c + a \cdot d + b \cdot d$. In Definition 2.8, for simplicity of notations we write sometimes xy instead of $x \cdot y$ and in (c), $0 + x = x$ instead of $0 + x = \{x\}$.

A Krasner hyperring $(R, +, \cdot)$ is called *commutative (with unit element)* if $(R, \cdot)$ is a commutative semigroup (with unit element).

Example 2.4.

(1) Let $R = \{0, 1, 2\}$ be a set with the hyperoperation $+$ and the binary operation $\cdot$ defined as follow:

+	0	1	2
0	0	1	2
1	1	1	R
2	2	R	2

and

$\cdot$	0	1	2
0	0	0	0
1	0	1	2
2	0	1	2

Then, $(R, +, \cdot)$ is a Krasner hyperring.

(2) The first construction of a hyperring appeared in Krasner's paper and it is the following one: Consider $(F, +, \cdot)$ a field, G a subgroup of $(F^*, \cdot)$ and take $F/G = \{aG \mid a \in F\}$ with the hyperaddition and the multiplication given by

$$aG \oplus bG = \{cG \mid c \in aG + bG\},$$
$$aG \odot bG = abG.$$

Then, $(F/G, \oplus, \odot)$ is a hyperring. If $(F, +, \cdot)$ is a unitary ring and G is a subgroup of the monoid $(F^*, \cdot)$ such that $xG = Gx$, for all $x \in F$, then $(F/G, \oplus, \odot)$ is a Krasner hyperring with identity.

(3) Let $(A, +, \cdot)$ be a ring and N a normal subgroup of its multiplicative semigroup. Then, the multiplicative classes $\overline{x} = xN$ $(x \in A)$ form a partition of R, and let $\overline{A} = A/N$ be the set of these classes. If for all $\overline{x}, \overline{y} \in \overline{A}$, we define

$$\overline{x} \oplus \overline{y} = \{\overline{z} \mid z \in \overline{x} + \overline{y}\}, \quad \text{and} \quad \overline{x} * \overline{y} = \overline{x \cdot y},$$

then the obtained structure is a Krasner hyperring.

(4) Let R be a commutative ring with identity. We set $\overline{R} = \{\overline{x} = \{x, -x\} \mid x \in R\}$. Then, $\overline{R}$ becomes a Krasner hyperring with respect to the hyperoperation $\overline{x} \oplus \overline{y} = \{\overline{x+y}, \ \overline{x-y}\}$ and multiplication $\overline{x} \otimes \overline{y} = \overline{x \cdot y}$.

Definition 2.9. Let $(R, +, \cdot)$ be a Krasner hyperring and A be a non-empty subset of R. Then, A is said to be a *subhyperring* of R if $(A, +, \cdot)$ is itself a Krasner hyperring.

The subhyperring A of R is *normal* in R if and only if $x + A - x \subseteq A$ for all $x \in R$.

Definition 2.10. A subhyperring A of a Krasner hyperring R is a *left (right) hyperideal* of R if $r \cdot a \in A$ $(a \cdot r \in A)$ for all $r \in R$, $a \in A$. A is called a *hyperideal* if A is both a left and a right hyperideal.

Lemma 2.2. *A non-empty subset A of a Krasner hyperring R is a left (right) hyperideal if and only if*

(1) $a, b \in A$ *implies* $a - b \subseteq A$,
(2) $a \in A$, $r \in R$ *imply* $r \cdot a \in A$ $(a \cdot r \in A)$.

Proof. It is straightforward. ∎

Definition 2.11. Let A and B be nonempty subsets of a Krasner hyperring R.

- The sum $A + B$ is defined by
$$A + B = \{x \mid x \in a + b \text{ for some } a \in A, \ b \in B\}.$$

- The product AB is defined by
$$AB = \left\{x \mid x \in \sum_{i=1}^{n} a_i b_i, a_i \in A, b_i \in B, n \in \mathbb{Z}^+\right\}.$$

If A and B are hyperideals of R, then $A + B$ and AB are also hyperideals of R.

The second type of a hyperring was introduced by Rota. The multiplication is a hyperoperation, while the addition is an operation, that is why she called it a multiplicative hyperring.

Definition 2.12. A triple $(R, +, \cdot)$ is called a *multiplicative hyperring* if

(1) $(R, +)$ is an abelian group.
(2) $(R, \cdot)$ is a semihypergroup.
(3) For all $a, b, c \in R$, we have $a \cdot (b + c) \subseteq a \cdot b + a \cdot c$ and $(b + c) \cdot a \subseteq b \cdot a + c \cdot a$.
(4) For all $a, b \in R$, we have $a \cdot (-b) = (-a) \cdot b = -(a \cdot b)$.

If in (3) we have equalities instead of inclusions, then we say that the multiplicative hyperring is *strongly distributive*.

An element e in R, such that for all $a \in R$, we have $a \in a \cdot e \cap e \cdot a$, is called a *weak identity* of R.

Example 2.5.

(1) Let $(R, +, \cdot)$ be a ring and I be an ideal of it. We define the following hyperoperation on R: For all $a, b \in R$, $a * b = a \cdot b + I$. Then, $(R, +, *)$ is a strongly distributive hyperring. Indeed, first of all, $(R, +)$ is an abelian group. Then, for all $a, b, c \in R$, we have
$$a*(b*c) = a*(b{\cdot}c+I) = \bigcup_{h \in I} a*(b{\cdot}c+h) = \bigcup_{h \in I} a{\cdot}(b{\cdot}c+h)+I = a{\cdot}b{\cdot}c+I$$
and similarly, we have $(a * b) * c = a \cdot b \cdot c + I$. Moreover, for all $a, b, c \in R$, we have $a*(b+c) = a{\cdot}(b+c)+I = a{\cdot}b+a{\cdot}c+I = a*b+a*c$ and similarly, we have $(b + c) * a = b * a + c * a$. Finally, for all $a, b \in R$, we have $a * (-b) = a \cdot (-b) + I = (-a) \cdot b + I = (-a) * b$ and $-(a * b) = (-a \cdot b) + I = (-a) \cdot b + I = a * (-b)$.
(2) Let $(R, +, \cdot)$ be a non-zero ring. For all $a, b \in R$ we define the hyperoperation $a * b = \{a \cdot b, 2a \cdot b, 3a \cdot b, \ldots\}$. Then, $(R, +, *)$ is a multiplicative hyperring, which is not strongly distributive. Notice that for all $a \in R$, we have $a * 0 = 0 * a = \{0\}$.

Proposition 2.1. *If $(R, +, \cdot)$ is a multiplicative hyperring, then for all $a, b, c \in R$,*
$$a \cdot (b - c) \subseteq a \cdot b - a \cdot c \quad and \quad (b - c) \cdot a \subseteq b \cdot a - c \cdot a.$$

If $(R, +, \cdot)$ is a strongly distributive, then for all $a, b, c \in R$,

$$a \cdot (b - c) = a \cdot b - a \cdot c \text{ and } (b - c) \cdot a = b \cdot a - c \cdot a.$$

Proof. The statement follows from the conditions (3) and (4) of Definition 2.12. ∎

Proposition 2.2. *In a strong distributive hyperring $(R, +, \cdot)$, we have $0 \in a \cdot 0$ and $0 \in 0 \cdot a$, for all $a \in R$.*

Proof. The statement follows from the above proposition, by considering $b = c$. ∎

Theorem 2.10. *For a strongly distributive hyperring $(R, +, \cdot)$, the following statements are equivalent:*

(1) *There exists $a \in R$ such that $|0 \cdot a| = 1$.*
(2) *There exists $a \in R$ such that $|a \cdot 0| = 1$.*
(3) *$|0 \cdot 0| = 1$.*
(4) *$|a \cdot b| = 1$, for all $a, b \in R$.*
(5) *$(R, +, \cdot)$ is a ring.*

Proof. (2⇒3): Suppose $a \neq 0$. For all $a \in R$ we have $0 \cdot 0 = (a - a) \cdot 0 = a \cdot 0 - a \cdot 0$ and so by (2), it follows that $0 \cdot 0 = \{0\}$, whence we obtain (3).

(3⇒4): For all $a \in R$, we have $0 \cdot 0 = a \cdot 0 - a \cdot 0$ and so by (3) it follows that $|a \cdot 0| = 1$, otherwise if we suppose that there exist $x \neq y$ elements of $a \cdot 0$, then $0 \cdot 0$ would contain $x - y \neq 0$ and 0, a contradiction. On the other hand, for all $a, b \in R$ we have $a \cdot 0 = a \cdot (b - b) = a \cdot b - a \cdot b$, whence it follows that $a \cdot b$ contains only an element. The other implications (4⇒5) and (5⇒2) are immediate. Similarly, the condition (1) is equivalent to (3), (4) and (5). ∎

Corollary 2.3. *A strongly distributive hyperring $(R, +, \cdot)$ is a ring if and only if there exist $a_0, b_0 \in R$ such that $|a_0 \cdot b_0| = 1$.*

Proof. According to the above theorem, it is sufficient to check that $|a_0 \cdot 0| = 1$. We have $a_0 \cdot 0 = a_0 \cdot (b_0 - b_0) = a_0 \cdot b_0 - a_0 \cdot b_0$, whence we obtain that $a_0 \cdot 0$ contains only 0. ∎

Notice that there exist multiplicative hyperrings, which are not strongly distributive and for which we have $a * 0 = \{0\}$ for all $a \in R$.

Definition 2.13. A hyperring $(R, +, \cdot)$ is called *unitary* if it contains an element u, such that $a \cdot u = u \cdot a = \{a\}$ for all $a \in R$.

We obtain the following result.

Theorem 2.11. *Every unitary strongly distributive hyperring $(R, +, \cdot)$ is a ring.*

Proof. If u is the unit element, then we have $u \cdot u = \{u\}$ and according to the above corollary, it follows that R is a ring. ∎

Theorem 2.12. *In any multiplicative hyperring $(R, +, \cdot)$, if there are $a, b \in R$ such that $|a \cdot b| = 1$, then $0 \cdot 0 = \{0\}$.*

Proof. We have $a \cdot 0 = a \cdot (b - b) \subseteq a \cdot b - a \cdot b = \{0\}$. On the other hand, $0 \cdot 0 = (a - a) \cdot 0 \subseteq a \cdot 0 - a \cdot 0$. But this must also be $\{0\}$, since $a \cdot 0$ is a singleton. ∎

Corollary 2.4. *In any unitary multiplicative hyperring $(R, +, \cdot)$, we have $0 \cdot 0 = \{0\}$.*

Definition 2.14. Let $(R, +, \cdot)$ be a multiplicative hyperring and H be a non-empty subset of R. We say that H is a *subhyperring* of $(R, +, \cdot)$ if $(H, +, \cdot)$ is a multiplicative hyperring.

In other words, H is a subhyperring of $(R, +, \cdot)$ if $H - H \subseteq H$ and for all $x, y \in H$, $x \cdot y \subseteq H$.

Definition 2.15. We say that H is a *hyperideal* of $(R, +, \cdot)$ if $H - H \subseteq H$ and for all $x, y \in H$, $r \in R$, $x \cdot r \cup r \cdot x \subseteq H$.

The intersection of two subhyperrings of a multiplicative hyperring $(R, +, \cdot)$ is a subhyperring of R. The intersection of two hyperideals of a multiplicative hyperring $(R, +, \cdot)$ is a hyperideal of R. Moreover, any intersection of subhyperrings of a multiplicative hyperring is a subhyperring, while any intersection of hyperideals of a multiplicative hyperring is a hyperideal. In this manner, we can consider the *hyperideal generated by any subset S* of $(R, +, \cdot)$, which is the intersection of all hyperideals of R, which contain S.

For each multiplicative hyperring $(R, +, \cdot)$, the *zero hyperideal* is the hyperideal generated by the additive identity 0. Contrary to what happens in ring theory, the zero hyperideal can contain other elements than 0. If we denote the zero hyperideal of R by $< 0 >$, then we have

$$\langle 0 \rangle = \Big\{ \sum_i x_i + \sum_j y_j + \sum_k z_k \mid \text{each sum is finite and for each } i, j, k \text{ there}$$

$$\text{exist } r_i, s_j, t_k, u_k \in R \text{ such that } x_i \in r_i \cdot 0, \; y_j \in 0 \cdot s_j, \; z_k \in t_k \cdot 0 \cdot u_k \Big\}.$$

Denote by $H \oplus K$ the hyperideal generated by $H \cup K$, where H and K are hyperideals of $(R, +, \cdot)$.

Theorem 2.13. *If H and K are hyperideals of R, then*

$$H \oplus K = \{h + k \mid h \in H, k \in K\}.$$

Proof. Denote the set $\{h + k \mid h \in H, k \in K\}$ by I. Then, I is a hyperideal of R, which contains H and K.

Moreover, if J is a hyperideal of R, containing H and K, then $I \subseteq J$. Hence, we have $I = H \oplus K$. ■

Notice that the above theorem can be extended to an whichever family of hyperideals.

If we denote by $\mathcal{I}$ the set of all hyperideals of a multiplicative hyperring $(R, +, \cdot)$, then $(\mathcal{I}, \subseteq)$ is a complete lattice. The infimum of any family of hyperideals is their intersection, while the supremum is the hyperideal generated by their union.

Definition 2.16. A *homomorphism (good homomorphism)* between two multiplicative hyperrings $(R, +, \circ)$ and $(R', +', \circ')$ is a map $f : R \to R'$ such that for all x, y of R, we have $f(x + y) = f(x) +' f(y)$ and $f(x \circ y) \subseteq f(x) \circ' f(y)$ $(f(x \circ y) = f(x) \circ' f(y)$ respectively).

The following definition introduces a hyperring in general form. Both addition and multiplication are hyperoperations, that satisfy a set of conditions.

Definition 2.17. A hyperringoid $(H, \oplus, \odot)$ is called a *hyperring* if the following conditions are satisfied.

    (1) $(H, \oplus)$ is a commutative hypergroup.
    (2) $(H, \odot)$ is a semihypergroup.
    (3) For all $x, y, z \in H, (x \oplus y) \odot z = (x \odot z) \oplus (y \odot z), z \odot (x \oplus y) = (z \odot x) \oplus (z \odot y)$.
    (4) For all $x \in H$ and all $u \in \omega_{(H, \oplus)}, x \odot u \subseteq \omega_{(H, \oplus)} \supseteq u \odot x$.

Definition 2.18. Let R be a hyperring. We define the relation γ as follows: $x \gamma y \Leftrightarrow \exists n \in \mathbb{N}, \exists k_i \in \mathbb{N}, \exists (x_{i1}, \ldots, x_{ik_i}) \in R^{k_i}, 1 \le i \le n$, such that

$$\{x, y\} \subseteq \sum_{i=1}^{n} \left(\prod_{j=1}^{k_i} x_{ij} \right).$$

Definition 2.19. Let R be a hyperring. We consider the relation α as follows:

$x \, \alpha \, y \Leftrightarrow \exists n \in \mathbb{N}, \, \exists(k_1, \ldots, k_n) \in \mathbb{N}^n, \, \exists \sigma \in \mathbb{S}_n$ and $[\exists(x_{i1}, \ldots, x_{ik_i}) \in R^{k_i}, \, \exists \sigma_i \in \mathbb{S}_{k_i}, \, (i = 1, \ldots, n)]$ such that

$$x \in \sum_{i=1}^{n} \left(\prod_{j=1}^{k_i} x_{ij} \right) \text{ and } y \in \sum_{i=1}^{n} A_{\sigma(i)},$$

where $A_i = \prod_{j=1}^{k_i} x_{i\sigma_i(j)}$.

The relation α and γ are reflexive and symmetric. Let α^* and γ^* be the transitive closure of α and γ.

Theorem 2.14. *Let* $(R, +, \cdot)$ *be a hyperring.*

(1) γ^* *is the smallest equivalence relation on* R *such that the quotient* R/γ^* *is a ring.* R/γ^* *is called the fundamental ring.*

(2) α^* *is the smallest equivalence relation on* R *such that the quotient* R/α^* *is a commutative ring.* R/α^* *is called the commutative fundamental ring.*

2.4 Hypermodules

Definition 2.20. A non-empty set M is a *hypermodule* over a hyperring R (R-hypermodule) if $(M, +)$ is a canonical hypergroup and there exists a map $\cdot : R \times M \to \mathcal{P}^*(M)$ by $(r, m) \mapsto rm$ such that for all $r_1, r_2 \in R$ and $m_1, m_2 \in M$, we have

(1) $r_1(m_1 + m_2) = r_1 m_1 + r_2 m_2$,
(2) $(r_1 + r_2)m_1 = r_1 m_1 + r_2 m_1$,
(3) $(r_1 r_2)m_1 = r_1(r_2 m_1)$.

Example 2.6.

(1) Let A be an ordinary ring and M be an R-module. If M' is a submodule of M and one defines the following scalar hyperoperation

$$\forall(a, x) \in A \times M, \, a \circ x = ax + M'$$

then M is an R-hypermodule.

(2) Let $M = \mathbb{R}^2$ with the following hyperoperation

$$a \cdot (x, y) = \{(u, v) \in \mathbb{R}^2 \mid xv = yv\}$$

for every $a \in \mathbb{R}$ and for every $(x, y) \in \mathbb{R}^2 - \{(0,0)\}$ and

$$a \cdot (0,0) = \{(0,0)\}.$$

Then, $\mathbb{R}^2$ is an $\mathbb{R}$-hypermodule.

Let A be a non-empty subset of an R-hypermodule M. Then A is called a *subhypermodule* of M if A is itself a hypermodule. A subhypermodule A of M is *normal* in M if $x + A - x \subseteq A$ for all $x \in M$.

Example 2.7. Let $R = \{0, 1, 2\}$ be a set with hyperoperation $+$ and binary operation $\cdot$ as follow:

+	0	1	2
0	0	1	2
1	1	1	R
2	2	R	2

and

$\cdot$	0	1	2
0	0	1	2
1	0	1	2
2	0	1	2

Then $(R, +, \cdot)$ is a hyperring. Let $M = \{0, 1, 2, 3, 4, 5, 6, 7, 8\}$ be a set with the hyperoperation as follows:

$\oplus$	0	1	2	3	4	5	6	7	8
0	0	1	2	3	4	5	6	7	8
1	1	1	{0,1,2}	4	4	{3,4,5}	7	7	{6,7,8}
2	2	{0,1,2}	2	5	{3,4,5}	5	8	{6,7,8}	8
3	3	4	5	3	4	5	{0,3,6}	{1,4,7}	{2,5,8}
4	4	4	{3,4,5}	4	4	{3,4,5}	{1,4,7}	{1,4,7}	M
5	5	{3,4,5}	5	5	{3,4,5}	5	{2,5,8}	M	{2,5,8}
6	6	7	8	{0,3,6}	{1,4,7}	{2,5,8}	6	7	8
7	7	7	{6,7,8}	{1,4,7}	{1,4,7}	M	7	7	{6,7,8}
8	8	{6,7,8}	8	{2,5,8}	M	{2,5,8}	8	{6,7,8}	8

Then $(M, \oplus)$ is a canonical hypergroup. Now, we define the external product from $R \times M \to M$ as follows:

$\otimes$	0	1	2	3	4	5	6	7	8
0	0	1	2	3	4	5	6	7	8
1	0	1	2	3	4	5	6	7	8
2	0	1	2	3	4	5	6	7	8

Then $(M, \oplus, \otimes)$ is an R-hypermodule. Clearly $A = \{0, 3, 6\}$ is a normal subhypermodule of M.

Let A be a subhypermodule of an R-hypermodule M. Then the quotient hypergroup $M/A = \{m + A \mid m \in M\}$, with external composition $R \times M/A \longrightarrow M/A$, $(r, m + A) \longrightarrow rm + A$ is an R-hypermodule, and it is called the *quotient R-hypermodule* of M by A.

Many authors worked on hypermodules. Here, we present some results from [49]. Let M be a hypermodule and N be a non-empty subset of M. Then, N is called a *subhypermodule* of M if $(N, +)$ is a canonical subhypergroup of $(M, +)$ and for every $r \in R$ and $n \in N$, $r \cdot n \subseteq N$. A subhypermodule N is called *normal* if for every $m \in M$, $m + N - m \subseteq N$. Let X be a subset of a hypermodule of M and $\{M_i \mid i \in I\}$ be the family of all subhypermodule of M which contain X. Then, $\bigcap_{i \in I} M_i$ is called the *hypermodule generated* by X. This hypermodule is denoted by $\langle X \rangle$. If $X = \{m_1, m_2, \ldots, m_n\}$, then the hypermodule $\langle X \rangle$ is denoted by $\langle m_1, m_2, \ldots, m_n \rangle$. Let M be an R-hypermodule, R_1 and M_1, M_2 be non-empty subsets of R and M, respectively. We define

$$R_1 \cdot M_1 = \Big\{ x \in M \mid x \in \sum_{i=1}^{n} r_i \cdot m_i, r_i \in R_1, m_i \in M_1, n \in \mathbb{N} \Big\},$$

$$M_1 + M_2 = \Big\{ x \in M \mid x \in m_1 + m_2, m_1 \in M_1, m_2 \in M_2 \Big\},$$

$$\mathbb{Z}X = \Big\{ m \in M \mid m \in \sum_{i=1}^{n} n_i x_i, n_i \in \mathbb{Z}, x_i \in X \Big\}.$$

Proposition 2.3. *Let M be an R-hypermodule and $X \subseteq M$. Then, $\langle X \rangle = \mathbb{Z}X + R \cdot X$.*

Definition 2.21. Let M be an R-hypermodule such that $(M, +)$ be an abelian group. Then, M is called *multiplicative hypermodule*.

If N is a subhypermodule of a hypermodule M, then we define the relation $m_1 \equiv m_2$ if and only if $m_1 \in m_2 + N$, for every $m_1, m_2 \in M$. This relation is denoted by $m_1 N^* m_2$.

Proposition 2.4. *Let N be a subhypermodule of hypermodule M. Then, N^* is an equivalence relation.*

Proof. Suppose that $m \in M$. Since 0 is neutral element and $0 \in N$, it follows that $m = m + 0 \in m + N$, the relation N^* is reflexive. Let $m_1, m_2 \in M$ and $m_1 N^* m_2$. Then, $m_1 \in m_2 + n$, for some $n \in N$. Hence, $m_2 \in m_1 - n \in m_1 + N$. So this relation is symmetric. Let $m_1, m_2, m_3 \in M$ such that $m_1 N^* m_2$ and $m_2 N^* m_3$. Then, for some $n_1, n_2 \in N$, $m_1 \in m_2 + n_1, m_2 \in m_3 + n_2$. So, $m_1 \in m_2 + n_1 \subseteq m_3 + n_1 + n_2 \subseteq m_3 + N$. Therefore, $m_1 N^* m_3$. This completes the proof. ∎

If I is a hyperideal of a hyperring R, then we define the relation with the following hyperoperations: $x \equiv y$ if and only if $x \in y + I$. This relation is denoted by xI^*y.

Proposition 2.5. *Let I be a hyperideal of R. Then, $[R : I^*]$ is a hyperring with the following hyperoperations:*

$$I^*(x) \oplus I^*(y) = \{I^*(z) \mid z \in I^*(x) + I^*(y)\},$$
$$I^*(x) \odot I^*(y) = \{I^*(z) \mid z \in I^*(x) \cdot I^*(y)\}.$$

Proof. The proof is straightforward. ∎

Theorem 2.15. *Let M be an R-hypermodule, I be an ideal of R and N be a subhypermodule of M. Then, $[M : N^*]$ is a $[R : I^*]$ hypermodule with the following hyperoperations:*

$$N^*(m_1) \oplus N^*(m_2) = \{N^*(m) \mid m \in N^*(m_1) + N^*(m_2)\},$$
$$I^*(r) \odot N^*(m) = \{N^*(m) \mid m \in I^*(r) \cdot N^*(m)\},$$

and $[M : N^*]$ is R-hypermodule with the following hyperoperations:

$$N^*(m_1) \oplus N^*(m_2) = \{N^*(m) \mid m \in N^*(m_1) + N^*(m_2)\},$$
$$r \odot N^*(m) = \{N^*(m) \mid m \in r \cdot N^*(m)\}.$$

Proof. The proof is straightforward. ∎

Let M be an R-hypermodule and N be a subhypermodule of M. Then, the zero element of $[M : N^*]$ is $\{N\}$ and $|\langle\{N\}\rangle| = 1$.

Proposition 2.6. *Let N be a normal subhypermodule of hypermodule M. Then, for every $m_1, m_2 \in M$ the following are equivalent:*

(1) $m_2 \in m_1 + N$,
(2) $m_1 - m_2 \subseteq N$,
(3) $(m_1 - m_2) \cap N \neq \emptyset$.

Proof. Suppose that $(m_1 - m_2) \cap N \neq \emptyset$. Then there exists $m \in (m_1 - m_2) \cap N$. So $-m_2 + m_1 \subseteq -m_2 + m + m_2 \subseteq N$. If $x \in -m_2 + m_1$, then $x \in N$. Hence, $-m_2 \in x - m_1$ and $m_2 \in m_1 - x \subseteq m_1 + N$. Therefore, (3) implies (1). It is easy to see that (1) implies (2) and (2) implies (3). ∎

Definition 2.22. Let M be an R-hypermodule and N be a subhypermodule of M. We denote $\Omega(N) = \{m \in M \mid m - m \subseteq N\}$.

Proposition 2.7. *Let M be an R-hypermodule and N be a subhypermodule of M. Then, $\Omega(N)$ is a subhypermodule of M and $N \subseteq \Omega(N)$.*

Proof. Since $N \neq \emptyset$, the set $\Omega(N)$ is non-empty. Let $m_1, m_2, m \in \Omega(N)$, $r \in R$, $x \in m_1 - m_2$ and $y \in r \cdot m$. Then,

$$x - x \subseteq (m_1 - m_2) - (m_1 - m_2) = (m_1 - m_1) + (m_2 - m_2) \subseteq N + N = N,$$
$$y - y \subseteq r \cdot m - r \cdot m = r \cdot (m - m) \subseteq N.$$

Hence, $m_1 - m_2 \subseteq \Omega(N)$ and $r \cdot m \subseteq \Omega(N)$. Moreover, for every $n \in N$, since N is a subhypermodule of M, $n - n \subseteq N$. Therefore, $\Omega(N)$ is a subhypermodule of M containing N. ∎

Proposition 2.8. *Let M be an R-hypermodule and $m_1, m_2 \in \Omega(\{0\})$. Then, $m_1 + m_2$ is a singleton set.*

Proof. The proof is straightforward. ∎

Proposition 2.9. *Let M be an hypermodule. Then, $\Omega(\{0\})$ is an abelian group and for every submodule M_1 of M, $M_1 \subseteq \Omega(\{0\})$.*

Proof. Suppose that $m_1, m_2 \in \Omega(\{0\})$ and $x, y \in m_1 + m_2$. Then

$$x - y \subseteq (m_1 + m_2) - (m_1 + m_2) = (m_1 - m_2) - (m_1 - m_2) = 0.$$

This implies that $m_1 + m_2$ is a singleton and $\Omega(\{0\})$ is a subgroup. Let M_1 be any subgroup of M and $x \in M_1$. Then, $x - x = \{0\}$. Hence $x \in \Omega(\{0\})$ and $M_1 \subseteq \Omega(\{0\})$. This completes the proof. ∎

Corollary 2.5. *Let M be an R-hypermodule and N be a subhypermodule of M. Then, N is normal if and only if $\Omega(N) = M$. Moreover, $(M, +)$ is an abelian group if and only if $\Omega(\{0\}) = M$.*

Let $H(M) = \{x \mid x \in m - m, \text{ for all } m \in M\}$.

Proposition 2.10. *Let M be an R-hypermodule and N be a subhypermodule of M. Then, N is normal if and only if $H(M) \subseteq N$.*

Proof. Suppose that N be a subhypermodule and $H(M) \subseteq N$. Then for every $m \in M$ and $n \in N$ we have $m + n - m = m - m + n \subseteq H(M) + n \subseteq N + N = N$. Hence, N is normal. Let N be a normal subhypermodule and $m \in M$. This implies that $m + 0 - m \subseteq m + N - m \subseteq N$. Hence $m - m \subseteq N$, for every $m \in M$. Therefore, $H(M) \subseteq N$. This completes the proof. ∎

Corollary 2.6. *Let M be an R-hypermodule. Then, $H(M)$ is the smallest normal subhypermodule of M.*

Corollary 2.7. *Let N_1 and N_2 be subhypermodules of M such that $N_1 \subseteq N_2$ and N_1 be normal subhypermodule. Then, N_2 is also normal.*

Corollary 2.8. *Let M be an R-hypermodule such that $\{0\}$ is normal. Then, all subhypermodules of M are normal.*

Theorem 2.16. *Let M be an R-hypermodule. Then, $(M, +)$ is abelian group if and only if $H(M) = \{0\}$.*

Proof.   We know that $(M, +)$ is abelian group if and only if $\Omega(\{0\}) = M$. Moreover, $\Omega(\{0\}) = M$ if and only if $\{0\}$ is a normal subhypermodule. Hence, $(M, +)$ is an abelian group if and only if $\{0\}$ is a normal subhypermodule of M. Since $H(M)$ is a smallest subhypermodule of M, then $\{0\}$ is normal if and only if $H(M) = \{0\}$. This completes the proof.   ■

Corollary 2.9. *Since $H(M)$ is the smallest normal subhypermodule of M, it follows that M is a module if and only if all subhypemodules of M are normal.*

Corollary 2.10. *Let N be a normal subhypermodule of hypermodule M. Then, the equivalence relation defined in Proposition 2.4, is a strongly regular relation. Hence $[M : N^*]$ is abelian group.*

Theorem 2.17. *Let M be an R-hypermodule and N be a normal subhypermodule of M. Then, $[M : N^*]$ is a multiplicative hypermodule.*

Proof.   Suppose that N is a normal subhypermodule of M. The zero element of this quotient hypermodule is $\{N\}$. Moreover, $\{N\}$ is normal. By Theorem 2.16, $[M : N^*]$ is a multiplicative hypermodule.   ■

Theorem 2.18. *Let M be a multiplicative hypermodule. Then, the following statements are equivalent:*

(1) *there exists $m \in M$ such that $|0 \cdot m| = 1$,*
(2) *there exists $r \in R$, such that $|r \cdot 0| = 1$,*
(3) *$|0 \cdot 0| = 1$,*
(4) *for all $r \in R, m \in M$, we have $|r \cdot m| = 1$.*

Proof.   ($2 \Rightarrow 3$) Suppose that $r \in R$. We have $0 \cdot 0 = (r - r) \cdot 0 = r \cdot 0 - r \cdot 0$ and by (2), it follows that $0 \cdot 0 = \{0\}$, whence we obtain (3).

($3 \Rightarrow 4$) Let $r \neq 0$ be an element of R. We have $0 \cdot 0 = (r - r) \cdot 0 = r \cdot 0 - r \cdot 0$. If there exists $x \neq y$ elements of $r \cdot 0$, then $0 \cdot 0$ would contain $x - y \neq 0$ and 0, and it is a contradiction. On the other hand, for every $r \in R$ and

$m \in M$, $r \cdot (m - m) = r \cdot m - r \cdot m$, whence it follows that $r \cdot m$ contains only an element. The other implications (4⇒1) are immediate. Similarly, the condition (1) is equivalent to (3) and (4). ∎

Proposition 2.11. *Let M be a multiplicative hypermodule. Then,*

(1) $0 \in r \cdot 0$, *for every $r \in R$,*

(2) $0 \in 0 \cdot m$, *for every $m \in M$,*

(3) *if there exist $r_0 \in R$ and $m_0 \in M$ such that $|r_0 \cdot m_0| = 1$, then $|0 \cdot 0| = 1$,*

(4) *if N is a subhypermodule of M, then for any element $N^*(m) \in [M : N^*]$, we have $|N^*(m) \odot N^*(0)| = 1$.*

Proof. By Theorem 2.18, we obtain (1), (2), (3) and (4). ∎

Proposition 2.12. *Let M be a multiplicative hypermodule over hyperring R. Then, the external hyperoperation $R \times M \to \mathcal{P}^*(M)$ is operation if and only if there exist $r_0 \in R$ and $m_0 \in M$ such that $|r_0 \cdot m_0| = 1$.*

Proof. By Theorem 2.18, it is sufficient to check that $|r_0 \cdot 0| = 1$. We have

$$r_0 \cdot 0 = r_0 \cdot (m_0 - m_0) = r_0 \cdot m_0 - r_0 \cdot m_0,$$

whence we obtain that $r_0 \cdot 0$ contain only 0. ∎

Corollary 2.11. *Let M be an R-hypermodule and N be a subhypermodule of M. Then, $[M : N^*]$ is also an R-hypermodule. Moreover, if N is a normal subhypermodule of M, then by the Corollary 2.10, $[M : N^*]$ is a multiplicative hypermodule and by Theorem 2.18, the external hyperoperation in this quotient is operation.*

Definition 2.23. A mapping $f : M \to M'$ is called a *homomorphism* if for all $a, b \in M$ and $r \in R$, we have

$$f(a + b) = f(a) + f(b), \quad f(ra) = rf(a) \text{ and } f(0) = 0.$$

Clearly, a homomorphism f is an isomorphism if f is both injective and surjective. We now write $M \cong M'$ if M is isomorphic to M'.

Let R be a hyperring and M be a hypermodule over R. We define the relation ε on M as follows:

$$x \varepsilon y \iff x, y \in \sum_{i=1}^{n} m_i'; \; m_i' = m_i \text{ or } m_i' = \sum_{j=1}^{n_i} \left(\prod_{k=1}^{k_{ij}} x_{ijk} \right) z_i,$$

$$m_i \in M, \quad x_{ijk} \in R, \quad z_i \in M.$$

The fundamental relation ε^* on M can be defined as the smallest equivalence relation such that the quotient M/ϵ^* be a module over the corresponding fundamental ring.

Chapter 3

H_v-groups

H_v-structures were introduced by Vougiouklis at the Fourth AHA congress (1990) [118]. The concept of an H_v-structure constitutes a generalization of the well-known algebraic hyperstructures (hypergroup, hyperring, hypermodule and so on). Actually some axioms concerning the above hyperstructures such as the associative law, the distributive law and so on are replaced by their corresponding weak axioms. Since the quotients of the H_v-structures with respect to the fundamental equivalence relations (β^*, γ^*, ϵ^*, etc.) are always ordinary structures.

The quotients of H_v-structure theory has been pursued in many directions by Vougiouklis, Davvaz, Spartalis, Dramalidis, Leoreanu-Fotea, Š. Hošková, and others. We invite the reader to consult the references for an in depth exposition of the theory and its applications.

3.1 H_v-groups and some examples

Definition 3.1. Let H be a non-empty set and $\circ : H \times H \to \mathcal{P}^*(H)$ be a hyperoperation. The " $\circ$ " in H is called *weak associative* if

$$x \circ (y \circ z) \cap (x \circ y) \circ z \neq \emptyset, \text{ for all } x, y, z \in H.$$

The " $\circ$ " is called *weak commutative* if

$$x \circ y \cap y \circ x \neq \emptyset, \text{ for all } x, y \in H.$$

The " $\circ$ " is called *strongly commutative* if

$$x \circ y = y \circ x, \text{ for all } x, y \in H.$$

The hyperstructure $(H, \circ)$ is called an H_v-*semigroup* if " $\circ$ " is weak associative. An H_v-semigroup is called an H_v-*group* if

$$a \circ H = H \circ a = H, \text{ for all } a \in H.$$

In an obvious way, the H_v-subgroup of an H_v-group is defined.

All the weak properties for hyperstructures can be applied for subsets. For example, if $(H, \circ)$ is a weak commutative H_v-group, then for all non-empty subsets A, B and C of H, we have

$$(A \circ B) \cap (B \circ A) \neq \emptyset \text{ and } A \circ (B \circ C) \cap (A \circ B) \circ C \neq \emptyset.$$

To prove this, one has simply to take one element of each set.

A motivation to study the above structures is given by the following examples.

Example 3.1. Let $(G, \cdot)$ be a group and ρ be an equivalence relation on G. In G/ρ, the set of quotient, consider the hyperoperation $\odot$ defined by $\rho(x) \odot \rho(y) = \{\rho(z) \mid z \in \rho(x) \cdot \rho(y)\}$, where $\rho(x)$ denotes the equivalence class of the element x. Then, $(G/\rho, \odot)$ is an H_v-group which is not always a hypergroup.

Example 3.2. On the set $\mathbb{Z}_{mn}$ consider the hyperoperation $\oplus$ defined by setting $0 \oplus m = \{0, m\}$ and $x \oplus y = x + y$ for all $(x, y) \in \mathbb{Z}_{mn}^2 - \{(0, m)\}$. Then $(\mathbb{Z}_{mn}, \oplus)$ is an H_v-group. $\oplus$ is weak associative but not associative, since taking $k \notin m\mathbb{Z}$ we have

$$(0 \oplus m) \oplus k = \{0, m\} \oplus k = \{k, m + k\},$$
$$0 \oplus (m \oplus k) = 0 \oplus (m + k) = \{m + k\}.$$

Moreover, it is weak commutative but not commutative.

Example 3.3. Consider the group $(\mathbb{Z}^n, +)$ and take $m_1, ..., m_n \in \mathbb{N}$. We define a hyperoperation $\oplus$ in $\mathbb{Z}^n$ as follows:

$$(m_1, 0, ..., 0) \oplus (0, 0, ..., 0) = \{(m_1, 0, ..., 0), (0, 0, ..., 0)\},$$
$$(0, m_2, ..., 0) \oplus (0, 0, ..., 0) = \{(0, m_1, ..., 0), (0, 0, ..., 0)\},$$
$$(0, 0, ..., m_n) \oplus (0, 0, ..., 0) = \{(0, 0, ..., m_n), (0, 0, ..., 0)\},$$

and $\oplus = +$ in the remaining cases. Then, $(\mathbb{Z}^n, \oplus)$ is an H_v-group.

Definition 3.2. Let $(H_1, \circ)$ and $(H_2, \star)$ be two H_v-groups. A map $f : H_1 \to H_2$ is called an H_v-*homomorphism* or a *weak homomorphism* if

$$f(x \circ y) \cap f(x) \star f(y) \neq \emptyset, \text{ for all } x, y \in H_1.$$

f is called an *inclusion homomorphism* if

$$f(x \circ y) \subseteq f(x) \star f(y), \text{ for all } x, y \in H_1.$$

Finally, f is called a *strong homomorphism* if

$$f(x \circ y) = f(x) \star f(y), \text{ for all } x, y \in H_1.$$

If f is onto, one to one and strong homomorphism, then it is called an *isomorphism*. Moreover, if the domain and the range of f are the same H_v-group, then the isomorphism is called an *automorphism*. We can easily verify that the set of all automorphisms of H, defined by $AutH$, is a group.

Several H_v-structures can be defined on a set H. A partial order on these hyperstructures can be introduced, as follows.

Definition 3.3. Let $(H, \circ)$ and $(H, \star)$ be two H_v-groups defined on the same set H. We say that $\circ$ *smaller than* $\star$, and $\star$ *greater than* $\circ$, and we write $\circ \leq \star$, if there is $f \in Aut(H, \star)$ such that $x \circ y \subseteq f(x \star y)$, for all $x, y \in H$.

If a hyperoperation is weak associative, then every greater hyperoperation, defined on the same set is also weak associative. In [122], the set of all H_v-groups with a scalar unit defined on a set with three elements is determined using this property.

The nth power of an element h, denoted h^s, is defined to be the union of all expressions of n times of h, in which the parentheses are put in all possible ways. An H_v-group $(H, \star)$ is called *cyclic* with finite period respect to $h \in H$, if there exists a positive integer s such that $H = h^1 \cup h^2 \cup ... \cup h^s$. The minimum of such s is called the *period of the generator h*. If all generators have the same period, then H is cyclic with period. If there exists $h \in H$ and s positive integer, the minimum one, such that $H = h^s$, then H is called *single-power cyclic* and h is a generator with single-power period s. The cyclicity in the infinite case is defined similarly. Thus, for example, the H_v-group $(H, \star)$ is called *single-power cyclic with infinite period* with generator h if every element of H belongs to a power of h and there exists $s_0 \geq 1$ such that for all $s \geq s_0$ we have $h^1 \cup h^2 \cup ... \cup h^{s-1} \subset h^s$. An element $a \in H$ is called *idempotent element* if $a^2 = a$.

3.2   Enumeration of H_v-groups

In [121], it is proved that all the quasi-hypergroups with two elements are H_v-groups. It is also proved that up to the isomorphism there are exactly 20 different H_v-groups.

Theorem 3.1. *All quasihypergroups with two elements are H_v-groups.*

Proof. One can see that on a set with two elements $H = \{a, b\}$, exactly eight hypergroups, up to isomorphism, can be defined. Besides the eight hypergroups, there are twelve more quasihypergroups, i.e., the reproduction axiom is valid. It is easy to see that the weak associativity is also valid for them, so they are H_v-groups. Therefore, the twenty H_v-groups, up to isomorphism, defined on a set $H = \{a, b\}$ are given in the following table.

H_1	H_2	H_3	H_4	H_5	H_6	H_7	H_8	H_9	H_{10}
a	H	H	H	a	H	H	H	H	a
b	a	H	a	H	H	H	H	b	H
b	a	a	H	H	H	H	H	b	b
a	b	b	b	b	a	b	H	a	a

H_{11}	H_{12}	H_{13}	H_{14}	H_{15}	H_{16}	H_{17}	H_{18}	H_{19}	H_{20}
b	H	H	H	H	a	H	H	H	b
H	H	b	a	a	H	H	H	b	H
a	b	H	a	b	H	a	b	a	H
b	a	a	H	H	a	H	H	H	a

In this table in each column the H_v-groups $H_1, H_2, \ldots, H_{20}$ are presented. The hyperproduct $a \circ a$, $a \circ b$, $b \circ a$ and $b \circ b$ are presented on the first, second, third and forth row, respectively.

In the above table, the first eight H_v-groups are hypergroups. Moreover, all the above, expect the H_{15} and H_{19} are weak commutative. ∎

Definition 3.4. An H_v-group is called H_b-group if it contains an operation which defines a group. In the reverse case, given a group, every greater hyperoperation defines an H_v-group or more precisely an H_b-group.

Example 3.4. Examples 3.2 and 3.3 are examples of H_b-groups.

In a given hypergroupoid, it is laborious to check the weak associativity. In the following, some properties are presented to reduce the cases of the triples of elements for which one needs to check the weak associativity.

Proposition 3.1. *Let $(H, \circ)$ be a weak commutative hypergroupoid. Then,*

$$a \circ (b \circ a) \cap (a \circ b) \circ a \neq \emptyset, \text{ for all } a, b \in H.$$

Proof. Since $(H, \circ)$ is weak commutative, consider $z \in a \circ b \cap b \circ a$. Then, $z \circ a \cap a \circ z \neq \emptyset$. Therefore $a \circ (b \circ a) \cap (a \circ b) \circ a \neq \emptyset$. ∎

Proposition 3.2. *Let $(H, \circ)$ be a commutative (strong) hypergroupoid. Then, for all $a, b \in H$ we have*

$$a \circ (a \circ b) \cap (a \circ a) \circ b \neq \emptyset \text{ implies } b \circ (a \circ a) \cap (b \circ a) \circ a \neq \emptyset.$$

Proof. It is straightforward. Since $b \circ (a \circ a) = (a \circ a) \circ b$ and $(b \circ a) \circ a = a \circ (b \circ a) = a \circ (a \circ b)$, it follows that $b \circ (a \circ a) \cap (b \circ a) \circ a \neq \emptyset$. ∎

Remark 3.1. From Propositions 3.1 and 3.2 we see that the cases one has to check with respect to the weak associativity are:

(1) If $(H, \circ)$ is weak commutative, then for $a, b \in H$ with $a \neq b$, check the triples (a, a, b), (a, b, b), (b, a, a), (b, b, a) and every triple with three different elements.
(2) If $(H, \circ)$ is commutative (strong), then for $a, b \in H$ with $a \neq b$, check the triples (a, a, b), (b, b, a) and every triple with three different elements.

Recall that an element e of a hypergroupoid $(H, \circ)$, is called a *unit* if $h \in e \circ h \cap h \circ e$, for all $h \in H$. If $h = e \circ h = h \circ e$, for all $h \in H$, then e is called a *scalar unit*.

Proposition 3.3. *Let e be a unit of a hypergroupoid $(H, \circ)$. If $x, y, z \in H$, such that $e \in \{x, y, z\}$, then $(x \circ y) \circ z \cap x \circ (y \circ z) \neq \emptyset$.*

Proof. Suppose that $x = e$. Then, we have $y \circ z \subseteq (e \circ y) \circ z$ and $y \circ z \subseteq e \circ (y \circ z)$. Thus, we obtain $(e \circ y) \circ z \cap e \circ (y \circ z) \neq \emptyset$. Similarly, one can check the other cases. ∎

In this paragraph we find all H_v-groups with three elements which contain a scalar unit element e. Therefore, the set of all H_v-groups defined on the set $H = \{e, a, b\}$ up to the isomorphism is determined. Notice that the number of all hypergroupoids (non-degenerate) with a unit element, on a set with three elements, is $(2^3 - 1)^4 = 2401$.

Problem 3.1. *Find all H_v-groups with three elements which contain a scalar unit. Therefore, for $H = \{e, a, b\}$, consider the multiplication table*

$\circ$	e	a	b
e	e	a	b
a	a	1	2
b	b	3	4

and then find all the fours, of non-empty subsets of H, in the place of the four $(1, 2, 3, 4)$, such that the reproduction axiom and the weak associativity are valid.

In the set H only one automorphism can be defined:

$$f : e \leftrightarrow e, \ a \leftrightarrow b.$$

So, two H_v-groups, $(H, \circ)$ and $(H, \star)$, are isomorphic if

$$a \star a = f(b \circ b), \ a \star b = f(b \circ a), \ b \star a = f(a \circ b) \text{ and } b \star b = f(a \circ a).$$

Thus, the fours $(a \circ a, a \circ b, b \circ a, b \circ b)$ and $(f(b \circ b), f(b \circ a), f(a \circ b), f(a \circ a))$ are isomorphic. In other words, instead of the table

$$\begin{array}{|c|c|} \hline 1 & 2 \\ \hline 3 & 4 \\ \hline \end{array}, \ \text{take the table} \ \begin{array}{|c|c|} \hline 4 & 3 \\ \hline 2 & 1 \\ \hline \end{array}$$

and then replace a by b and b by a. The result is the isomorphic H_v-group to the first one.

Theorem 3.2. *Let $H = \{e, a, b\}$ be a set with three elements. Consider the set $\mathcal{U}$ of all quasihypergroups $(H, \circ)$ defined in H, with e to be the scalar unit. Then the subset of all with all not weak associative quasihypergroups is described as follows. Denote by a four $(a \circ a, a \circ b, b \circ a, b \circ b)$ the only one product needed to be defined. Then the fours which are not weak associative, with the corresponding isomorphic to the fours, are the following:*

$$(e, \{e, b\}, \{e, b\}, a) \cong (b, \{e, a\}, \{e, a\}, e),$$
$$(e, \{e, b\}, H, a) \cong (b, H, \{e, a\}, e),$$
$$(e, H, \{e, b\}, a) \cong (b, \{e, a\}, H, e),$$
$$(\{e, b\}, e, a, \{e, a\}) \cong (\{e, b\}, b, e, \{e, a\}),$$
$$(\{e, b\}, e, a, H) \cong (H, b, e, \{e, a\}),$$
$$(\{e, b\}, e, \{a, b\}, \{e, a\}) \cong (\{e, b\}, \{a, b\}, e, \{e, a\}),$$
$$(\{e, b\}, e, \{a, b\}, H) \cong (H, \{a, b\}, e, \{e, a\}),$$
$$(\{e, b\}, \{e, b\}, a, \{e, a\}) \cong (\{e, b\}, b, \{e, a\}, \{e, a\}),$$
$$(\{e, b\}, \{e, b\}, a, H) \cong (H, b, \{e, a\}, \{e, a\}),$$
$$(\{e, b\}, e, b, \{e, a\}) \cong (\{e, b\}, a, e, \{e, a\}),$$
$$(\{e, b\}, e, b, H) \cong (H, a, e, \{e, a\}),$$
$$(\{e, b\}, \{e, a\}, b, \{e, a\}) \cong (\{e, b\}, a, \{e, b\}, \{e, a\}),$$
$$(\{e, b\}, a, b, \{e, a\})$$
$$(\{e, b\}, a, b, H) \cong (H, a, b, \{e, a\}),$$
$$(\{e, b\}, a, \{e, b\}, H) \cong (H, \{e, a\}, b, \{e, a\}),$$
$$(\{e, b\}, b, a, \{e, a\})$$
$$(\{e, b\}, b, a, H) \cong (H, b, a, \{e, a\}).$$

Proof. To prove this theorem the following algorithm was applied: In position 1 of the four $(1, 2, 3, 4)$, the sets $\{e\}, \{a\}, \{b\}, \{e, a\}, \{e, b\}, \{a, b\}$ and H, were consequently put. Then for each case, positions $2, 3, 4$ were completed in order to obtain small hyperoperations such that the reproduction axiom is valid. In each case it was checked if the weak associativity was valid taking into account Remark 3.1. If the weak associativity was valid, then the result was considered a "minimal" H_v-group. If the weak associativity was not valid, then the sets in positions $2, 3, 4$ were enlarged by adding new elements. Again the weak associativity was checked and this algorithm was continued. The procedure was stopped if the weak associativity was valid. In each result, whether it was weak associative or not, the isomorphic, reproductive, hyperoperation was obtained and they were used in the next steps. In each enlargement of the hyperoperations it was checked if a weak associative hyperoperation from the ones obtained previously, was already contained. The results are the ones written in the theorem.   ∎

Within the same procedure the following, "complement", theorem is proved.

Theorem 3.3. *The set of all H_v-groups, with a scalar unit e, defined on a set with three $H = \{e, a, b\}$ is the following: Every greater four $(a \circ a, a \circ b, b \circ a, b \circ b)$ than the ones, with the corresponding isomorphic ones, defines an H_v-group in H.*

M_1 : $(e, b, b, \{e, a\}) \cong (\{e, b\}, a, a, e)$,
M_2 : $(e, \{e, b\}, \{e, b\}, \{a, b\}) \cong (\{a, b\}, \{e, a\}, \{e, a\}, e)$,
M_3 : $(e, H, H, a) \cong (b, H, H, e)$,
M_4 : $(e, \{a, b\}, \{a, b\}, e)$,
M_5 : $(e, H, H, b) \cong (a, H, H, e)$,
M_6 : $(a, \{e, b\}, \{e, b\}, a) \cong (b, \{e, a\}, \{e, a\}, b)$,
M_7 : (a, H, H, b),
M_8 : (b, e, e, a),
M_9 : $(\{e, b\}, \{e, a\}, b, H) \cong (H, a, \{e, b\}, \{e, a\})$,
M_{10} : $(\{e, b\}, b, \{e, a\}, H) \cong (H, \{e, b\}, a, \{e, a\})$.

An H_v-group is called *minimal* if it contains no other H_v-group defined on the same set H. For a given H_v-group, not minimal, one can remove an element from one product $x \circ y$, with $x, y \in H$, and then to check if the remaining hyperproduct gives an H_v-group. So, a series can be obtained, and if H is finite, leads to a minimal H_v-group. Therefore, the above set of fours, is the set of minimal ones of all H_v-groups with three elements which have a scalar unit.

Remark 3.2. One can classify all the minimals according to their fundamental group H/β^*. For this classification we obtain the following: The H_v-group M_8 has $H/\beta^* = \{\{e\}, \{a\}, \{b\}\}$, so it has three elements in the fundamental group. The H_v-group M_1 has $H/\beta^* = \{\{e, a\}, \{b\}\}$, so it has two elements in the fundamental groups. The rest of the minimals have one element in the fundamental group.

Remark 3.3. The H_v-group M_1 is a very thin hypergroup. There is also the very thin H_v-group of the second kind. This has the four (e, b, b, H) and contains the M_1. The other very thin H_v-groups can be obtained from the group M_8 by setting in one of the positions $1, 2, 3$ or 4 greater subsets. These are H_v-groups and, under the isomorphism, we have the fours:

$$(\{e, b\}, e, e, a), \ (\{a, b\}, e, e, a), \ (H, e, e, a),$$
$$(b, \{e, a\}, e, a), \ (b, \{e, b\}, e, a), \ (b, H, e, a).$$

Therefore, in the set of all H_v-groups on $H = \{e, a, b\}$, there are exactly 8 very thin hypergroups.

Remark 3.4. From the minimals only the M_9 and M_{10} are not commutative. Enlarging them, one can obtain only two more non-commutative in the set $H = \{e, a, b\}$. Therefore, in the set of all H_v-groups in H, with e scalar unit, there are only four non-commutative H_v-groups: the M_9, M_{10} and $(H, \{e, a\}, b, H)$, $(H, b, \{e, a\}, H)$.

3.3   Fundamental relation on H_v-groups

Let $(H, \circ)$ be an H_v-group. The relation β^* is the smallest equivalence relation on H such that the quotient H/β^* is a group. β^* is called the *fundamental equivalence relation* on H. The relation β^* was introduced on hypergroups by Koskas [64] and studied mainly by Corsini [16] and Davvaz [30, 29]. Vougiouklis in [120] studied the relation β^* on H_v-groups.

If $\mathcal{U}$ denotes the set of all finite products of elements of H, then a relation β can be defined on H whose transitive closure is the fundamental relation β^*. The relation β is defined as follows: for x and y in H we write $x\beta y$ if and only if $\{x, y\} \subseteq u$ for some $u \in \mathcal{U}$. The following theorem was proved by Vougiouklis [120].

Theorem 3.4. *The fundamental relation* β^* *is the transitive closure of the relation* β.

Proof. Suppose that $\hat{\beta}$ is the transitive closure of β and $\hat{\beta}(a)$ is the class of the element $a \in H$. First, we prove that the quotient set $H/\hat{\beta}$ is a group. The product $\odot$ in $H/\hat{\beta}$ is defined in the usual manner

$$\hat{\beta}(a) \odot \hat{\beta}(b) = \{\hat{\beta}(c) \mid c \in \hat{\beta}(a) \circ \hat{\beta}(b)\},$$

for all $a, b \in H$. Take $a' \in \hat{\beta}(a)$ and $b' \in \hat{\beta}(b)$. Then, we have

$a'\hat{\beta}a \Leftrightarrow$ there exist $x_1, ..., x_{m+1}$ with $x_1 = a'$, $x_{m+1} = a$ and $u_1, ..., u_m \in \mathcal{U}$
   such that $\{x_i, x_{i+1}\} \subseteq u_i$, for $i = 1, ..., m$

and

$b'\hat{\beta}b \Leftrightarrow$ there exist $y_1, ..., y_{n+1}$ with $y_1 = b'$, $y_{n+1} = b$ and $v_1, ..., v_n \in \mathcal{U}$
   such that $\{y_j, y_{j+1}\} \subseteq v_j$, for $j = 1, ..., n$.

Thus, we obtain

$$\{x_i, x_{i+1}\} \circ y_1 \subseteq u_i \circ v_1, \text{ for } i = 1, ..., m-1, \tag{3.1}$$

$$x_{m+1} \circ \{y_j, y_{j+1}\} \subseteq u_m \circ v_j, \text{ for } j = 1, ..., n. \tag{3.2}$$

Therefore, we obtain

$u_i \circ v_1 = t_i \in \mathcal{U}$, for $i = 1, ..., m-1$ and $u_m \circ v_j = t_{m+j-1} \in \mathcal{U}$, for $j = 1, ..., n$.

So, $t_k \in \mathcal{U}$, for all $k \in \{1, ..., m + n - 1\}$. Now, pick up any elements $z_1, ..., z_{m+n}$ such that

$z_i \in x_i \circ y_1$, for $i = 1, ..., m$ and $z_{m+j} \in x_{m+1} \circ y_{j+1}$, for $j = 1, ..., n$.

By using Equations 3.1 and 3.2, we have

$$\{z_k, z_{k+1}\} \subseteq t_k, \text{ for } k = 1, ..., m + n - 1.$$

So, every element $z_1 \in x_1 \circ y_1 = a' \circ b'$ is $\hat{\beta}$ equivalent to every element $z_{m+n} \in x_{m+1} \circ y_{n+1} = a \circ b$. Therefore, $\hat{\beta}(a) \odot \hat{\beta}(b)$ is singleton. Hence, we can write

$$\hat{\beta}(a) \odot \hat{\beta}(b) = \hat{\beta}(c), \text{ for all } c \in \hat{\beta}(a) \circ \hat{\beta}(b).$$

Moreover, since $\circ$ is weak associative, it follows that $\odot$ is associative, and consequently, $H/\hat{\beta}$ is a group.

Now, let σ be an equivalence relation in H such that H/σ is a group. Denote $\sigma(a)$ the class of a. Then, $\sigma(a) \odot \sigma(b)$ is singleton for all $a, b \in H$, i.e.,

$$\sigma(a) \odot \sigma(b) = \sigma(c), \text{ for all } c \in \sigma(a) \circ \sigma(b).$$

But also, for every $a, b \in H$ and $A \subseteq \sigma(a)$, $B \subseteq \sigma(b)$, we have

$$\sigma(a) \odot \sigma(b) = \sigma(a \circ b) = \sigma(A \circ B).$$

So, the above relation is valid for all finite products. That means that

$$\sigma(x) = \sigma(u), \text{ for all } u \in \mathcal{U} \text{ and } x \in u.$$

Thus, for every $a \in H$, we obtain

$$x \in \beta(a) \;\Rightarrow\; x \in \sigma(a).$$

But σ is transitively closed, so we have

$$x \in \hat{\beta}(a) \;\Rightarrow\; x \in \sigma(a).$$

Thus, the relation $\hat{\beta}$ is the smallest equivalence relation in H such that $H/\hat{\beta}$ is a group, i.e., $\hat{\beta} = \beta^*$.   ∎

We can rewrite the definition of β^* on H as follows.

  $a\beta^*b$ if and only if there exist $z_1, ..., z_{n+1} \in H$ with $z_1 = a$, $z_{n+1} = b$ and $u_1, ..., u_n \in \mathcal{U}$ such that $\{z_i, z_{i+1}\} \subseteq u_i$ $(i = 1, ..., n)$.

Suppose that $\beta^*(a)$ is the equivalence class containing $a \in H$. Then the product $\odot$ on H/β^* is defined as follows:

$$\beta^*(a) \odot \beta^*(b) = \beta^*(c) \text{ for all } c \in \beta^*(a) \circ \beta^*(b).$$

Since H/β^* is a group, one can say that $(H, \odot)$ is by virtue a group. This is why the hyperstructure H is called H_v-group.

Problem 3.2. *Freni in [55] proved that for hypergroups $\beta^* = \beta$. We do not know yet if this equality is valid for H_v-groups.*

Example 3.5. Consider the H_v-group defined in Example 3.2. From the sum

$$[\ldots (0 \oplus \underbrace{m) \oplus \ldots \oplus m]}_{n-1 \text{ times}} \oplus m = \{0, m, 2m, \ldots, (n-1)m\},$$

it is obtained that $\beta(0) = \{0, m, 2m, \ldots, (n-1)m\}$. Similarly, for every $0 < k < n-1$ we have

$$k \oplus [\ldots (0 \oplus \underbrace{m) \oplus \ldots \oplus m]}_{n-1 \text{ times}} \oplus m = \{k, k+m, k+2m, \ldots, k+(n-1)m\}.$$

So $\beta(k) = k + m\mathbb{Z}$. That means that $\beta^* = \beta$ and $\mathbb{Z}_{mn}/\beta^* \cong \mathbb{Z}_m$.

Example 3.6. Consider the H_v-group defined in Example 3.3. We have

$$\mathbb{Z}^n / \beta^* \cong \mathbb{Z}_{m_1} \times \mathbb{Z}_{m_2} \times \ldots \times \mathbb{Z}_{m_n}.$$

Let $(H, \circ)$ be an H_v-group. An element $x \in H$ is called *single* if its fundamental class is singleton, i.e., $\beta^*(x) = \{x\}$. Denote by S_H the set of all single elements of H.

Theorem 3.5. *Let $(H, \circ)$ be an H_v-group and $x \in S_H$. Let $a \in H$ and take any element $v \in H$ such that $x \in a \circ v$. Then,*

$$\beta^*(a) = \{h \in H \mid h \circ v = x\}.$$

Proof. We have $x \in a \circ v$. So, $x = a \circ v$ which means $x = \beta^*(a) \circ \beta^*(v)$. Thus, for all $h \in \beta^*(a)$ we have $h \circ v = x$. Conversely, let $x = h \circ v$. Then, $x = \beta^*(h) \circ \beta^*(v)$. Since H/β^* is a group, we have $\beta^*(h) = x \circ (\beta^*(v))^{-1} = \beta^*(a)$, so $h \in \beta^*(a)$. ∎

Theorem 3.6. *Let $(H, \circ)$ be an H_v-group and $x \in S_H$. Then, the core of H is $\omega_H = \{u \mid u \circ x = x\} = \{u \mid x \circ u = x\}$.*

Proof. It is obvious. ∎

Theorem 3.7. *Let $(H, \circ)$ be an H_v-group and $x \in S_H$. Then, $x \circ y = \beta^*(x \circ y)$ and $y \circ x = \beta^*(y \circ x)$, for all $y \in H$.*

Proof. Suppose that for some y there exist $t \in x \circ y$ and $t' \in \beta^*(t)$ such that $t' \notin x \circ y$. From the reproductivity, there exists $v \neq x$ in H such that $t' \in v \circ y$. So, we have $\beta^*(v) \circ \beta^*(y) = \beta^*(t')$. On the other hand, $\beta^*(x) \circ \beta^*(y) = \beta^*(t)$. Thus, $\beta^*(v) \circ \beta^*(y) = \beta^*(x) \circ \beta^*(y)$. Hence, $\beta^*(v) = \beta^*(x)$. Since $x \in S_H$, $\beta^*(x) = \{x\}$. Thus, $v = x$ which is a contradiction. ∎

The previous theorem proves that the product of a single element with any arbitrary element is always a whole fundamental class.

Suppose that $(H, \circ)$ is an H_v-group such that S_H is non-empty. Then, the only greater hyperoperations $\circ \; < \; \star$ for which the H_v-groups $(H, \star)$ contain single elements are the ones with the same fundamental group, since the fundamental classes are determined from the products of a single element with the elements of the group. On the other hand, a less hyperoperation $\diamond < \circ$ can have the same set S_H if only in the products of non-single elements the $\diamond$ is less than $\circ$. Finally, if ρ and σ are equivalence relations with $\rho < \sigma$ such that H/ρ and H/σ are non-equal groups, then they can not have both single elements.

Corollary 3.1. *Let* $(H, \circ)$ *be an* H_v-*group. If* S_H *is non-empty, then* $\beta^* = \beta$.

Proof. It is obvious, since all the β^*-classes can be obtained as products of two elements one of which is single.   ■

Using the fundamental equivalence relation, one can define semidirect hyperproducts of H_v-groups, see [120].

Theorem 3.8. *Let* $(A, \circ)$ *and* $(B, \star)$ *be two* H_v-*groups,* β^* *be the fundamental relation on* B *and let's take the group* $\text{Aut}A$*. Consider any homomorphism*

$$\widehat{} : B/\beta^* \to \text{Aut}A$$
$$\beta^*(b) \mapsto \widehat{\beta^*(b)}.$$

We denote $\widehat{\beta^*(b)} = \widehat{b}$*. In* $A \times B$*, we define a hyperproduct as follows:*

$$(a, b) \cdot (c, d) = \{(x, y) \mid x \in a \circ \widehat{b}(c), \ y \in b \star d\} = (a \circ \widehat{b}(c), b \star d).$$

Then, the set $A \times B$ *becomes an* H_v-*group.*

Proof. From the fundamental property, for all $x, y \in B$ we have

$$\widehat{x \star y} = \widehat{xy} = \widehat{z}, \text{ for all } z \in x \star y.$$

Now, suppose that $(a_1, b_1), (a_2, b_2), (a_3, b_3) \in A \times B$. Then,

$$\Big((a_1, b_1) \cdot (a_2, b_2)\Big) \cdot (a_3, b_3)$$
$$= \bigcup_{\substack{x \in a_1 \circ \widehat{b_1}(a_2) \\ y \in b_1 \star b_2}} (x, y) \cdot (a_3, b_3)$$
$$= \bigcup_{\substack{x \in a_1 \circ \widehat{b_1}(a_2) \\ y \in b_1 \star b_2}} \{(z, w) \mid z \in x \circ \widehat{y}(a_3), \ w \in y \star b_3\}$$
$$= \{(z, w) \mid z \in (a_1 \circ \widehat{b_1}(a_2)) \circ \widehat{b_1 \star b_2}(a_3), w \in (b_1 \star b_2) \star b_3\}$$
$$= \Big((a_1 \circ \widehat{b_1}(a_2)) \circ \widehat{b_1 \star b_2}(a_3), \ (b_1 \star b_2) \star b_3\Big).$$

On the other hand, we have

$$(a_1, b_1) \cdot \Big((a_2, b_2) \cdot (a_3, b_3)\Big)$$
$$= \bigcup_{\substack{x \in a_2 \circ \widehat{b_2}(a_3) \\ y \in b_2 \star b_3}} (a_1, b_1) \cdot (x, y)$$
$$= \bigcup_{\substack{x \in a_2 \circ \widehat{b_2}(a_3) \\ y \in b_2 \star b_3}} \{(z, w) \mid z \in a_1 \circ \widehat{b_1}(x), \ w \in b_1 \star y\}$$
$$= \{(z, w) \mid z \in a_1 \circ \widehat{b_1}(a_2 \circ \widehat{b_2}(a_3)), \ w \in b_1 \star (b_2 \star b_3)\}$$
$$= \{(z, w) \mid z \in a_1 \circ (\widehat{b_1}(a_2) \circ \widehat{b_1}(\widehat{b_2}(a_3))), \ w \in b_1 \star (b_2 \star b_3)\}$$
$$= \{(z, w) \mid z \in a_1 \circ (\widehat{b_1}(a_2) \circ \widehat{b_1 \star b_2}(a_3)), \ w \in b_1 \star (b_2 \star b_3)\}$$
$$= \Big(a_1 \circ (\widehat{b_1}(a_2) \circ \widehat{b_1 \star b_2}(a_3)), \ w \in b_1 \star (b_2 \star b_3)\Big)$$

But A and B are H_v-groups, so

$$\left(a_1 \circ \widehat{b_1}(a_2)\right) \circ \widehat{b_1 \star b_2}(a_3) \cap a_1 \circ \left(\widehat{b_1}(a_2) \circ \widehat{b_1 \star b_2}(a_3)\right) \neq \emptyset$$

and
$$(b_1 \star b_2) \star b_3 \cap b_1 \star (b_2 \star b_3) \neq \emptyset.$$

Therefore, we obtain

$$(a_1, b_1) \cdot \left((a_2, b_2) \cdot (a_3, b_3)\right) \cap (a_1, b_1) \cdot \left((a_2, b_2) \cdot (a_3, b_3)\right) \neq \emptyset,$$

i.e., the weak associativity is valid.

It is also easy to see that the reproduction axiom is valid too. ∎

The H_v-group defined in Theorem 3.8 is called *semidirect hyperproduct* of A and B corresponding to $\widehat{}$ and it is denoted by $A \widehat{\times} B$.

Lemma 3.1. *Let $(A, \circ)$ and $(B, \star)$ be two H_v-groups, β^* the fundamental relation on B and $\widehat{} : B/\beta^* \to AutA$. Then,*

$$\widehat{b}(\beta^*(a)) = \beta^*(\widehat{b}(a)), \text{ for all } a \in A, b \in B.$$

Proof. It is straightforward. ∎

3.4 Reversible H_v-groups

Let $(H, \circ)$ be an H_v-group with (left, right) identity elements. Then, H is called (*left, right*) *reversible* in itself when any relation $c \in a \circ b$ implies the existence of a left inverse a' of a and a right inverse b' of b such that $b \in a' \circ c$ and $a \in c \circ b'$.

An H_v-group $(H, \circ)$ is called *feebly quasi-canonical* if it is regular, reversible and satisfies the following conditions:

> For each $a \in H$, if a', a'' are inverses of a, then for each $x \in H$, we have $a' \circ x = a'' \circ x$ and $x \circ a' = x \circ a''$.

A feebly quasi-canonical H_v-group H is called *feebly canonical* if it is strongly commutative.

In the rest of this section, we study a wide class of reversible H_v-groups that investigated by Spartalis [102].

Let $(H, \circ)$ be an H_v-group with left or right identity elements. We denote by E_l (respectively, E_r) the set of left (respectively, right) identities. Hence, the set of all identities of the H_v-group H is $E = E_l \cup E_r$. We also denote by $i_l(x, e)$ the set of all left inverses of an element x with respect to the identity e of E, i.e., $i_l(x, e) = \{x' \in H \mid e \in x' \circ x\}$. Consequently,

$i_l(x) = \bigcup_{e \in E} i_l(x, e)$ is the set of all left inverses of the element x. If A is a non-empty subset of H, then $i_l(A) = \bigcup_{a \in A} i_l(a)$. Similar notations hold for the right inverses, too.

Definition 3.5. Let $(H, \circ)$ be an H_v-group. Then, H is called *left completely reversible* in itself if, for all $a, b \in H$, it satisfies the following condition:

(1) $c \in a \circ b$ implies $b \in u \circ c$, for all $u \in i_l(a)$.

Similarly, H is called *right completely reversible* in itself if

(2) $c \in a \circ b$ implies $a \in c \circ v$, for all $v \in i_r(b)$.

Lemma 3.2. *If H is left completely reversible, then for each $e_l \in E_l$, $e_r \in E_r$, $e \in E = E_l \cup E_r$ and $a \in H$, we have*

(1) $i_l(e_l) = E_l$;
(2) $i_l(a) = i_l(a, e_r)$;
(3) $i_l(i_l(i_l(a, e))) \subseteq i_l(a, e)$.

Proof.   (1) Suppose that $e_l \in E_l$. Clearly, $E_l \subseteq i_l(e_l, e_l) \subseteq i_l(e_l)$. Moreover, let $u \in i_l(e_l)$. Since for all $x \in H$, $x \in e_l \circ x$, it follows that $x \in u \circ x$. Therefore, $u \in E_l$, i.e., $i_l(e_l) \subseteq E_l$ and so $i_l(e_l) = E_l$.

(2) Suppose that $a \in H$ and $e_r \in E_r$. Obviously, $i_l(a, e_r) \subseteq i_l(a)$. Conversely, assume that $u \in i_l(a)$. From the relation $a \in a \circ e_r$ it follows that $e_r \in u \circ a$, that is $u \in i_l(a, e_r)$ and so $i_l(a) = i_l(a, e_r)$.

(3) Suppose that $a \in H$, $e \in E$ and $u \in i_l(a, e)$. Since $e \in u \circ a$, we have that $a \in w \circ e$ for all $w \in i_l(u)$ and $e \in v \circ a$ for all $v \in i_l(i_l(u))$. Consequently, $v \in i_l(a, e)$ and so $i_l(i_l(i_l(a, e))) \subseteq i_l(a, e)$.   ∎

Proposition 3.4. *Let H be left completely reversible, $a \in H$ and $e \in E$ such that $a \in i_l(i_l(a, e))$. Then, the following conditions hold:*

(1) $i_l(a, e) = i_l(a) = i_l(i_l(i_l(a, e)))$;
(2) If u is an inverse of a with respect to e for which the hypothesis holds, i.e., $a \in i_l(u)$, then $i_l(i_l(a)) = i_l(u)$ and, moreover, for all $x \in H$, $v \in i_l(a)$ we have $u \circ x \subseteq v \circ x$.

Proof.   (1) Suppose that $a \in i_l(i_l(a, e))$. Then, $i_l(a) \subseteq i_l(i_l(i_l(a, e)))$. According to Lemma 3.2(3), we have

$$i_l(a) \subseteq i_l(i_l(i_l(a, e))) \subseteq i_l(a, e) \subseteq i_l(a).$$

Thus, $i_l(a,e) = i_l(a) = i_l(i_l(i_l(a,e)))$.

(2) Suppose that u is an inverse of the element a with respect to e such that $a \in i_l(u)$. Then, there exists an identity $e' \in E$ such that $a \in i_l(u, e')$. Therefore, $u \in i_l(a) \subseteq i_l(i_l(u, e'))$. According to (1), we have $i_l(u, e') = i_l(u) = i_l(i_l(i_l(u)))$. Consequently, $u \in i_l(a) \subseteq i_l(i_l(u))$. So, $i_l(u) \subseteq i_l(i_l(a)) \subseteq i_l(u)$. Hence, $i_l(i_l(a)) = i_l(u)$. Finally, let $x \in H$ and $y \in u \circ x$. From the reversibility of H, it follows that $x \in a \circ y$ and $y \in v \circ x$, for all $v \in i_l(a)$. Thus, $u \circ x \subseteq v \circ x$. ∎

Notice that if $a \in H$, $a \in E$ and $a \in i_l(a,e)$, then the following relation is satisfied:

$$i_l(a,e) = i_l(a) = i_l(i_l(i_l(a,e))) = i_l(i_l(a)).$$

Proposition 3.5. *Let H be left completely reversible and $a \in H$ such that $a \in \bigcap_{u \in i_l(a)} i_l(u)$. Then, for all $x \in H$, $u, v \in i_l(a)$ and $e_r \in E_r$ we have*

(1) $u \circ x = v \circ x$;

(2) If E_r is non-empty, then $i_l(a) \subseteq u \circ e_r \subseteq i_r(a, e_r)$.

Proof. (1) We observe that the assumption of Proposition 3.33 are satisfied for all $e \in E$ and for all $u \in i_l(a)$. Therefore, for all $x \in H$ and $u, v \in i_l(a)$, we have $u \circ x \subseteq v \circ x$ and so $u \circ x = v \circ x$.

(2) Suppose that $e_r \in E_r$. Then, we have $i_l(a) \circ e_r = \bigcup_{u \in i_l(a)} u \circ e_r$ and because of (1), it follows that $i_l(a) \circ e_r = u \circ e_r$, for all $u \in i_l(a)$. Thus, $i_l(a) \subseteq u \circ e_r$. Moreover, let $y \in u \circ e_r$. Since $a \in i_l(u)$, it follows that $e_r \in a \circ y$, that is, $y \in i_r(a, e_r)$. ∎

If $(H, \circ)$ is a strongly commutative H_v-group, then from the first condition of previous proposition we conclude that the concepts of left (respectively, right) completely reversible H_v-group and the feebly canonical H_v-group are identical.

Theorem 3.9. *If H is completely reversible, then it is a feebly quasi-canonical H_v-group.*

Proof. At the first, we prove that all identities and inverses in H are two sided. Let $a \in H$, $e \in E$, $u \in i_l(a,e)$ and $v \in i_r(a,e)$. Since $e \in a \circ v$, it follows that $v \in u \circ e$. Since $e \in i_r(e)$, $u \in v \circ e$. Since $a \in i_l(v,e)$, we have $e \in a \circ u$. Therefore, $i_l(a,e) \subseteq i_r(a,e)$. Similarly, we obtain $i_r(a,e) \subseteq i_l(a,e)$. Hence, $i_l(a,e) = i_r(a,e)$. Now, suppose that $e_l \in E_l$

and $a \in H$. Then, there exists $u \in H$ such that $e_l \in u \circ a \cap a \circ u$ and so $a \in a \circ e_l$. Thus, $E_l \subseteq E_r$. In the same manner, every right identity is also a left identity and hence $E_l = E_r$. Consequently, H is regular and reversible. Moreover, by Proposition 3.5(1) for the left (right) completely reversible H_v-groups, we have that, for $a \in H$, if u, v are inverses of a, then for all $x \in H$, $u \circ x = v \circ x$ and $x \circ u = x \circ v$. Therefore, H is a feebly quasi-canonical H_v-group.   ∎

In what follows, we consider H_v-groups with only two sided identities and inverses. The following relation introduced by De Salvo [50] and studied by Spartalis [102].

Consider the binary relation $\sim$ on H as follows:

$$x \sim y \ \Leftrightarrow \ \text{there exists } z \in H \text{ such that } \{x, y\} \subseteq i(z),$$

for all $x, y \in H$. If H is left completely reversible, then by using Proposition 3.5(1) we obtain that $\sim$ is an equivalence relation. In the quotient set $\hat{H} = H/\sim$ we define the following hyperoperation between classes in the usual manner

$$\hat{x} \odot \hat{y} = \{\hat{w} \mid w \in \hat{x} \circ \hat{y}\},$$

for all $x, y \in H$, where $\hat{x}$ is the equivalence class containing x. According to Proposition 3.5(1), this hyperoperation is equivalent to the following one

$$\hat{x} \odot \hat{y} = \{\hat{w} \mid w \in x \circ y\}.$$

Lemma 3.3. $(\hat{H}, \odot)$ *is an H_v-group.*

Proof.   It is straightforward.   ∎

Proposition 3.6. *Let H be left completely reversible and $e \in E$. Then, $\hat{H}$ is a left reversible in itself H_v-group and $\hat{e} = E$ is the unique identity of $\hat{H}$. Moreover, $\hat{e}$ is a right scalar, $\hat{H}$ is regular and each $\hat{x} \in \hat{H}$ has a unique inverse.*

Proof.   Suppose that $a \in E$. By Lemma 3.2 (1) we have $i(e) = E$. So, $E \subseteq \hat{e}$. Further, if $x \in \hat{e}$, then there exists $y \in H$ such that $\{x, e\} \subseteq i(y)$. Since $y \in i(e)$, it follows that $x \in E$. Hence, $\hat{e} \subseteq E$. Thus, $\hat{e} = E$. Obviously, for all $\hat{x} \in \hat{H}$, $\hat{x} \in \hat{x} \odot \hat{e} \cap \hat{e} \odot \hat{x}$. Now, suppose that $\hat{s}$ is a left identity or $\hat{t}$ is a right identity of $\hat{H}$. Then, $\hat{e} \in \hat{s} \odot \hat{e} \cap \hat{e} \odot \hat{t}$ and so there exist $e' \in E$ and $t' \in \hat{t}$ such that $e \in s \circ e' \cap e \circ t'$. Therefore, $s, t' \in i(e) = i(e')$ and hence $\hat{s} = \hat{e} = \hat{t}$. Finally, for all $\hat{x} \in \hat{H}$,

$$\hat{x} \odot \hat{e} = \{\hat{w} \mid w \in x \circ e = x \circ E\}.$$

If $w \in x \circ e'$ and $e' \in E$, then for all $u \in i(x)$, $e' \in u \circ w$. Thus, $\{w, x\} \subseteq i(u)$ and so $x \sim w$. Consequently, $\hat{x} \odot \hat{e} = \{\hat{x}\}$.

It is easy to prove that each $\hat{x} \in \hat{H}$ has a unique inverse, i.e., $i(\hat{x}) = \{\hat{a}\}$, where $a \in i(x)$. Moreover, $\hat{H}$ is a regular H_v-group. Finally, we show that $\hat{H}$ is left reversible in itself. Suppose that $\hat{x}, \hat{y} \in \hat{H}$ and $\hat{z} \in \hat{x} \odot \hat{y}$. Then, there exists $y' \in \hat{y}$ such that $z \in x \circ y'$ and hence for all $a \in i(x)$, $y' \in a \circ z$. Since $i(\hat{x}) = \{\hat{a}\}$, it follows that $\hat{y} \in \hat{a} \odot \hat{z}$.   ∎

Now, let K be an H_v-subgroup of H. Let the left coset expansion $H/K = \{x \circ K \mid x \in H\}$ satisfies the following conditions:

(1*) for all $x \in H$, $x \in x \circ K$,
(2*) for all $x, y \in H$, $x \circ K \cap y \circ K \neq \emptyset$ implies $x \circ K = y \circ K$.

It is easy to see that H/K becomes an H_v-group with respect to the usual hyperoperation:

$$x \circ K \odot y \circ K = \{z \circ K \mid z \in (x \circ K) \circ (y \circ K)\},$$

for all x, $y \in H$. A similar remark holds for the right coset expansion. Moreover, for the right coset expansion, we have the following proposition.

Proposition 3.7. *Let H be left completely reversible and K be an H_v-subgroup of H. If $K \cap E \neq \emptyset$ and for all $x \in H$, $K \circ (K \circ x) \subseteq K \circ x$, then $H/K = \{K \circ x \mid x \in H\}$ is an H_v-group.*

Proof. It suffices to prove the following conditions:

(1) for all $x \in H$, $x \in K \circ x$,
(2) for all $x, y \in H$, $K \circ x \cap K \circ y \neq \emptyset$ implies $K \circ x = K \circ y$.

Obviously, for all $x \in H$, $x \in K \circ x$. Moreover, suppose that $x, y \in H$ and $z \in K \circ x \cap K \circ y$. Then, $K \circ z \subseteq K \circ (K \circ x) \subseteq K \circ x$. From $z \in K \circ x$, it follows that $z \in u \circ x$, where $u \in K$. Therefore, for all $a \in i(u)$, $x \in a \circ z$. Since $i(u) \cap K \neq \emptyset$, we have $x \in K \circ z$. Thus, $K \circ x \subseteq K \circ z$. Consequently, $K \circ z = K \circ x$. Similarly, $K \circ z = K \circ y$.   ∎

Theorem 3.10. *Let H be left completely reversible and E be the set of identities and H/E be the left coset expansion of H with respect to E. Then the following conditions hold:*

(1) E is a total H_v-subgroup of H;

(2) for all $x \in H$, $x \circ E = \hat{x}$, that is H/E is identical with the H_v-group $\hat{H}$;

(3) E is the smallest of the H_v-subgroup K of H such that H/K satisfies *(1*)* and *(2*)*.

Proof. By using Proposition 3.5(2) and Lemma 3.2(2), for all $a \in H$, $e \in E$ and $u \in i(a)$, we obtain

$$i(a) = u \circ a = i(a, e). \tag{3.3}$$

(1) Suppose that $e \in E$. Then, from the previous relation, we obtain

$$i(e) = u \circ e', \text{ for all } u \in i(e) \text{ and } e' \in E.$$

Moreover, according to Lemma 3.2(1), $i(e) = E$ and hence $E = e'' \circ e'$, for all $e', e'' \in E$. Therefore, E is a total H_v-subgroup of H.

(2) By hypothesis $H/E = \{x \circ E \mid x \in H\}$ and $x \circ E = \bigcup_{e \in E} x \circ e$. Suppose that $x \in H$, $e \in E$ and $u \in i(x)$. Then, applying (3.3) we obtain $x \circ E = i(u) = x \circ e$ and hence $H/E = \{x \circ e \mid x \in H\}$. Furthermore, for all $x, y \in H$ we have the following

$$x \sim y \Rightarrow \text{there exists } z \in H \text{ such that } \{x, y\} \subseteq i(z)$$
$$\Leftrightarrow \text{there exists } z \in H, \ e \in E \text{ such that } y \in x \circ e = i(z)$$
$$\Leftrightarrow y \in x \circ e.$$

Consequently, H/E is identical with the H_v-group $\hat{H}$.

(3) Suppose that K is an H_v-subgroup of H such that the left coset expansion H/K satisfies (1). Let $e \in E$. Then, $e \in e \circ K$ and so $e \in e \circ u$, for some $u \in K$. Therefore, $u \in i(e)$ and since $i(e) = E$ we have $K \cap E \neq \emptyset$. Finally, since E is a total H_v-subgroup of H, it follows that $E \subseteq K$. ∎

3.5   A sequence of finite H_v-groups

This section deals with a sequence of H_v-structures. The results are obtained by Antampoufis and Dramalidis [7]. Firstly, they defined a hyperoperation on a set and studied, in the general case, the hyperstructure resulting. The hyperstructure is an H_v-group. The hyperoperation is defined in every finite hyperstructure using indices of the cyclic group $\mathbb{Z}_n$. The case of infinite order is separately studied with indices in $\mathbb{N}$. They also studied the existence of identities, inverses elements and powers of the elements of the H_v-group. The hyperstructures of small order, provided with the particular hyperoperation are groups, those of greater order are hypergroups and then those of the greatest order lead to H_v-groups. A sequence of finite H_v-groups with common properties is created. Finally, they presented the motivating example. The hyperoperation is defined in a geometrical figure on $\mathbb{R}^2$ [6, 51], which is partitioned into a finite or infinite number of parts. The hyperoperation is defined in the sense of the *boundary* among the parts. A lot of properties of this geometrical H_v-structures are studied.

Consider the set $H = \{\alpha\} \cup \{z \mid z = a_r, r \in \mathbb{Z}_n\}$. On H we define the hyperoperation $\circ$ as follows:

Definition 3.6. For every $x, y \in H$ define

$$\circ : H \times H \to \mathcal{P}^*(H) \colon (x, y) \mapsto x \circ y$$

such that

$x \circ y = \{\alpha, a_{p-1}, a_{p+1}, a_{k-1}, a_{k+1}\}$, $x = a_p$, $y = a_k$, for all $p, k \in \mathbb{Z}_n$
$x \circ \alpha = \alpha \circ x = x$, for all $x \in H$.

Some properties of the hyperoperation $\circ$:

(1) Obviously, $\circ$ is commutative, i.e., $x \circ y = y \circ x$, for all $x, y \in H$.
(2) According to Definition 3.6, $\alpha \circ \alpha = \alpha$ and $\alpha \circ a_p = a_p \circ \alpha = a_p$, for all $p \in \mathbb{Z}_n$. That means that $\alpha \circ x = x \circ \alpha = x$, for all $x \in H$, so the element α is scalar unit element.
(3) $\alpha^m = \alpha$, $m \in \mathbb{N}^*$.
(4) Since $\alpha^2 = \alpha$, it follows that the element α is also an idempotent element and obviously $I_\circ(\alpha, \alpha) = \alpha$.
(5) $|H| = n + 1$.
(6) $H/\beta^* = H$.
(7) $x \cdot y = x^2 \cup y^2$, for all $x, y \in H$ with $(x, y) \neq (\alpha, y)$ and $(x, y) \neq (x, \alpha)$.

Proposition 3.8. $I_\circ(a_p, \alpha) = H \setminus \{\alpha\}$, *for all* $p \in \mathbb{Z}_n$.

Proof. Suppose that $a_k \in I_\circ(a_p, \alpha)$, where $k, p \in \mathbb{Z}_n$. Then, $\alpha \in a_k \circ a_p$ and $\alpha \in a_p \circ a_k$. So, $\alpha \in \{\alpha, a_{k-1}, a_{k+1}, a_{p-1}, a_{p+1}\}$ and $\alpha \in \{\alpha, a_{p-1}, a_{p+1}, a_{k-1}, a_{k+1}\}$. Since the previous relations are both true for every $k, p \in \mathbb{Z}_n$, we obtain $I_\circ(a_p, \alpha) = H \setminus \{\alpha\}$, for all $p \in \mathbb{Z}_n$. ∎

Proposition 3.9. $\displaystyle\bigcup_{x \in H} x^2 = \bigcup_{k \in \mathbb{Z}_n} a_k^2 = H$.

Proof. Notice that

$$\bigcup_{k \in \mathbb{Z}_n} a_k^2 = \bigcup_{k \in \mathbb{Z}_n} \{\alpha, a_{k-1}, a_{k+1}\} = \{\alpha, a_{-1}, a_1, a_0, a_2, a_3, \ldots, a_{n-1}, a_n\}$$
$$= \{\alpha, a_0, a_1, a_2, a_3, \ldots, a_{n-1}\} = H,$$

since $a_{-1} = a_{n-1}$ and $a_n = a_0$.

Also, since $\alpha^2 = \alpha$, it follows that $\bigcup_{x \in H} x^2 = \bigcup_{k \in \mathbb{Z}_n} a_k^2 = H$. ∎

Proposition 3.10. $a_k^2 \subseteq a_p \circ a_k$, for all $k, p \in \mathbb{Z}_n$.

Proof. For $k, p \in \mathbb{Z}_n$,

$$a_p \circ a_k = \{\alpha, a_{p-1}, a_{p+1}, a_{k-1}, a_{k+1}\} \supset \{\alpha, a_{k-1}, a_{k+1}\} = a_k^2.$$

When $p = k$ we get that $a_k \circ a_k = \{\alpha, a_{k-1}, a_{k+1}\} = a_k^2$. So, generally, $a_k^2 \subseteq a_p \circ a_k$, for all $k, p \in \mathbb{Z}_n$. ∎

Proposition 3.11. *The hyperstructure* $(H, \circ)$ *is a commutative* H_v-*group.*

Proof. For the reproduction axiom, if $x = \alpha$, then

$$x \circ H = \alpha \circ H = \bigcup_{h \in H} (\alpha \circ h) = \bigcup_{h \in H} h = H = H \circ \alpha = H \circ x.$$

If $x = a_k, k \in \mathbb{Z}_n$, then

$$x \circ H = a_k \circ H = \bigcup_{h \in H} (a_k \circ h) = (a_k \circ \alpha) \cup [\bigcup_{k \in \mathbb{Z}_n} (a_k \circ a_p)]$$
$$= \{a_k\} \cup \{\alpha, a_{k-1}, a_{k+1}, a_{-1}, a_0, a_1, a_2, a_3, \ldots, a_{k-1}, a_{k+1}, \ldots, a_{n-2}, a_n\}$$
$$= \{\alpha, a_0, a_1, a_2, a_3, \ldots, a_{n-2}, a_{n-1}\} = H,$$

since $a_{-1} = a_{n-1}$ and $a_n = a_0$.

Obviously $H \circ a_k = H \circ x = H$, then

$$x \circ H = H \circ x = H, \text{ for all } x \in H.$$

Since $\circ$ is commutative, we have to check only the following cases for the associativity: For $p, k \in \mathbb{Z}_n$,

- $\alpha \circ (\alpha \circ a_k) = \alpha \circ a_k = a_k$ and $(\alpha \circ \alpha) \circ a_k = \alpha \circ a_k = a_k$.
- $\alpha \circ (a_p \circ a_k) = \{\alpha, a_{p-1}, a_{p+1}, a_{k-1}, a_{k+1}\} = a_p \circ a_k$ and $(\alpha \circ a_p) \circ a_k = a_p \circ a_k$.
- $a_k \circ (a_k \circ \alpha) = a_k \circ a_k = \{\alpha, a_{k-1}, a_{k+1}\}$ and $(a_k \circ a_k) \circ \alpha = \{\alpha, a_{k-1}, a_{k+1}\}$.

So far, notice that if one or two elements α appear in the triples (x, y, z), then the equality appears for the associativity, i.e. $x \circ (y \circ z) = (x \circ y) \circ z$. Furthermore, we have to check the following two cases: For $p, k, m \in \mathbb{Z}_n$,

- $a_k \circ (a_k \circ a_p) = a_k \circ \{\alpha, a_{k-1}, a_{k+1}, a_{p-1}, a_{p+1}\}$
  $= \{a_k, a_{k-1}, a_{k+1}, a_{k-2}, a_{k+2}, a_p, a_{p-2}, a_{p+2}\}$
  and $(a_k \circ a_k) \circ a_p = \{\alpha, a_{k-1}, a_{k+1}\} \circ a_p$
  $= \{a_p, a_{k-2}, a_k, a_{p-1}, a_{p+1}, a_{k+2}\}.$

So,

$$a_k \circ (a_k \circ a_p) \cap (a_k \circ a_k) \circ a_p = \{a_p, a_k, a_{k-2}, a_{k+2}\} \neq \emptyset.$$

- $a_k \circ (a_p \circ a_m) = \{a_k, a_{k-1}, a_{k+1}, a_p, a_{p-2}, a_{p+2}, a_m, a_{m-2}, a_{m+2}\}$
  and
  $(a_k \circ a_p) \circ a_m = \{a_m, a_{k-2}, a_k, a_{m-1}, a_{m+1}, a_{k+2}, a_p, a_{p-2}, a_{p+2}\}.$

Thus,

$$a_k \circ (a_p \circ a_m) \cap (a_k \circ a_p) \circ a_m = \{a_p, a_k, a_m, a_{p-2}, a_{p+2}\} \neq \emptyset.$$

Therefore, generally $x \circ (y \circ z) \cap (x \circ y) \circ z \neq \emptyset$, for all $x, y \in H$. ∎

Proposition 3.12. $a_k^p \subseteq a_k^m$, for all $k \in \mathbb{Z}_n, p, m \in \mathbb{N}, p, m > 1 \Leftrightarrow m \geqq p$.

Proof. Since $\circ$ is commutative, for $k \in \mathbb{Z}_n$,

$a_k^2 = \{\alpha, a_{k-1}, a_{k+1}\};$

$a_k^3 = a_k^2 \circ a_k = \{\alpha, a_{k-2}, a_{k-1}, a_k, a_{k+1}, a_{k+2}\};$

$a_k^4 = a_k^3 \circ a_k \cup a_k^2 \circ a_k^2$

$= \{\alpha, a_{k-3}, a_{k-2}, a_{k-1}, a_k, a_{k+1}, a_{k+2}, a_{k+3}\} \cup \{\alpha, a_{k-2}, a_{k-1}, a_k, a_{k+1}, a_{k+2}\}$

$= \{\alpha, a_{k-3}, a_{k-2}, a_{k-1}, a_k, a_{k+1}, a_{k+2}, a_{k+3}\}.$

So, by induction, for $p, m \in \mathbb{N}, p, m > 1$ we obtain

$$a_k^p = \{\alpha, a_{k-p+1}, a_{k-p+2}, \ldots, a_{k+p-2}, a_{k+p-1}\}$$
$$a_k^m = \{\alpha, a_{k-m+1}, a_{k-m+2}, \ldots, a_{k+m-2}, a_{k+m-1}\}.$$

Let $a_k^p \subseteq a_k^m \Rightarrow k - m + 1 \leqq k - p + 1$ and $k + m - 1 \geqq k + p - 1 \Rightarrow m \geqq p$.
Now, let $m \geqq p \Rightarrow k - m + 1 \leqq k - p + 1$ and $k + m - 1 \geqq k + p - 1, k \in \mathbb{Z}_n$.
Since $k - p + 1, k + p - 1, k - m + 1, k + m - 1 \in \mathbb{Z}_n$, it follows that

$$\{a_{k-p+1}, a_{k-p+2}, \ldots, a_{k+p-1}\} \subseteq \{a_{k-m+1}, a_{k-m+2}, \ldots, a_{k+m-1}\}.$$

Hence, we have

$$\{\alpha, a_{k-p+1}, a_{k-p+2}, \ldots, a_{k+p-1}\} \subseteq \{\alpha, a_{k-m+1}, a_{k-m+2}, \ldots, a_{k+m-1}\}$$

So, $a_k^p \subseteq a_k^m$. ∎

Corollary 3.2. $a_k^m = a_k^{m-1} \circ a_k$, *for all* $k \in \mathbb{Z}_n, m \in \mathbb{N}, m > 1$.

Proof.   For $k \in \mathbb{Z}_n, m \in \mathbb{N}, m > 1$, since $(\circ)$ is commutative:

$a_k^m = a_k^{m-1} \circ a_k \cup a_k^{m-2} \circ a_k^2 \cup \ldots \cup a_k^{m/2} \circ a_k^{m/2}$, if $m = 2p, p \in \mathbb{N}$;

$a_k^m = a_k^{m-1} \circ a_k \cup a_k^{m-2} \circ a_k^2 \cup \ldots \cup a_k^{m+1/2} \circ a_k^{m-1/2}$, if $m = 2p+1, p \in \mathbb{N}$

According to Proposition 3.12, the power $m-1$ is the greatest one in both cases, so

$$a_k^m = a_k^{m-1} \circ a_k, \text{ for all } k \in \mathbb{Z}_n, m \in \mathbb{N}, m > 1.$$ ∎

Proposition 3.13. *The* $(H, \circ)$ *is a single-power cyclic* H_v*-group and each element* $a_k \in H, k \in \mathbb{Z}_n, n = 2, 3$ *is a generator with period 3.*

Proof.   Let $x = a_k, k \in \mathbb{Z}_n$ be any element of H. Then,

$$a_k^2 = \{\alpha, a_{k-1}, a_{k+1}\},$$
$$a_k^3 = a_k^2 \circ a_k = \{\alpha, a_{k-2}, a_{k-1}, a_k, a_{k+1}, a_{k+2}\}.$$

For $n = 2, H = \{\alpha, a_0, a_1\}$:

Notice that $|H| = 3$ and $|a_k^2| = 3$ but $a_k \notin a_k^2$, so $a_k^2 \neq H$. Since $|a_k^3| = 6$ and $a_k \in a_k^3$, it follows that $a_k^3 = H$. That means that $(\{\alpha, a_0, a_1\}, \circ)$ is single-power cyclic H_v-group and each element $a_k \in H, k \in \mathbb{Z}_2$ is generator with period 3.

For $n = 3, H = \{\alpha, a_0, a_1, a_2\}$:

Notice that $|H| = 4$, $|a_k^3| = 6$ and $a_k \in a_k^3 \Rightarrow a_k^3 = H$. That means that $(\{\alpha, a_0, a_1, a_2\}, \circ)$ is single-power cyclic H_v-group and each element $a_k \in H, k \in \mathbb{Z}_3$ is generator with period 3.   ∎

Proposition 3.14. *The* $(H, \circ)$ *is a single-power cyclic* H_v*-group and each element* $a_k \in H, k \in \mathbb{Z}_n$ *is a generator with period the minimum* $m \in \mathbb{N}, m > 1$ *such that* $m \geq \frac{n+1}{2}, n \geq 4$.

Proof.   Let $x = a_k, k \in \mathbb{Z}_n, n \geq 4$ be any element of H. Then, from Proposition 3.12,

$$a_k^m = \{\alpha, a_{k-m+1}, a_{k-m+2}, \ldots, a_{k+m-2}, a_{k+m-1}\}, m \in \mathbb{N}, m > 1.$$

Notice that $|a_k^m| = 1 + [(k + m - 1) - (k - m + 1) + 1] = 2m$.

Since $|H| = n + 1$, we obtain

$$a_k^m = H \Rightarrow 2m \geq n + 1 \Rightarrow m \geq \frac{n+1}{2}.$$ ∎

Consider now the infinite set $H' = \{\alpha\} \cup \{z/z = a_m, m \in \mathbb{N}\}$. Then $(H', \circ)$ is also a commutative H_v-group.

Proposition 3.15. *The* $(H', \circ)$ *is a single-power cyclic* H_v-*group with infinite period and each element* $a_m \in H', m \in \mathbb{N}$ *is a generator.*

Proof. Let $x = a_m, a_m \in H', m \in \mathbb{N}$ be any element of H'. Then,

$a_m^1 = a_m$,

$a_m^2 = \{\alpha, a_{m-1}, a_{m+1}\}$,

$a_m^3 = \{\alpha, a_{m-2}, a_{m-1}, a_m, a_{m+1}, a_{m+2}\}$,

$a_m^4 = \{\alpha, a_{m-3}, a_{m-2}, a_{m-1}, a_m, a_{m+1} a_{m+2}, a_{m+3}\}$,

...

$a_m^{n-1} = \{\alpha, a_{m-n+2}, a_{m-n+3}, \ldots, a_{m+n-3}, a_{m+n-2}\}, n \in \mathbb{N}^*$,

$a_m^n = \{\alpha, a_{m-n+1}, a_{m-n+2}, \ldots, a_{m+n-2}, a_{m+n-1}\}, n \in \mathbb{N}^*$,

...

So, every element of H' belongs to a power of a_m and there exists $n_0 \geqq 1$ such that for every $n \geqq n_0$:

$$a_m^1 \cup a_m^2 \cup a_m^3 \cup \ldots \cup a_m^{n-1} \subset a_m^n.$$ ∎

Let us consider the geometrical shape of the figure below. It is partitioned into $n + 1$ parts such that:

(1) There is a part, denoted by α, which borders all the rest.
(2) In addition, each of the remaining parts, denoted by $a_i, i \in \mathbb{Z}_n$ borders the two others (its adjacent ones).

Thus, the set H of all parts of the figure consists of one central and n peripheral parts.

Definition 3.7. On H we introduce a hyperoperation $(*)$ such that

(1) the hyperproduct between the central part α and any other x is equal to x, i.e. $\alpha * x = x * \alpha = x$.

(2) The hyperproduct of two peripheral parts is the set of all parts border to them (their adjacent ones), i.e., $a_i * a_j = \{\alpha, a_{i-1}, a_{i+1}, a_{j-1}, a_{j+1}\}, i, j \in \mathbb{Z}_n$.

We call $*$, *boundary hyperoperation*. Notice that $* \equiv \circ$, for example

$$\alpha * \alpha = \alpha, \alpha * a_2 = a_2, a_1 * a_2 = \{\alpha, a_0, a_1, a_2, a_3\}, a_2 * a_5 = \{\alpha, a_1, a_3, a_4, a_6\}.$$

We study below some cases, depending on n, of the boundary hyperoperation, $n = 1, 2, 3, 4, 5$ then a sequence of hyperstructures is created and in each one we present the corresponding figure, the Cayley table of the hyperoperation and the kind of the hyperstructure resulting.

Case 1

If $n = 1$ then $H = \{\alpha, a_0\}$

$*$	α	a_0
α	α	a_0
a_0	a_0	α

Notice that $(\{\alpha, a_0\}, *)$ is a group and $H \cong \mathbb{Z}_2$.

Case 2

If $n = 2$ then $H = \{\alpha, a_0, a_1\}$

$*$	α	a_0	a_1
α	α	a_0	a_1
a_0	a_0	α, a_1	H
a_1	a_1	H	α, a_0

Notice that $(\{\alpha, a_0, a_1\}, *)$ is a hypergroup. It is also a H_b-group, greater than a group, isomorphic to $\mathbb{Z}_3$. We say that H contains $\mathbb{Z}_3$ up to isomorphism.

Case 3

If $n = 3$ then $H = \{\alpha, a_0, a_1, a_2\}$

$*$	α	a_0	a_1	a_2
α	α	a_0	a_1	a_2
a_0	a_0	α, a_1, a_2	H	H
a_1	a_1	H	α, a_0, a_2	H
a_2	a_2	H	H	α, a_0, a_1

Notice that $(\{\alpha, a_0, a_1, a_2\}, *)$ is a hypergroup. It is also a H_b-group which contains $\mathbb{Z}_4$ up to isomorphism.

Case 4

If $n = 4$ then $H = \{\alpha, a_0, a_1, a_2, a_3\}$

$*$	α	a_0	a_1	a_2	a_3
α	α	a_0	a_1	a_2	a_3
a_0	a_0	α, a_1, a_3	H	α, a_1, a_3	H
a_1	a_1	H	α, a_0, a_2	H	α, a_0, a_2
a_2	a_2	α, a_1, a_3	H	α, a_1, a_3	H
a_3	a_3	H	α, a_0, a_2	H	α, a_0, a_2

Notice that $(\{\alpha, a_0, a_1, a_2, a_3\}, *)$ is a H_v-group.

Case 5

If $n = 5$ then $H = \{\alpha, a_0, a_1, a_2, a_3, a_4\}$

$*$	α	a_0	a_1	a_2	a_3	a_4
α	α	a_0	a_1	a_2	a_3	a_4
a_0	a_0	α, a_1, a_4	$\alpha, a_0, a_1, a_2, a_4$	α, a_1, a_3, a_4	α, a_1, a_2, a_4	$\alpha, a_0, a_1, a_3, a_4$
a_1	a_1	$\alpha, a_0, a_1, a_2, a_4$	α, a_0, a_2	$\alpha, a_0, a_1, a_2, a_3$	α, a_0, a_2, a_4	α, a_0, a_2, a_3
a_2	a_2	α, a_1, a_3, a_4	$\alpha, a_0, a_1, a_2, a_3$	α, a_1, a_3	$\alpha, a_1, a_2, a_3, a_4$	α, a_0, a_1, a_3
a_3	a_3	α, a_1, a_2, a_4	α, a_0, a_2, a_4	$\alpha, a_1, a_2, a_3, a_4$	α, a_2, a_4	$\alpha, a_0, a_2, a_3, a_4$
a_4	a_4	$\alpha, a_0, a_1, a_3, a_4$	α, a_0, a_2, a_3	α, a_0, a_1, a_3	$\alpha, a_0, a_2, a_3, a_4$	α, a_0, a_3

Notice that $(\{\alpha, a_0, a_1, a_2, a_3, a_4\}, *)$ is a H_v-group.

3.6   Fuzzy H_v-groups

It is well known that the concept of fuzzy sets, introduced by Zadeh [141], has been extensively applied to many scientific fields.

Definition 3.8. Let X be a set. A *fuzzy subset A* in X is characterized by a membership function $\mu_A : X \to [0,1]$ which associates with each point $x \in X$ its *grade* or *degree of membership* $\mu_A(x) \in [0,1]$.

Let A and B be fuzzy sets in X. Then

- $A = B$ if and only if $\mu_A(x) = \mu_B(x)$ for all $x \in X$,
- $A \subseteq B$ if and only if $\mu_A(x) \leq \mu_B(x)$ for all $x \in X$,
- $C = A \cup B$ if and only if $\mu_C(x) = max\{\mu_A(x), \mu_B(x)\}$ for all $x \in X$,
- $D = A \cap B$ if and only if $\mu_D(x) = min\{\mu_A(x), \mu_B(x)\}$ for all $x \in X$.

The *complement* of A, denoted by A^c, is defined by

$$\mu_{A^c}(x) = 1 - \mu_A(x) \text{ for all } x \in X.$$

For the sake of simplicity, we shall show every fuzzy set by its membership function.

Definition 3.9. Let f be a mapping from a set X to a set Y. Let μ be a fuzzy subset of X and λ be a fuzzy subset of Y. Then the *inverse image* $f^{-1}(\lambda)$ of λ is the fuzzy subset of X defined by $f^{-1}(\lambda)(x) = \lambda(f(x))$ for all $x \in X$. The *image* $f(\mu)$ of μ is the fuzzy subset of Y defined by

$$f(\mu)(y) = \begin{cases} \sup\{\mu(t) \mid t \in f^{-1}(y)\} & \text{if } f^{-1}(y) \neq \emptyset \\ 0 & \text{otherwise} \end{cases}$$

for all $y \in Y$.

It is not difficult to see that the following assertions hold:

(1) If $\{\lambda_i\}_{i \in I}$ be a family of fuzzy subsets of Y, then

$$f^{-1}\left(\bigcup_{i \in I} \lambda_i\right) = \bigcup_{i \in I} f^{-1}(\lambda_i) \text{ and } f^{-1}\left(\bigcap_{i \in I} \lambda_i\right) = \bigcap_{i \in I} f^{-1}(\lambda_i).$$

(2) If μ is a fuzzy subset of X, then $\mu \subseteq f^{-1}(f(\mu))$. Moreover, if f is one to one, then $f^{-1}(f(\mu)) = \mu$.

(3) If λ is a fuzzy subset of Y, then $f(f^{-1}(\lambda)) \subseteq \lambda$. Moreover, if f is onto, then $f(f^{-1}(\lambda)) = \lambda$.

The concept of a fuzzy subgroup of a group $(G, \cdot)$ is introduced in [89]. If G is a group and $\mu : G \to [0, 1]$ is a fuzzy subset of G, then μ is called a *fuzzy subgroup* if it satisfies,

(1) $\min\{\mu(x), \mu(y)\} \leq \mu(x \cdot y)$, for all $x, y \in G$,
(2) $\mu(x) \leq \mu(x^{-1})$, for all $x \in G$.

In [23, 26], Davvaz applied the concept of fuzzy sets to the theory of algebraic hyperstructures and defined fuzzy subhypergroup (respectively, H_v-subgroup) of a hypergroup (resp. H_v-group) which is a generalization of the concept of Rosenfeld's fuzzy subgroup of a group. In this section first we define fuzzy H_v-subgroup of an H_v-group and then we obtain the relation between a fuzzy H_v-subgroup and level H_v-groups. This relation is expressed in terms of a necessary and sufficient condition.

Definition 3.10. Let $(H, \circ)$ be an H_v-group and let μ be a fuzzy subset of H. Then, μ is said to be a *fuzzy H_v-subgroup* of H if the following axioms hold.

(1) $\min\{\mu(x), \mu(y)\} \leq \inf\limits_{\alpha \in x \circ y} \{\mu(\alpha)\}$, for all $x, y \in H$.

(2) For all $x, a \in H$ there exists $y \in H$ such that $x \in a \circ y$ and

$$\min\{\mu(a), \mu(x)\} \leq \mu(y).$$

(3) For all $x, a \in H$ there exists $z \in H$ such that $x \in z \circ a$ and

$$\min\{\mu(a), \mu(x)\} \leq \mu(z).$$

Condition (2) is called the *left fuzzy reproduction axiom*, while (3) is called the *right fuzzy reproduction axiom*.

Example 3.7. Let $(G, \cdot)$ be a group and μ be a fuzzy subgroup of G. If we define the following hyperoperation on G, $\star : G \times G \to \mathcal{P}^*(G)$ with $x \star y = \{t \mid \mu(t) = \mu(x \cdot y)\}$, then $(G, \star)$ is an H_v-group and μ is a fuzzy H_v-subgroup of G.

Example 3.8. Suppose that H is a set and μ is a fuzzy subset of H. We define the hyperoperation $\star : H \times H \to \mathcal{P}^*(H)$ as follows: Assume that $x, y \in H$, if $\mu(x) \leq \mu(y)$, then $y \star x = x \star y = \{t \mid t \in H, \mu(x) \leq \mu(t) \leq \mu(y)\}$. Then, $(H, \star)$ is a hypergroup as well as a join space. If $(H, \cdot)$ is a group and μ is a fuzzy subgroup of H, then μ is a subhypergroup of $(H, \star)$.

Lemma 3.4. *Let $(H, \circ)$ be an H_v-group and μ be a fuzzy H_v-subgroup of H. Then,*

$$\min\{\mu(x_1), \ldots, \mu(x_n)\} \leq \inf\limits_{\alpha \in (\ldots((x_1 \circ x_2) \circ x_3) \ldots) \circ x_n} \{\mu(\alpha)\},$$

for all $x_1, x_2, \ldots, x_n \in H$.

Proof. We shall prove the validity of this lemma by mathematical induction. First, the lemma is clearly true for $n = 2$. To complete the proof we assume the validity of the lemma for $n = k - 1$, that is, we assume that

$$\min\{\mu(x_1), \ldots, \mu(x_{k-1})\} \leq \inf\limits_{r \in (\ldots((x_1 \circ x_2) \circ x_3) \ldots) \circ x_{k-1}} \{\mu(r)\},$$

for all $x_1, x_2, \ldots, x_{k-1} \in H$. Then,

$$\min\{\mu(x_1), \ldots, \mu(x_{k-1}), \mu(x_k)\} = \min\{\min\{\mu(x_1), \ldots, \mu(x_{k-1})\}, \mu(x_k)\}$$
$$\leq \min\left\{ \inf_{r \in (\ldots((x_1 \circ x_2) \circ x_3)\ldots)\circ x_{k-1}} \{\mu(r), \mu(x_k)\}\right\}$$
$$= \inf_{r \in (\ldots((x_1 \circ x_2) \circ x_3)\ldots)\circ x_{k-1}} \{\min\{\mu(r), \mu(x_k)\}$$
$$\leq \inf_{r \in (\ldots((x_1 \circ x_2) \circ x_3)\ldots)\circ x_{k-1}} \left\{ \inf_{\alpha \in r \circ x_k} \{\mu(\alpha)\}\right\}$$
$$\leq \inf_{\alpha \in ((\ldots((x_1 \circ x_2) \circ x_3)\ldots)\circ x_{k-1})\circ x_k} \{\mu(\alpha)\}. \quad \blacksquare$$

Let $(H, \circ)$ be an H_v-group. An n-ary hyperproduct can be defined, induced by $\circ$, by inserting $n - 2$ parentheses in the sequence of elements $x_1, \ldots, x_n$ in a standard position. Let us denote by $p(x_1, \ldots, x_n)$ such a pattern of $n - 2$ parentheses and by, P_n the set of all such patterns.

Corollary 3.3. *Let $(H, \circ)$ be an H_v-group and μ be a fuzzy H_v-subgroup of H. Then,*

$$\min\{\mu(x_1), \ldots, \mu(x_n)\} \leq \inf\{\mu(\alpha) \mid \alpha \in \bigcup_{p(x_1, \ldots, x_n) \in P_n(x_1, \ldots, x_n)} p(x_1, \ldots, x_n)\}.$$

for all $x_1, x_2, \ldots, x_{k-1} \in H$.

Theorem 3.11. *Let $(H, \circ)$ be an H_v-group and μ be a fuzzy subset of H. Then, μ is a fuzzy H_v-subgroup of H if and only if for every t, $0 \leq t \leq 1$, $\mu_t \neq \emptyset$ is an H_v-subgroup of H.*

Proof. Let μ be a fuzzy H_v-subgroup of H. For every x, y in μ_t we have $\min\{\mu(x), \mu(y)\} \geq t$ and so $\inf_{\alpha \in x \circ y} \{\mu(\alpha)\} \geq t$. Therefore, for every $\alpha \in x \circ y$ we have $\alpha \in \mu_t$, so $x \circ y \subseteq \mu_t$. Hence, for every $a \in \mu_t$ we have $a \circ \mu_t \subseteq \mu_t$. Now, let $x \in \mu_t$. Then, there exists $y \in H$ such that $x \in a \circ y$ and $\min\{\mu(a), \mu(x)\} \leq \mu(y)$. From $x \in \mu_t$ and $a \in \mu_t$ we get $\min\{\mu(x), \mu(a)\} \geq t$. So, $y \in \mu_t$, and this proves $\mu_t \subseteq a \circ \mu_t$.

Conversely, assume that for every t, $0 \leq t \leq 1$, $\mu_t \neq \emptyset$ is an H_v-subgroup of H. For every x, y in H we can write $\mu(x) \geq \min\{\mu(x), \mu(y)\}$ and $\mu(y) \geq \min\{\mu(x), \mu(y)\}$. If we put $t_0 = \min\{\mu(x), \mu(y)\}$, then $x \in \mu_{t_0}$ and $y \in \mu_{t_0}$, so $x \circ y \subseteq \mu_{t_0}$. Therefore, for every $\alpha \in x \circ y$ we have $\mu(\alpha) \geq t_0$ implying $\inf_{\alpha \in x \circ y} \{\mu(\alpha)\} \geq \min\{\mu(x), \mu(y)\}$ and in this way the condition (1) of Definition 3.10 is verified. To verify the second condition, if for every $a, x \in H$ we put $t_1 = \min\{\mu(a), \mu(x)\}$, then $x \in \mu_{t_1}$ and $a \in \mu_{t_1}$. So, there exists $y \in \mu_{t_1}$ such that $x \in a \circ y$. On the other hand, since $y \in \mu_{t_1}$,

then $t_1 \leq \mu(y)$ and hence $\min\{\mu(a), \mu(x)\} \leq \mu(y)$. The proof of the third condition of Definition 3.10 is similar to the proof of the second condition.
∎

The following two corollaries are exactly obtained from Theorem 3.11.

Corollary 3.4. *Let $(H, \circ)$ be an H_v-group and μ be a fuzzy H_v-subgroup of H. If $0 \leq t_1 < t_2 \leq 1$, then $\mu_{t_1} = \mu_{t_2}$ if and only if there is no x in H such that $t_1 \leq \mu(x) < t_2$.*

Corollary 3.5. *Let $(H, \circ)$ be an H_v-group and μ be a fuzzy H_v-subgroup of H. If the range of μ is the finite set $\{t_1, t_2, \ldots, t_n\}$, then the set $\{\mu_{t_i} \mid 1 \leq i \leq n\}$ contains all the level H_v-subgroups of μ. Moreover, if $t_1 > t_2 > \ldots > t_n$, then all the level H_v-subgroups μ_{t_i} form the following chain $\mu_{t_1} \subseteq \mu_{t_2} \subseteq \ldots \subseteq \mu_{t_n}$.*

Proposition 3.16. *Let $(H, \circ)$ be an H_v-group and μ be a fuzzy subset of H. Then, μ is a fuzzy H_v-subgroup of H if and only if for every t, $0 \leq t \leq 1$,*

(1) $\mu_t \circ \mu_t \subseteq \mu_t$,

(2) $a \circ (H - \mu_t) - (H - \mu_t) \subseteq a \circ \mu_t$, *for all* $a \in \mu_t$.

Proof. Let μ be a fuzzy H_v-subgroup of H. Then, by Theorem 3.11, μ_t is an H_v-subgroup of H, and then it is clear that the three conditions of Definition 3.10 are valid.

Conversely, suppose that the three conditions of Definition 3.10 hold. Then, by Theorem 3.11 it is enough to prove that μ_t is an H_v-subgroup of H. For the proof of the left reproduction axiom, it is enough to show that $\mu_t \subseteq a \circ \mu_t$, for every $a \in \mu_t$. Assume that there exists $x \in \mu_t$ such that $x \notin a \circ \mu_t$. Since $x \in \mu_t$, then there exists $b \in H$ such that $x \in a \circ b$. If $b \in \mu_t$, then $x \in a \circ b \subseteq a \circ \mu_t$, which is a contradiction. If $b \in H - \mu_t$, then we have $x \in \{x\} - (H - \mu_t) \subseteq a \circ b - (H - \mu_t) \subseteq a \circ (H - \mu_t) - (H - \mu_t) \subseteq a \circ \mu_t$, again a contradiction. Therefore, $\mu_t - a \circ \mu_t = \emptyset$ which implies that $\mu_t \subseteq a \circ \mu_t$. The proof of the right reproduction axiom is similar. ∎

Theorem 3.12. *Let $(H, \circ)$ be an H_v-group. Then, every H_v-subgroup of H is a level H_v-subgroup of a fuzzy H_v-subgroup of H.*

Proof. Let A be an H_v-subgroup of H. For a fixed real number c, $0 < c \leq 1$, the fuzzy subset μ is defined as follows

$$\mu(x) = \begin{cases} c \text{ if } x \in A, \\ 0 \text{ otherwise.} \end{cases}$$

We have $A = \mu_c$ and by Theorem 3.11, it is enough to prove that μ is a fuzzy H_v-subgroup. This is straightforward and we omit it. ∎

Corollary 3.6. *Let $(H, \circ)$ be an H_v-group and A be a non-empty subset of H. Then, a necessary and sufficient condition for A to be an H_v-subgroup is that $A = \mu_{t_0}$, where μ is a fuzzy H_v-subgroup and $0 < t_0 \leq 1$.*

Definition 3.11. Let $(H, \circ)$ be an H_v-group and μ be a fuzzy H_v-subgroup of H. Then, μ is called *right fuzzy closed with respect to H* if for every a, b in H all the x in $b \in a \circ x$ satisfy $\min\{\mu(b), \mu(a)\} \leq \mu(x)$. We call μ *left fuzzy closed with respect to H* if for all a, b in H all the y in $b \in y \circ a$ satisfy $\min\{\mu(b), \mu(a)\} \leq \mu(y)$. If μ is left and right fuzzy closed, then μ is called *fuzzy closed*.

Theorem 3.13. *If the fuzzy H_v-subgroup μ is right fuzzy closed, then $\mu_t.(H - \mu_t) = H - \mu_t$.*

Proof. If $b \in \mu_t \circ (H - \mu_t)$, then there exists $a \in \mu_t$ and $x \in H - \mu_t$ such that $b \in a \circ x$. Therefore, $\mu(x) < t \leq \mu(a)$ and since μ is right fuzzy closed we get $\min\{\mu(a), \mu(b)\} \leq \mu(x)$. Hence, $\mu(b) \leq \mu(x) < t$ which implies that $b \in H - \mu_t$. So, we have proved $\mu_t \circ (H - \mu_t) \subseteq H - \mu_t$.

On the other hand, if $x \in H - \mu_t$, then for every $a \in \mu_t$ by the reproduction axiom there exists $y \in H$ such that $x \in a \circ y$ and so it is enough to prove $y \in H - \mu_t$. Since μ is a fuzzy H_v-subgroup of H, by the definition we have $\min\{\mu(a), \mu(y)\} \leq \inf_{\alpha \in a \circ y}\{\mu(\alpha)\}$ which implies that

$$\min\{\mu(a), \mu(y)\} \leq \mu(x). \tag{3.4}$$

Since μ is right fuzzy closed so

$$\min\{\mu(x), \mu(a)\} \leq \mu(y). \tag{3.5}$$

Now, from $x \in H - \mu_t$ we get $a \in \mu_t$ and so $\mu(x) < t \leq \mu(a)$. Using 3.5 we obtain $\mu(x) \leq \mu(y)$. Therefore, $\mu(x) \leq \min\{\mu(a), \mu(y)\}$ and by 3.4 the relation $\min\{\mu(a), \mu(y)\} = \mu(x)$ is obtained. But $\mu(x) < \mu(a)$ and hence $\min\{\mu(a), \mu(y)\} = \mu(y)$. So, $\mu(x) = \mu(y)$. Since $x \in H - \mu_t$ we get $y \in H - \mu_t$ and the theorem is proved. ∎

Now, we define anti fuzzy H_v-subgroup of an H_v-group and then we present some results in this connection.

Definition 3.12. Let $(H, \circ)$ be an H_v-group and let μ be a fuzzy subset of H. Then, μ is said to be an *anti fuzzy H_v-subgroup* of H if the following axioms hold.

(1) $\sup\limits_{\alpha \in x \circ y} \{\mu(\alpha)\} \leq \max\{\mu(x), \mu(y)\}$, for all $x, y \in H$.

(2) For all $x, a \in H$ there exists $y \in H$ such that $x \in a \circ y$ and

$$\mu(y) \leq \max\{\mu(a), \mu(x)\}.$$

(3) For all $x, a \in H$ there exists $z \in H$ such that $x \in z \circ a$ and

$$\mu(z) \leq \max\{\mu(a), \mu(x)\}.$$

Condition (2) is called the *left anti fuzzy reproduction axiom*, while (3) is called the *right anti fuzzy reproduction axiom*.

For the sake of similarity, only left reproduction axiom for the H_v-groups is verified throughout this section.

Lemma 3.5. *Let $(H, \circ)$ be an H_v-group and μ be an anti fuzzy H_v-subgroup of H. Then,*

(1) $\sup\limits_{\alpha \in (\dots((x_1 \circ x_2) \circ x_3) \dots) \circ x_n} \{\mu(\alpha)\} \leq \max\{\mu(x_1), \dots, \mu(x_n)\}$, *for all $x_1, x_2, \dots, x_n \in H$,*

(2) $\sup\{\mu(\alpha) \mid \alpha \in \bigcap\limits_{p(x_1, \dots, x_n) \in P_n(x_1, \dots, x_n)} p(x_1, \dots, x_n)\} \leq \max\{\mu(x_1), \dots, \mu(x_n)\}.$

Proof. The proof is similar to the proof of Lemma 3.4. ∎

Proposition 3.17. *Let H be an H_v-group and μ be a fuzzy H_v-subgroup of H. Then, the set $\overline{\mu} = \{x \in H \mid \mu(x) = 1\}$ is empty or an H_v-subgroup of H.*

Proof. Let $\overline{\mu} \neq \emptyset$. Then, for all x, y in $\overline{\mu}$ we have $1 = \min\{\mu(x), \mu(y)\} \leq \inf\limits_{\alpha \in x \circ y} \{\mu(\alpha)\}$. Therefore, for every $\alpha \in x \circ y$ we have $\mu(\alpha) = 1$ which implies that $\alpha \in \overline{\mu}$. So, $x \circ y \subseteq \overline{\mu}$ implying $x \circ y \in \mathcal{P}^*(\overline{\mu})$. Hence, for every $a \in \overline{\mu}$, we have $a \circ \overline{\mu} \subseteq \overline{\mu}$ and to prove the left reproduction axiom it is enough to prove $\overline{\mu} \subseteq a \circ \overline{\mu}$.

Since μ is a fuzzy H_v-subgroup of H, for every $x \in \overline{\mu}$ there exists $y \in H$ such that $x \in a \circ y$ and $\min\{\mu(a), \mu(x)\} \leq \mu(y)$. Since $x \in \overline{\mu}$ and $a \in \overline{\mu}$, we have $\min\{\mu(a), \mu(x)\} = 1$. Therefore, $\mu(y) = 1$ which implies that $y \in \overline{\mu}$ and the proposition is proved. ∎

Example 3.9. Let $(G, \circ)$ be a group and μ be a fuzzy subset of G. We define the hyperoperation $\circ : G \times G \to \mathcal{P}^*(G)$ as follows: $x \circ y = \{t \mid \mu(t) \leq \mu(x \circ y)\}$. Then,

(1) $(G, \circ)$ is an H_v-group.

(2) If μ is an anti fuzzy subgroup of $(G, \circ)$, then μ is an anti fuzzy H_v-subgroup of $(G, \circ)$.

Theorem 3.14. *Let H be an H_v-group and μ be a fuzzy subset of H. Then μ is a fuzzy H_v-subgroup of H if and only if it's complement μ^c is an anti fuzzy H_v-subgroup of H.*

Proof.   Let μ be a fuzzy H_v-subgroup of H, for every x, y in H, we have $\min\{\mu(x), \mu(y)\} \leq \inf_{\alpha \in x \circ y}\{\mu(\alpha)\}$, or $\min\{1 - \mu^c(x), 1 - \mu^c(y)\} \leq \inf_{\alpha \in x \circ y}\{1 - \mu^c(\alpha)\}$, or $\min\{1 - \mu^c(x), 1 - \mu^c(y)\} \leq 1 - \sup_{\alpha \in x \circ y}\{\mu^c(\alpha)\}$, or $\sup_{\alpha \in x \circ y}\{\mu^c(\alpha)\} \leq 1 - \min\{1 - \mu^c(x), 1 - \mu^c(y)\}$, or $\sup_{\alpha \in x \circ y}\{\mu^c(\alpha)\} \leq \max\{\mu^c(x), \mu^c(y)\}$, and in this way the first condition is verified for μ^c.

Since μ is a fuzzy H_v-subgroup of H, for every a, x in H, there exists $y \in H$ such that $x \in a \circ y$ and $\min\{\mu(a), \mu(x)\} \leq \mu(y)$, or $\min\{1 - \mu^c(a), 1 - \mu^c(x)\} \leq 1 - \mu^c(y)$, or $\mu^c(y) \leq 1 - \min\{1 - \mu^c(a), 1 - \mu^c(x)\}$, or $\mu^c(y) \leq \max\{\mu^c(a), \mu^c(x)\}$ and the second condition is satisfied. Thus, μ^c is an anti fuzzy H_v-subgroup. The converse also can be proved similarly. ∎

Now, let H be a non-empty set and μ be a fuzzy subset of H. Then, for $0 \leq t \leq 1$, the lower level subset of μ is the set $\overline{\mu_t} = \{x \in H \mid \mu(x) \leq t\}$. Clearly $\overline{\mu_1} = H$ and if $t_1 < t_2$, then $\overline{\mu_{t_1}} \subseteq \overline{\mu_{t_2}}$.

Theorem 3.15. *Let H be an H_v-group and μ be a fuzzy subset of H. Then, μ is an anti fuzzy H_v-subgroup of H if and only if for every t, $0 \leq t \leq 1$, $\overline{\mu_t} \neq \emptyset$ is an H_v-subgroup of H.*

Proof.   Let μ be an anti fuzzy H_v-subgroup of H. For every x, y in $\overline{\mu_t}$ we have $\mu(x) \leq t$ and $\mu(y) \leq t$. Hence, $\max\{\mu(x), \mu(y)\} \leq t$ and so $\sup_{\alpha \in x \circ y}\{\mu(\alpha)\} \leq t$. Therefore, for every $\alpha \in x \circ y$ we have $\mu(\alpha) \leq t$ which implies that $\alpha \in \overline{\mu_t}$, so $x \circ y \subseteq \overline{\mu_t}$ implying $x \circ y \in \mathcal{P}^*(\overline{\mu_t})$. Hence, for every $a \in \overline{\mu_t}$ we have $a \circ \overline{\mu_t} \subseteq \overline{\mu_t}$ and to prove this part of the theorem it is enough to prove that $\overline{\mu_t} \subseteq a \circ \overline{\mu_t}$.

Since μ is an anti fuzzy H_v-subgroup of H, for every $x \in \overline{\mu_t}$ there exists $y \in H$ such that $x \in a \circ y$ and $\mu(y) \leq \max\{\mu(a), \mu(x)\}$. From $x \in \overline{\mu_t}$ and $a \in \overline{\mu_t}$ we get $\max\{\mu(x), \mu(a)\} \leq t$ and so $y \in \overline{\mu_t}$. Therefore, we have proved that for every $x \in \overline{\mu_t}$ there exists $y \in \overline{\mu_t}$ such that $x \in a \circ y$ implying that $x \in a \circ \overline{\mu_t}$ and this proves $\overline{\mu_t} \subseteq a \circ \overline{\mu_t}$.

Conversely, assume that for every t, $0 \leq t \leq 1$, $\overline{\mu_t} \neq \emptyset$ is an H_v-subgroup of H. For every x, y in H we can write $\mu(x) \leq \max\{\mu(x), \mu(y)\}$

and $\mu(y) \leq \max\{\mu(x), \mu(y)\}$ and if we put $t_0 = \max\{\mu(x), \mu(y)\}$, then $x \in \overline{\mu_{t_0}}$ and $y \in \overline{\mu_{t_0}}$. Since $\overline{\mu_{t_0}}$ is an H_v-subgroup, so $x \circ y \subseteq \overline{\mu_{t_0}}$. Therefore, for every $\alpha \in x \circ y$ we have $\mu(\alpha) \leq t_o$ implying $\sup\limits_{\alpha \in x \circ y} \{\mu(\alpha)\} \leq t_0$ and so $\sup\limits_{\alpha \in x \circ y} \{\mu(\alpha)\} \leq \max\{\mu(x), \mu(y)\}$ and in this way the first condition of Definition 3.12 is verified. To verify the second condition, if for every $a, x \in H$ we put $t_1 = \max\{\mu(a), \mu(x)\}$, then $x \in \overline{\mu_{t_1}}$ and $a \in \overline{\mu_{t_1}}$. Since $a \circ \overline{\mu_{t_1}} = \overline{\mu_{t_1}}$, so there exists $y \in \overline{\mu_{t_1}}$ such that $x \in a \circ y$. On the other hand, since $y \in \overline{\mu_{t_1}}$, then $\mu(y) \leq t_1$ and hence $\mu(y) \leq \max\{\mu(a), \mu(x)\}$ and the second condition of Definition 3.12 is satisfied. ∎

Definition 3.13. Let X be a non-empty set and x_t, with $t \in [0, 1]$, be a fuzzy point of X characterized by the fuzzy subset μ defined by $\mu(x) = t$ and $\mu(y) = 0$, for $y \in X - \{x\}$. We define the *height* of x_t by $hgt(\mu) = t$. Moreover,

$$\widetilde{\mu} = \{x_t \mid \mu(x) \geq t, \ x \in X\}.$$

and $\widetilde{X}$ the family of all fuzzy points in X. The *support* of fuzzy subset μ of X is the set $supp(\mu) = \{x \in X \mid \mu(x) > 0\}$.

Proposition 3.18. *Let H be an H_v-group and μ be a fuzzy H_v-subgroup of H. Then, the set $supp(\mu)$ is an H_v-subgroup of H.*

Proof. For every x, y in $supp(\mu)$, we have $\mu(x) > 0, \mu(y) > 0$ and so $\min\{\mu(x), \mu(y)\} > 0$ which implies that $\inf\limits_{\alpha \in x \circ y} > 0$. Therefore, for every $\alpha \in x \circ y$ we have $\mu(\alpha) > 0$ which implies that $\alpha \in supp(\mu)$. Hence, for every $a \in supp(\mu)$, we have $a \circ supp(\mu) \subseteq supp(\mu)$ and to prove the left reproduction axiom it is enough to prove $supp(\mu) \subseteq a \circ supp(\mu)$.

Since μ is a fuzzy H_v-subgroup of H, for every $x \in supp(\mu)$ there exists $y \in H$ such that $x \in a \circ y$ and $\min\{\mu(a), \mu(x)\} \leq \mu(y)$. Since $x \in supp(\mu)$ and $a \in supp(\mu)$ we have $\min\{\mu(a), \mu(x)\} > 0$. Therefore, $\mu(y) > 0$ which implies that $y \in supp(\mu)$ and the proposition is proved. ∎

Definition 3.14. Let H be an H_v-group and μ be a fuzzy H_v-subgroup of H. We define the following hyperoperation on $\widetilde{\mu}$,

$$\circ : \widetilde{\mu} \times \widetilde{\mu} \to \mathcal{P}^*(\widetilde{\mu})$$
$$x_t \circ y_s = \{\alpha_{t \wedge s} \mid \alpha \in x \circ y\},$$

where $t \wedge s = \min\{t, s\}$.

Suppose that $x_t, \ y_s \in \widetilde{\mu}$. Then, $\mu(x) \geq t, \ \mu(y) \geq s$ and so

$\min \{\mu(x), \mu(y)\} \geq t \wedge s$ which implies that $\inf\limits_{\alpha \in x \circ y} \{\mu(\alpha)\} \geq t \wedge s$. Therefore, for every $\alpha \in x \circ y$, we have $\alpha_{t \wedge s} \in \widetilde{\mu}$.

Lemma 3.6. $(\widetilde{\mu}, \circ)$ *is an* H_v-*semigroup.*

Proof. For every $x_t, y_s, z_r \in \widetilde{\mu}$, we have

$$(x_t \circ y_s) \circ z_r = \{\alpha_{(t \wedge s) \wedge r} \mid \alpha \in (x \circ y) \circ z\},$$
$$x_t \circ (y_s \circ z_r) = \{\alpha_{t \wedge (s \wedge r)} \mid \alpha \in x \circ (y \circ z)\}.$$

Since $(H, \circ)$ is weak associative, $(H, \circ)$ is weak associative. ∎

Definition 3.15. Let H_1, H_2 be two H_v-groups and μ_1, μ_2 be fuzzy H_v-subgroups of H_1, H_2, respectively. Let $\widetilde{f}$ be a mapping from $\widetilde{\mu}_1$ into $\widetilde{\mu}_2$ such that $supp\, \widetilde{f}\, (x_t) = supp\, \widetilde{f}\, (x_s)$, for all $x_t, x_s \in \widetilde{\mu}_1$. Then, f is called

(1) a *strong fuzzy homomorphism* if

$$\widetilde{f}\, (x_t \circ y_s) = \widetilde{f}\, (x_t) \circ \widetilde{f}\, (y_s), \text{ for all } x_t, y_s \in \widetilde{\mu}_1,$$

(2) an *inclusion fuzzy homomorphism* if

$$\widetilde{f}\, (x_t \circ y_s) \subseteq \widetilde{f}\, (x_t) \circ \widetilde{f}\, (y_s), \text{ for all } x_t, y_s \in \widetilde{\mu}_1,$$

(3) a *fuzzy* H_v-*homomorphism* if

$$\widetilde{f}\, (x_t \circ y_s) \cap \widetilde{f}\, (x_t) \circ \widetilde{f}\, (y_s) \neq \emptyset, \text{ for all } x_t, y_s \in \widetilde{\mu}_1 .$$

A mapping $\widetilde{f} : \widetilde{\mu}_1 \rightarrow \widetilde{\mu}_2$ is called a *fuzzy isomorphism* if it is bijective and strong fuzzy homomorphism. Two fuzzy H_v-subgroups μ_1 and μ_2 are said to be *fuzzy isomorphic*, denoted by $\mu_1 \cong \mu_2$, if there exists a fuzzy isomorphism from $\widetilde{\mu}_1$ onto $\widetilde{\mu}_2$.

Theorem 3.16. *Let* H_1, H_2 *be two* H_v-*groups and* μ_1, μ_2 *be fuzzy* H_v-*subgroups of* H_1, H_2, *respectively. Let* $\widetilde{f} : \widetilde{\mu}_1 \rightarrow \widetilde{\mu}_2$ *be a fuzzy inclusion homomorphism. Then,*

(1) $hgt\, \widetilde{f}\, (x_t) = hgt\, \widetilde{f}\, (y_t)$.
(2) $hgt\, \widetilde{f}\, (x_t) \leq hgt\, \widetilde{f}\, (x_s)$, *whenever* $t \leq s$.

Proof. (1) For every x_t and y_t in $\widetilde{\mu}_1$, there exists $z \in H_1$ such that $y \in x \circ z$ and $\min \{\mu_1(x), \mu_1(y)\} \leq \mu_1(z)$. Since $\mu_1(x) \geq t$ and $\mu_1(y) \geq t$, we have $\mu_1(z) \geq t$ which implies that $z_t \in \tilde{\mu}_1$. From $y_t \in x_t \circ z_t$ we get $\tilde{f}(y_t) \in \tilde{f}(x_t \circ z_t)$ or $\tilde{f}(y_t) \in \tilde{f}(x_t) \circ \tilde{f}(z_t)$ and so

$$hgt\, \tilde{f}(y_t) = \min\{hgt\, \tilde{f}(x_t), hgt\, \tilde{f}(z_t)\} \leq hgt\, \tilde{f}(x_t).$$

Similarly, we obtain $hgt\, \tilde{f}(x_t) \leq hgt\, \tilde{f}(y_t)$. Therefore, $hgt\, \tilde{f}(x_t) = hgt\, \tilde{f}(y_t)$.

(2) Suppose that $t \leq s$. If $x_t \in \tilde{\mu}_1$, then there exists $y \in H_1$ such that $x \in x \circ y$ and $\mu_1(x) \leq \mu_1(y)$, so $y_t \in \tilde{\mu}_1$. From $x_t \in x_t \circ y_t$ we have $\tilde{f}(x_t) \in \tilde{f}(x_t \circ y_t)$ which implies $\tilde{f}(x_t) \in \tilde{f}(x_s \circ y_t)$ or $\tilde{f}(x_t) \in \tilde{f}(x_s) \circ \tilde{f}(y_t)$. Therefore,

$$hgt\, \tilde{f}(x_t) = \min\{hgt\, \tilde{f}(x_s), hgt\, \tilde{f}(y_t)\} \leq hgt\, \tilde{f}(x_s). \quad \blacksquare$$

Theorem 3.17. *Let H_1, H_2 be two H_v-groups and μ_1, μ_2 be fuzzy H_v-subgroups of H_1, H_2, respectively. A mapping $\tilde{f} : \tilde{\mu}_1 \to \tilde{\mu}_2$ is a fuzzy strong homomorphism if and only if there exists an ordinary strong homomorphism of H_v-groups $f : supp(\mu_1) \to supp(\mu_2)$ and an increasing function $\varphi : (0,1] \to (0,1]$ such that*

$$\tilde{f}(x_t) = [f(x)]_{\varphi(t)}, \text{ for all } x_t \in \tilde{\mu}_1.$$

Proof. Suppose that $\tilde{f} : \tilde{\mu}_1 \to \tilde{\mu}_2$ is a fuzzy strong homomorphism. We define a mapping $f : supp(\mu_1) \to supp(\mu_2)$ and a function $\varphi : (0,1] \to (0,1]$ as follows:

$$f(x) = supp\, \tilde{f}(x_{\mu_1(x)}), \text{ for all } x \in supp(\mu_1)$$

and

$$\varphi(t) = hgt\, \tilde{f}(x_t), \text{ for all } t \in (0,1].$$

Since $\tilde{f}$ is a fuzzy strong homomorphism, then $supp\, \tilde{f}(x_t) = supp\, \tilde{f}(x_{\mu_1(x)})$ and so $supp\, \tilde{f}(x_t) = f(x)$ which implies that $\tilde{f}(x_t) = [f(x)]_{\varphi(t)}$. By definition of φ and Theorem 3.16, it is easy to see that φ is increasing. Therefore, it remains to show that f is a strong homomorphism from the H_v-group $supp(\mu_1)$ into the H_v-group $supp(\mu_2)$.

For every $x, y \in supp(\mu_1)$, we put $\mu_1(x) = t$ and $\mu_1(y) = s$. Then, we have

$$[f(x \circ y)]_{\varphi(t \wedge s)} = \bigcup_{\alpha \in x \circ y} [f(\alpha)]_{\varphi(t \wedge s)}$$

$$= \bigcup_{\alpha \in x \circ y} \tilde{f}\ (\alpha_{t \wedge s})$$

$$= \tilde{f}\ (\bigcup_{\alpha \in x \circ y} \alpha_{t \wedge s})$$

$$= \tilde{f}\ (x_t \circ y_s) = \tilde{f}\ (x_t) \circ \tilde{f}\ (y_s)$$

$$= [f(x)]_{\varphi(t)} \circ [f(y)]_{\varphi(s)}$$

$$= \bigcup_{z \in f(x) \circ f(y)} z_{\varphi(t) \wedge \varphi(s)}$$

$$= [f(x) \circ f(y)]_{\varphi(t) \wedge \varphi(s)}.$$

Therefore, $f(x \circ y) = f(x) \circ f(y)$, i.e., f is a strong homomorphism.

Conversely, we consider a mapping $\tilde{f} : \tilde{\mu}_1 \to \tilde{\mu}_2$ such that $\tilde{f}\ (x_t) = [f(x)]_{\varphi(t)}$. It is enough to show that $\tilde{f}$ is a strong fuzzy homomorphism. For every $x_t, y_s \in \tilde{\mu}_1$ $(t \leq s)$, we have

$$\tilde{f}\ (x_t \circ y_s) = \bigcup_{\alpha \in x \circ y} \tilde{f}\ (\alpha_{t \wedge s})$$

$$= \bigcup_{\alpha \in x \circ y} [f(\alpha)]_{\varphi(t \wedge s)}$$

$$= \bigcup_{\alpha \in x \circ y} [f(\alpha)]_{\varphi(t)}$$

$$= [f(x \circ y)]_{\varphi(t)} = [f(x) \circ f(y)]_{\varphi(t) \wedge \varphi(s)}$$

$$= [f(x)]_{\varphi(t)} \circ [f(y)]_{\varphi(s)}$$

$$= \tilde{f}\ (x_t) \circ \tilde{f}\ (y_s).$$ ∎

Let f be a strong homomorphism from H_1 into H_2. We can define a mapping $\tilde{f} : \tilde{H}_1 \to \tilde{H}_2$ as follows: $\tilde{f}\ (x_t) = [f(x)]_t$. Obviously, $\tilde{f}$ is a strong fuzzy homomorphism from $\tilde{H}_1$ into $\tilde{H}_2$, where $\varphi(\lambda) = \lambda, \forall \lambda \in (0, 1]$. Therefore, the concept of strong fuzzy homomorphism between two H_v-groups can be seen as an extension of the concept of strong homomorphism between two H_v-groups.

Remark 3.5. Let $f : X \to Y$ and let $\varphi : (0, 1] \to (0, 1]$ be an increasing mapping. We define the mapping $f_\varphi : \tilde{X} \to \tilde{Y}$ by $f_\varphi(x_t) = [f(x)]_{\varphi(t)}$. Then, for every fuzzy subset μ of X we have

$$f_\varphi(\mu)(y) = \sup_{x \in f^{-1}(y)} \{\varphi(\mu(x))\}.$$

Theorem 3.18. *Let φ be bijective and $f : H_1 \to H_2$ be a surjective strong homomorphism and let μ be a fuzzy H_v-subgroup of H_1. Then, $f_\varphi(\mu)$ is a fuzzy H_v-subgroup of H_2.*

Proof.   Let μ be a fuzzy H_v-subgroup of H_1. By Theorem 3.11, for every t, $0 \le t \le 1$, level subset $\mu_t (\mu_t \neq \emptyset)$ is an H_v-subgroup of H_1 and so $f(\mu_{\varphi^{-1}(t)})$ is an H_v-subgroup of H_2. Now, it is enough to show that

$$f(\mu_{\varphi^{-1}(t)}) = (f_\varphi(\mu))_t.$$

For every y in $(f_\varphi(\mu))_t$ we have $f_\varphi(\mu)(y) \ge t$ which implies that

$$\sup_{x \in f^{-1}(y)} \{\varphi(\mu(x))\} \ge t.$$

Therefore, there exists $x_0 \in f^{-1}(y)$ such that $\varphi(\mu(x_0)) \ge t$ which implies that $\mu(x_0) \ge \varphi^{-1}(t)$ or $x_0 \in \mu_{\varphi^{-1}(t)}$ and so $f(x_0) \in f(\mu_{\varphi^{-1}(t)})$ implying $y \in f(\mu_{\varphi^{-1}(t)})$. Now, for every y in $f(\mu_{\varphi^{-1}(t)})$, there exists $x \in \mu_{\varphi^{-1}(t)}$ such that $f(x) = y$. Since $x \in \mu_{\varphi^{-1}(t)}$, we have $\mu(x) \ge \varphi^{-1}(t)$ or $\varphi(\mu(x)) \ge t$ and so $\sup_{x \in f^{-1}(y)} \{\varphi(\mu(x))\} \ge t$ which implies that $f_\varphi(\mu)(y) \ge t$. Therefore, $y \in (f_\varphi(\mu))_t$. ∎

3.7 H_v-semigroups and noise problem

Noise pollution is displeasing human, animal or machine-created sound that disrupts the activity or balance of human or animal life. The source of most outdoor noise worldwide is not only transportation systems (including motor vehicle noise, aircraft noise and rail noise), but, noise caused by people as well (audio entertainment systems, electric megaphones and loud people) [54]. The fact that noise pollution is also a cause of annoyance, is that, a 2005 study by Spanish researchers found that in urban areas households are willing to pay approximately 4 Euros per decibel per year for noise reduction [11]. Poor urban planning may give rise to noise pollution, since side-by-side industrial and residential buildings can result in noise pollution in the residential area. We set the following problem: The noise pollution that comes from a certain block of flats in urban areas, obviously annoys not only the block of flats itself but possibly neighboring blocks of flats or buildings, as well. If every city is considered as a set Ω with elements the blocks of flats or the buildings, then the above situation could be described with an algebraic hyperstructure and its properties. In this section, the main reference is [53] and we present the right reproductive

H_v-semigroup, as a tool to study the noise pollution problem in urban areas.

Definition 3.16. A hypergroupoid $(H, \circ)$ such that the weak associativity holds and $H \circ x = H$, for all $x \in H$, is called *right reproductive H_v-semigroup*. A hypergroupoid $(H, \circ)$ such that the weak associativity holds and $x \circ H = H$, for all $x \in H$, is called *left reproductive H_v-semigroup*.

Now we give the following definition.

Definition 3.17. Let $\Omega \neq \varnothing$ and $f : \Omega \to \mathcal{P}(\Omega)$ be a map, then we define a hyperoperation $r_L \colon \Omega \times \Omega \to \mathcal{P}(\Omega)$, on Ω as follows: for all $x, y \in \Omega$, we set

$$x r_L y = f(x) \cup \{x\}.$$

We call the hyperoperation r_L, *noise hyperoperation*. Note that the noise hyperoperation, always contains the element $x \in \Omega$. That means that the element $x \in \Omega$ could be considered as the representative of the elements of the set $x r_L y$. So, we symbolize:

$$x r_L y = f(x) \cup \{x\} = [x].$$

If $x \in f(x)$, for all $x \in \Omega$, then the hyperoperation is simplified as

$$x r_L y = f(x) = [x].$$

Therefore, the noise hyperoperation r_L depends only on the left element. That means that if one composes an element x, on the left, with any other element y, on the right, then the result is always the same set $[x]$.

Example 3.10. Consider a set $\Omega = \{x_1, x_2, x_3, x_4, x_5, x_6, x_7, x_8, x_9\}$ and a map $f : \Omega \to \mathcal{P}(\Omega)$ such that

$$
\begin{aligned}
f(x_1) &= \{x_2\}, \\
f(x_2) &= \{x_2, x_3\}, \\
f(x_3) &= \{x_2\}, \\
f(x_4) &= \{x_4\} \\
f(x_5) &= \{x_5, x_6, x_7\}, \\
f(x_6) &= \{x_6, x_7\}, \\
f(x_7) &= \{x_5\}, \\
f(x_8) &= \{x_8, x_9\}, \\
f(x_9) &= \emptyset.
\end{aligned}
$$

Then, as in the defined above noise hyperoperation, we have

$$[x_1] = f(x_1) \cup \{x_1\} = \{x_1, x_2\},$$
$$[x_2] = f(x_2) = \{x_2, x_3\},$$
$$[x_3] = f(x_3) \cup \{x_3\} = \{x_2, x_3\}$$
$$[x_4] = f(x_4) = \{x_4\},$$
$$[x_5] = f(x_5) = \{x_5, x_6, x_7\},$$
$$[x_6] = f(x_6) = \{x_6, x_7\}$$
$$[x_7] = f(x_7) \cup \{x_7\} = \{x_5, x_7\},$$
$$[x_8] = f(x_8) = \{x_8, x_9\},$$
$$[x_9] = f(x_9) \cup \{x_9\} = \{x_9\}.$$

Then, the "multiplication" table of (r_L) is given by:

r_L	x_1	x_2	x_3	x_4	x_5	x_6	x_7	x_8	x_9
x_1	x_1, x_2	x_1, x_2	x_1, x_2	x_1, x_2	x_1, x_2	x_1, x_2	x_1, x_2	x_1, x_2	x_1, x_2
x_2	x_2, x_3	x_2, x_3	x_2, x_3	x_2, x_3	x_2, x_3	x_2, x_3	x_2, x_2	x_2, x_3	x_2, x_3
x_3	x_2, x_3	x_2, x_3	x_2, x_3	x_2, x_3	x_2, x_3	x_2, x_3	x_2, x_2	x_2, x_3	x_2, x_3
x_4	x_4	x_4	x_4	x_4	x_4	x_4	x_4	x_4	x_4
x_5	$x_5, x_6,$ x_7	$x_5, x_6,$ x_7	x_5, x_6, x_7	$x_5, x_6,$ x_7	$x_5, x_6,$ x_7	$x_5, x_6,$ x_7	$x_5, x_6,$ x_7	$x_5, x_6,$ x_7	$x_5, x_6,$ x_7
x_6	x_6, x_7	x_6, x_7	x_6, x_7	x_6, x_7	x_6, x_7	x_6, x_7	x_6, x_7	x_6, x_7	x_6, x_7
x_7	x_5, x_7	x_5, x_7	x_5, x_7	x_5, x_7	x_5, x_7	x_5, x_7	x_5, x_7	x_5, x_7	x_5, x_7
x_8	x_8, x_9	x_8, x_9	x_8, x_9	x_8, x_9	x_8, x_9	x_8, x_9	x_8, x_9	x_8, x_9	x_8, x_9
x_9	x_9	x_9	x_9	x_9	x_9	x_9	x_9	x_9	x_9

Example 3.11. Let $X \neq \varnothing$ and $\mu : X \to [0, 1]$ be a fuzzy subset of X. We define the hyperoperation $(\odot)$ on X as follows: for all $x, y \in X$, $\odot : X \times X \to \mathcal{P}(X)$ such that

$$x \odot y = \{z \in X \mid \mu(z) = \mu(x)\}.$$

Then, consider the map $f(x) = \{z \in X \mid \mu(z) = \mu(x)\}$. Since $x \in f(x)$, for all $x \in X$, as above we have the hyperoperation r_L on X as follows: $r_L : X \times X \to \mathcal{P}(X)$, for all $x, y \in X$, such that $x r_L y = f(x)$.

Some properties of (r_L) are as follows.

(1) $x r_L \Omega = [x]$, for all $x \in \Omega$.
(2) $[x] r_L y = [x] r_L [y] \supseteq [x]$, for all $y \in \Omega$.
(3) $x^2 = x r_L x = [x]$, for all $x \in \Omega$.

Proposition 3.19. *The hypergroupoid (Ω, r_L) is an H_v-semigroup.*

Proof. We have to prove that the weak associativity holds. Indeed, for all $x, y, z \in \Omega$,

$$xr_L(yr_Lz) = \bigcup_{v \in yr_Lz} (xr_Lv) = \bigcup_{v \in [y]} (xr_Lv) = [x],$$

$$(xr_Ly)r_Lz = \bigcup_{w \in xr_Ly} (wr_Lz) = \bigcup_{w \in [x]} (wr_Lz) = \bigcup_{w \in [x]} [w] \supseteq [x].$$

Therefore, $(xr_Ly)r_Lz \supseteq xr_L(yr_Lz)$. Thus, we have

$$(xr_Ly)r_Lz \cap xr_L(yr_Lz) \neq \varnothing,$$

for all $x, y, z \in \Omega$. ∎

Proposition 3.20. *For all $x \in \Omega$, $\Omega r_Lx = \Omega$ and $xr_L\Omega = [x]$.*

Proof. For all $x \in \Omega$, we have

$$\Omega r_Lx = \bigcup_{\omega \in \Omega} (\omega r_Lx) = \bigcup_{\omega \in \Omega} [\omega] = \Omega.$$

On the other hand,

$$xr_L\Omega = \bigcup_{\omega \in \Omega} (xr_L\omega) = [x].$$ ∎

Proposition 3.21. *The hypergoupoid (Ω, r_L) is a right reproductive H_v-semigroup.*

Proof. It is straightforward. ∎

Note that the right reproductive H_v-semigroup (Ω, r_L) is an H_v-group if, for all $x \in \Omega$, we have $xr_L\Omega = \Omega$.

Proposition 3.22. *The strong associativity of (r_L) is valid if and only if we have*

$$\bigcup_{w \in [x]} (wr_Lz) = [x], \text{ for all } x, z \in \Omega.$$

Proof. Suppose that $(x, y, z) \in \Omega^3$ such that $(xr_Ly)r_Lz = xr_L(yr_Lz)$. Then,

$$(xr_Ly)r_Lz = xr_L(yr_Lz) \Rightarrow [x]r_Lz = xr_L[y]$$
$$\Rightarrow [x]r_Lz = [x]$$
$$\Rightarrow \bigcup_{w \in [x]} (wr_Lz) = [x].$$

Now, let $(x, y, z) \in \Omega^3$ such that

$$\bigcup_{w \in [x]} (wr_Lz) = [x].$$

Then,

$$(xr_Ly)r_Lz = [x]r_Lz = \bigcup_{w \in [x]} (wr_Lz) = [x],$$
$$xr_L(yr_Lz) = xr_L[y] = [x].$$ ∎

For the hyperoperation r_L, we shall check the conditions such that the strong or the weak commutativity is valid.

Proposition 3.23. *If $y \in [x]$ and $x \in [y]$, for all $x, y \in \Omega$, then the weak commutativity of (r_L) is valid. The strong commutativity of (r_L) is valid, iff $[x] = [y]$, for all $x, y \in \Omega$.*

Proof.   If $y \in [x]$ and $x \in [y]$, for all $x, y \in \Omega$, then

$$y \in [x] \text{ and } x \in [y] \Rightarrow x, y \in [x] \text{ and } x, y \in [y]$$
$$\Rightarrow [x] \cap [y] \neq \varnothing$$
$$\Rightarrow (xr_Ly) \cap (yr_Lx) \neq \varnothing.$$

The proof for the strong commutativity is straightforward.   ∎

Proposition 3.24. *Let $(\Omega, +, r_L)$ be an H_v-ring. If $xr_L\Omega = \Omega$, for all $x \in \Omega$ then the hyperstructure $(\Omega, +, r_L)$ is a dual H_v-ring, i.e. both $(\Omega, +, r_L)$ and $(\Omega, r_L, +)$ are H_v-rings.*

Proof.   We have that the (Ω, r_L) is an H_v-group. For the weak distributivity of $(+)$ with respect to (r_L) we have for all $x, y, z \in \Omega$,

$$x + (yr_Lz) \supseteq x + y$$

and

$$(x + y)r_L(x + z) = \bigcup_{s \in x+y, t \in x+z} (sr_Lt) \supseteq \bigcup_{s \in x+y} s = x + y.$$

So, we have

$$[x + (yr_Lz)] \cap [(x + y)r_L(x + z)] \neq \varnothing, \text{ for all } x, y, z \in \Omega.$$

Similarly, the weak distributivity of $+$ with respect to r_L from the right side.   ∎

Proposition 3.25.

(1) *All the elements of Ω are right unit elements with respect to r_L.*
(2) *All the elements of Ω are left absorbing-like elements with respect to r_L.*

Proof.   Since $x \in xr_Ly$, for all $x, y \in \Omega$, the proof is straightforward.   ∎

Proposition 3.26. *The left scalar elements of the H_v-semigroup (Ω, r_L), are left absorbing elements.*

Proof. Let $\beta \in \Omega$ be a left scalar unit element, then $\beta r_L x = x$, for all $x \in \Omega$. But since $\beta \in \beta r_L x$, for all $x \in \Omega$, we obtain that $\beta r_L x = \beta$, for all $x \in \Omega$. ∎

Proposition 3.27. *The right scalar unit elements of the H_v-semigroup (Ω, r_L), are idempotent elements.*

Proof. If $\alpha \in \Omega$ is a right scalar unit element, then $x r_L \alpha = x$, for all $x \in \Omega$. So, $\alpha r_L \alpha = \alpha$. This implies that $\alpha^2 = \alpha$. ∎

Proposition 3.28. *If there exists $x \in \Omega$ such that $f(x) = x$ or $[x] = x$, then x is left absorbing element and every element of (Ω, r_L) is right scalar unit of x.*

Proof. Suppose there exists $x \in \Omega$ such that $[x] = x$, then for all $y \in \Omega$, we have

$$x r_L y = [x] = x.$$

That means that x is left absorbing element and every element of (Ω, r_L) is right scalar unit of x. ∎

Since all the elements of the H_v-semigroup (Ω, r_L) are right unit elements, let us denote by $I^l_{r_L}(x, y)$ the set of the left inverses of the element $x \in \Omega$, associated with the right unit $y \in \Omega$, with respect to the hyperoperation (r_L). The set of the right inverses of the element $x \in \Omega$, associated with the right unit $y \in \Omega$, with respect to the hyperoperation (r_L), is denoted by $I^r_{r_L}(x, y)$.

Proposition 3.29. $y \in I^l_{r_L}(x, y)$.

Proof. Suppose that $x' \in \Omega$ such that $x' \in I^l_{r_L}(x, y) \Rightarrow y \in x' r_L x$. But, for all $x \in \Omega$ the relation $y \in y r_L x$ is valid. That means that $y \in I^l_{r_L}(x, y)$. ∎

Proposition 3.30. $I^r_{r_L}(x, y) = \Omega$ *if and only if* $y \in [x]$.

Proof. Let $y \in \Omega$ be right unit element and $x \in \Omega$, then
$$y \in [x] \Leftrightarrow y \in x r_L x', \text{ for all } x' \in \Omega$$
$$\Leftrightarrow x' \in I^r_{r_L}(x, y), \text{ for all } x'\Omega$$
$$\Leftrightarrow I^r_{r_L}(x, y) = \Omega.$$
∎

Since $x \in [x]$, for all $x \in \Omega$, the following corollary is obvious.

Corollary 3.7. $I^r_{r_L}(x, x) = \Omega$.

Remark 3.6. Notice that, according to Example 3.10, the elements x_4 and x_9 are idempotent elements, since $x_4^2 = x_4$ and $x_9^2 = x_9$. They are, also, left absorbing elements, since $x_4 r_L x = x_4$ and $x_9 r_L x = x_9$, for all $x \in \Omega$. Also, taking for instance, the element x_2 of Ω, notice that $I_{r_L}^l (x, x_2) = \{x_1, x_2, x_3\}$, for all $x \in \Omega$. Even more, since $x_2 \in [x_1]$ we obtain that $I_{r_L}^r (x_1, x_2) = \Omega$ and $I_{r_L}^r (x_1, x_1) = \Omega$.

Definition 3.18. The inclusion associativity in an H_v-semigroup can be held in two ways, either $(xy)z \subset x(yz)$ and then is called *inclusion on the right parenthesis* or $(xy)z \supseteq x(yz)$ and then is called *inclusion on the left parenthesis*.

Proposition 3.31. *Let $(H, \circ)$ be an H_v-semigroup where the inclusion on the left parenthesis holds. Let $a_1, a_2, , a_n$ be n elements (distinct or not) of H, then we set:*

$$p = a_1 \circ a_2 \circ, ..., \circ a_n = [...[(a_1 \circ a_2) \circ a_3] a_4 \circ ...] a_n.$$

Then, every other possible meaningful expression p' obtained by inserting parentheses in the finite sequence $a_1, a_2, ..., a_n$ is subset to p. Remark that for the above H_v-semigroup and for any positive integer n, the nth power a^n, of an element a is

$$a^n = [...[(a \circ a) \circ a] a \circ ...] a.$$

Proof.   It is straightforward.   ∎

Proposition 3.32. *The right reproductive H_v-semigroup (Ω, r_L) is single-power cyclic with infinite period with generator $x \in \Omega$ if and only if there exists $s_0 \geq 2$ such that for all $s \geq s_0$, we have $\cup [x]^{s-2} = \Omega$.*

Proof.   For all $x \in \Omega$, we have

$$x^2 = x r_L x = [x],$$
$$x^3 = x^2 r_L x = [x] r_L [x] = [x]^2,$$
$$x^4 = [x]^3,$$
$$...$$
$$x^{s-1} = [x]^{s-2}.$$

Suppose that there exists $s_0 \geq 2$, such that for all $s \geq s_0$, we have

$$\cup [x]^{s-2} = \Omega \Leftrightarrow [x] \cup [x]^2 \cup ... \cup [x]^{s-2} = \Omega \Leftrightarrow x^2 \cup x^3 \cup ... \cup x^{s-1} = \Omega$$

and since $x \in x^2$, it follows that there exists $s_0 \geq 1$ such that for all $s \geq s_0$

$$x^1 \cup x^2 \cup x^3 \cup ... \cup x^{s-1} = \Omega.$$

So, every element of Ω belongs to a power of x.

But, since the next one is valid, i.e., $x^2 = [x] \ni x$, $x^3 = x^2 r_L x \supseteq x^2$ and by induction $x^1 \subset x^2 \subset ... \subset x^{s-1} \subset x^s$, for all $x \in \Omega$, then there exists $s_0 \geq 1$ such that for all $s \geq s_0$, $x^s = \Omega$. So, there exists $s_0 \geq 1$ such that for all $s \geq s_0$

$$x^1 \cup x^2 \cup x^3 \cup ... \cup x^{s-1} \subset x^s. \qquad \blacksquare$$

As we mentioned above, the noise pollution in urban areas coming from a spot, annoys a certain area in which the noisy spot belongs to. That was the motivation which led to the mathematical expression $x r_L y = [x]$, for all $x, y \in \Omega$. That means that if a city is considered as a set Ω with elements its buildings (or spots which could produce noise pollution), then every building (or a spot) x, which is a source of noise pollution, together with any other building (or a spot) y of the city, will affect anyhow the noise pollution area $[x]$, where $x \in [x]$ and maybe y. It is clear, that the source of the noise pollution x, could not be seen as the center of a cyclic disk, but as any spot of a certain area which is affected by x. We shall try to explain some of the properties of the noise hyperoperation (r_L) developed above, in terms of noise pollution problems in urban areas. The property $x \in x r_L y$, for all $y \in \Omega$ means that the building x, as a source of noise pollution, first of all, annoys the residents of the building x. The property $r_L[y] = [x]$ means that the source of noise pollution x together with any region $[y]$ is not only independent on the spots of the region $[y]$ but the noise pollution region remains $[x]$, as well. The property $[x] r_L y = [x] r_L [y] \supseteq [x]$ means that the noise pollution region that results when either the noise pollution region $[x]$ operates with the spot y or with the region $[y]$, is the same and anyhow this noise pollution region is bigger than $[x]$. The property $x r_L y = x r_L z$ means that $[x]$ remains the noise pollution region when x as a source of noise pollution affects any other spot of the city Ω. Continuously, the relation $x r_L \Omega \neq \Omega$ means that, the noise pollution region coming from spot x, can't affect the whole city Ω. The weak associativity which is expressed by the inclusion on the left parenthesis, i.e., $(x r_L y) r_L z \supseteq x r_L (y r_L z)$ actually means that, the noise pollution region coming from the noise pollution region $[x]$ together with any spot, is not only bigger than that one which comes from the noise pollution spot x together with any other region but includes it, as well. An absorbing element, as in the relation $\alpha r_L x = \alpha$, could be considered as a spot surrounded by a wall or a forest, which doesn't annoy any other spot of the city Ω.

Since the weak associativity is valid, the concept of transitive closure can be applied here, in order to obtain the fundamental β^* classes. The actual meaning of this situation is that the city Ω can be divided, using the noise hyperoperation, in a partition, where every fundamental class does not annoy any other blocks of flats from other fundamental classes. The next example gives an idea:

Example 3.12. According to Example 3.10, consider now that Ω is a city where $\Omega = \{x_1, x_2, x_3, x_4, x_5, x_6, x_7, x_8, x_9\}$. From "multiplication" table of (r_L), we obtain that $\beta^*(x_1) = \{x_1, x_2, x_3\}$,

$$\beta^*(x_4) = \{x_4\}, \beta^*(x_5) = \{x_5, x_6, x_7\}, \beta^*(x_8) = \{x_8, x_9\}.$$

So, the fundamental semigroup Ω/β^* is

$$\Omega/\beta^* = \big\{ \{x_1, x_2, x_3\}, \{x_4\}, \{x_5, x_6, x_7\}, \{x_8, x_9\} \big\}$$

and the "multiplication" table is

$\circ$	$\underline{x}_1$	$\underline{x}_4$	$\underline{x}_5$	$\underline{x}_8$
$\underline{x}_1$	$\underline{x}_1$	$\underline{x}_1$	$\underline{x}_1$	$\underline{x}_1$
$\underline{x}_4$	$\underline{x}_4$	$\underline{x}_4$	$\underline{x}_4$	$\underline{x}_4$
$\underline{x}_5$	$\underline{x}_5$	$\underline{x}_5$	$\underline{x}_5$	$\underline{x}_5$
$\underline{x}_8$	$\underline{x}_8$	$\underline{x}_8$	$\underline{x}_8$	$\underline{x}_8$

In other words and beyond the mathematical content of the present example, the city Ω was divided into four regions, where every region (fundamental class) does not annoy any other spot belonging to the rest regions. So, one could consider that among the four regions there exists a green park, full of trees, which absorbs the possible noise pollution caused by any of the four regions. Since $\beta^*(x_4) = \{x_4\}$, the element $x_4 \in \Omega$ (spot or building of the city) is a single element and that means that it doesn't annoy any other spot of the city Ω , so it can be considered as the remotest spot of the city.

Chapter 4

H_v-rings

4.1 H_v-rings and some examples

Definition 4.1. A multi-valued system $(R, +, \cdot)$ is an H_v-*ring* if

(1) $(R, +)$ is an H_v-group;

(2) $(R, \cdot)$ is an H_v-semigroup;

(3) $\cdot$ is weak distributive with respect to $+$, i.e., for all x, y, z in R we have

$$(x \cdot (y + z)) \cap (x \cdot y + x \cdot z) \neq \emptyset \text{ and } ((x + y) \cdot z) \cap (x \cdot z + y \cdot z) \neq \emptyset.$$

An H_v-ring may be commutative with respect either to $+$ or $\cdot$. If H is commutative with respect to both $+$ and $\cdot$, then we call it a *commutative* H_v-*ring*. If there exists $u \in R$ such that $x \cdot u = u \cdot x = \{x\}$ for all $x \in R$, then u is called the *scalar unit* of R and it is denoted by 1.

Example 4.1. Let $(R, +, \cdot)$ be a ring and $\mu : R \to [0, 1]$ be a function. We define the hyperoperations $\uplus, \otimes, *$ on R as follows:

$$x \uplus y = \{t \mid \mu(t) = \mu(x + y)\},$$
$$x \otimes y = \{t \mid \mu(t) = \mu(x \cdot y)\},$$
$$x \star y = y \star x = \{t \mid \mu(x) \leq \mu(t) \leq \mu(y)\}, \quad (\text{if } \mu(x) \leq \mu(y)).$$

Then, $(R, \star, \star)$, $(R, \star, \otimes)$, $(R, \star, \uplus)$, $(R, \uplus, \star)$, and $(R, \uplus, \otimes)$ are H_v-rings.

Definition 4.2. An H_v-ring $(R, +, \cdot)$ is called a *dual* H_v-*ring* if $(R, \cdot, +)$ is an H_v-ring. If both $+$ and $\cdot$ are weak commutative, then R is called a *weak commutative dual* H_v-*ring*.

Proposition 4.1. *If* $(H, \star)$ *is an* H_v-*group, then for every hyperoperation* $\circ$ *such that* $\{x, y\} \subseteq x \circ y$ *for all* $x, y \in H$, *the hyperstructure* $(H, \star, \circ)$ *is a dual* H_v-*ring.*

Proof. First we prove that $(H, \star, \circ)$ is an H_v-ring. For every x, y, z in H, we have

$$\{x\} \cup (y \star z) \subseteq x \circ (y \star z)$$
$$(x \star x) \cup (x \star z) \cup (y \star x) \cup (y \star z) = \{x, y\} \star \{x, z\} \subseteq (x \circ y) \star (x \circ z).$$

Thus, $y \star z \subseteq (x \circ (y \star z)) \cap ((x \circ y) \star (x \circ z)) \neq \emptyset$. Therefore, the left and similarly the right weak distributivity are valid and the rest axioms can be easily verified.

Now, we prove that $(H, \circ, \star)$ is an H_v-ring. For every x, y, z in H, we have

$$(x \star y) \cup (x \star z) = x \star \{y, z\} \subseteq x \star (y \circ z)$$
$$(x \star y) \cup (x \star z) \subseteq (x \star y) \circ (x \star z).$$

So, we have $(x \star y) \cup (x \star z) \subseteq (x \star (y \circ z)) \cap ((x \star y) \circ (x \star z)) \neq \emptyset$. Thus, $\star$ is a left weak distributive with respect to $\circ$ and the rest axioms are easily verified. ∎

Proposition 4.2. *Let $(H, +)$ be an H_v-group with a scalar zero element 0. Then, for every hyperoperation $\odot$ such that*

$$\{x, y\} \subseteq x \odot y \text{ for all } x, y \text{ in } H \setminus \{0\}, \ x \odot 0 = 0 \odot x = 0 \text{ for all } x \text{ in } H,$$

the hyperstructure $(H, +, \odot)$ is an H_v-ring.

Proof. For every non-zero elements $x, y, z \in H$, we have

$$y + z \subseteq (x \odot (y + z)) \cap ((x \odot y) + (x \odot z)) \neq \emptyset.$$

Moreover, if one of the elements x, y, z is zero, then the strong distributivity is valid. The rest of the weak axioms are also valid. ∎

Proposition 4.3. [51] *We define the following three hyperoperations on the set $\mathbb{R}^n$, where $\mathbb{R}$ is the set of real numbers:*

$$x \oplus y = \{r(x + y) | \ r \in [0, 1]\},$$
$$x \odot y = \{x + r(y - x) | \ r \in [0, 1]\},$$
$$x \bullet y = \{x + ry | \ r \in [0, 1]\}.$$

Then the hyperstructure $(\mathbb{R}^n, \star, \circ)$ is a weak commutative dual H_v-ring where $\star, \circ \in \{\oplus, \odot, \bullet\}$.

Proof. The associativity:

(1) We have

$$x \oplus (y \oplus z) = \{rx + rmy + rmz \mid r, m \in [0,1]\}$$
$$(x \oplus y) \oplus z = \{tnx + tny + tz \mid t, n \in [0,1]\}.$$

If $r = t = 0$ then $\{0\} \subseteq (x \oplus (y \oplus z)) \cap ((x \oplus y) \oplus z)$. If $m = n = 1$ then

$$\{r(x + y + z) \mid r \in [0,1]\} = \{t(x + y + z) \mid t \in [0,1]\}$$
$$\subseteq (x \oplus (y \oplus z)) \cap ((x \oplus y) \oplus z).$$

We claim that for all $m, n \in [0,1)$ and for all $r, t \in (0,1]$ the following assertion is valid:

$$x \oplus (y \oplus z) \neq (x \oplus y) \oplus z.$$

Indeed, if there exist $m, n \in [0,1)$ and $r, t \in (0,1]$ such that we have the equality in the above condition to be hold, then $r = tn$, $rm = tn$, $rm = t$ which imply $rm = r$, $tn = t$. So $m = n = 1$, which is a contradiction.

(2) We have

$$x \odot (y \odot z) = \{(1-r)x + r(1-m)y + rmz \mid r, m \in [0,1]\}$$
$$(x \odot y) \odot z = \{(1-t)(1-n)x + t(1-n)y + nz \mid t, n \in [0,1]\}.$$

We claim that the above two sets are equal. Let $r, m \in [0,1]$. Then we have

$$(1-n)(1-t) = 1-r, \ t(1-n) = r(1-m), \ n = rm$$

if and only if

$$(1-rm)(1-t) = 1-r, \ t(1-rm) = r(1-m), \ n = rm.$$

It is obvious that $n = rm \in [0,1]$. Now, we shall prove that $t \in [0,1]$. If $rm = 1$ then $r = m = 1$, so we obtain $n = 1$ and $0t = 0$ which is valid for all $t \in [0,1]$. If $rm \neq 1$ then we have $t = \dfrac{r(1-m)}{1-rm}$ and $r \neq 1$ or $m \neq 1$. Let $m \neq 1$, then $0 \leq rm < 1$, so we have $1 - rm > 0$. Now, from $r \leq 1$ we obtain

$$r - rm \leq 1 - rm \ \Leftrightarrow \ \frac{r(1-m)}{1-rm} \leq 1 \ \Leftrightarrow \ t \leq 1.$$

Obviously, $r(1-m) \geq 0$, so $t \geq 0$.

Let $n, t \in [0,1]$. Using the same technique, we can easily show that $r, m \in [0,1]$.

(3) We have

$$x \bullet (y \bullet z) = \{x + ry + rmz \mid r, m \in [0, 1]\}$$
$$(x \bullet y) \bullet z = \{x + ny + tz \mid t, n \in [0, 1]\}.$$

By setting $r = n$ and $rm = t$ we obtain $x \bullet (y \bullet z) \subseteq (x \bullet y) \bullet z$.

The commutativity: For the first two hyperoperations we have:

$$x \oplus y = [0, x + y] = [0, y + x] = y \oplus x \text{ and } x \odot y = [x, y] = [y, x] = y \odot x.$$

For the third one, we have

$$x \bullet y = \{x + ry \mid r \in [0, 1]\} \text{ and } y \bullet x = \{mx + y \mid m \in [0, 1]\}.$$

The above two sets have common elements for all $x, y \in \mathbb{R}^n$, only in the case $r = m = 1$.

The reproduction axiom: We can see easily that, for all $x \in \mathbb{R}^n$

$$x \oplus \mathbb{R}^n = \mathbb{R}^n \oplus x = \mathbb{R}^n, \quad x \odot \mathbb{R}^n = \mathbb{R}^n \odot x = \mathbb{R}^n, \quad x \bullet \mathbb{R}^n = \mathbb{R}^n \bullet x = \mathbb{R}^n.$$

The distributivity: The following assertions hold for all $x, y, z \in \mathbb{R}^n$.

$$(x \oplus (y \oplus z)) \cap ((x \oplus y) \oplus (x \oplus z)) \neq \emptyset,$$
$$((x \oplus y) \oplus z) \cap ((x \oplus z) \oplus (y \oplus z)) \neq \emptyset,$$
$$x \oplus (y \odot z) = (x \oplus y) \odot (x \oplus z),$$
$$(x \odot y) \oplus z = (x \odot z) \oplus (y \odot z),$$
$$(x \oplus (y \bullet z)) \cap ((x \oplus y) \bullet (x \oplus z)) \neq \emptyset,$$
$$((x \bullet y) \oplus z) \cap ((x \oplus z) \bullet (y \oplus z)) \neq \emptyset,$$
$$x \odot (y \odot z) = (x \odot y) \odot (x \odot z),$$
$$(x \odot y) \odot z = (x \odot z) \odot (y \odot z),$$
$$(x \odot (y \oplus z)) \cap ((x \odot y) \oplus (x \odot z)) \neq \emptyset,$$
$$((x \oplus y) \odot z) \cap ((x \odot z) \oplus (y \odot z)) \neq \emptyset,$$
$$(x \odot (y \bullet z)) \cap ((x \odot y) \bullet (x \odot z)) \neq \emptyset,$$
$$((x \bullet y) \odot z) \cap ((x \odot z) \bullet (y \odot z)) \neq \emptyset,$$
$$(x \bullet (y \bullet z)) \cap ((x \bullet y) \bullet (x \bullet z)) \neq \emptyset,$$
$$((x \bullet y) \bullet z) \cap ((x \bullet z) \bullet (y \bullet z)) \neq \emptyset,$$
$$x \bullet (y \odot z) = (x \bullet y) \odot (x \bullet z),$$
$$(x \odot y) \bullet z = (x \bullet z) \odot (y \bullet z),$$
$$(x \bullet (y \oplus z)) \cap ((x \bullet y) \oplus (x \bullet z)) \neq \emptyset,$$
$$((x \oplus y) \bullet z) \cap ((x \bullet z) \oplus (y \bullet z)) \neq \emptyset.$$

Among the 9 cases, we shall prove here the last one. The rest of them can be proved in a similar way

$$x \bullet (y \oplus z) = \{x + mty + mtz \mid m, t \in [0, 1]\},$$
$$(x \bullet z) \oplus (y \bullet z) = \{2rx + rny + rkz \mid r, n, k \in [0, 1]\}.$$

If $m = n = k = 0$ and $r = \frac{1}{2}$ then $\{x\} \subseteq (x \bullet (y \oplus z)) \cap ((x \bullet y) \oplus (x \bullet z))$. Also, $((x \oplus y) \bullet z) \cap ((x \bullet z) \oplus (y \bullet z)) \neq \emptyset$, for all $x, y, z \in \mathbb{R}^n$. ∎

Now, we present some general constructions which can be useful in the theory of representations of several classes of H_v-groups.

Let $(H, \circ)$ be a hypergroupoid; by $\triangle_H$ we mean the diagonal of the Cartesian product $H \times H$, i.e., $\triangle_H = \{[x, x] \mid x \in H\}$. Let us define a mapping $D : H \to H \times H$ by $D(x) = [x, x]$ for all $x \in H$, i.e., $\triangle_H = D(H)$.

Lemma 4.1. *Let* $(H, \circ)$ *be a hypergroupoid. Define a hyperoperation* $\star$ *on the diagonal* $\triangle_H$ *as follows:*

$$[x, x] \star [y, y] = D(x \circ y \cup y \circ x) = \{[u, u] \mid u \in x \circ y \cup y \circ x\}$$

for any pair $[x, x], [y, y] \in \triangle_H$. *Then the following assertions hold:*

(1) *For any hypergroupoid* $(H, \circ)$ *we have that* $(\triangle_H, \star)$ *is a commutative hypergroupoid.*

(2) *If* $(H, \circ)$ *is a weakly associative hypergroupoid, then the hypergroupoid* $(\triangle_H, \star)$ *is weakly associative as well.*

(3) *If* $(H, \circ)$ *is a quasihypergroup, the the hypergroupoid* $(\triangle_H, \star)$ *satisfies also the reproduction axiom, i.e., it is a quasihypergroup.*

(4) *If* $(H, \circ)$ *is associative, then the hypergroupoid* $(\triangle_H, \star)$ *is weakly associative (but not associative in general).*

Proof. The assertion (1) follows immediately from the above definition of the hyperoperation $\star$.

(2) Suppose that $[x, x], [y, y], [z, z] \in \triangle_H$. Then

$$([x, x] \star [y, y]) \star [z, z]$$

$$= D(x \circ y \cup y \circ x) \star [z, z] = (D(x \circ y) \cup D(y \circ x)) \star [z, z]$$

$$= (D(x \circ y) \star [z, z]) \cup (D(y \circ x) \star [z, z])$$

$$= \Big(\bigcup_{u \in x \circ y} [u, u] \star [z, z] \Big) \cup \Big(\bigcup_{v \in y \circ x} [v, v] \star [z, z] \Big)$$

$$= \Big(\bigcup_{u \in x \circ y} D(u \circ z \cup z \circ u) \Big) \cup \Big(\bigcup_{v \in y \circ x} D(v \circ z \cup z \circ v) \Big)$$

$$= \Big(\bigcup_{u \in x \circ y} D(u \circ z) \Big) \cup \Big(\bigcup_{u \in x \circ y} D(z \circ u) \Big) \cup \Big(\bigcup_{v \in y \circ x} D(v \circ z) \Big) \cup \Big(\bigcup_{v \in y \circ x} D(z \circ u) \Big)$$

$$= D\Big(\bigcup_{u \in x \circ y} u \circ z \Big) \cup D\Big(\bigcup_{u \in x \circ y} z \circ u \Big) \cup D\Big(\bigcup_{v \in y \circ x} v \circ z \Big) \cup D\Big(\bigcup_{v \in y \circ x} z \circ u \Big)$$

$$= D(x \circ y \circ z) \cup D(z \circ x \circ y) \cup D(y \circ x \circ z) \cup D(z \circ y \circ x)$$

$$= D(x \circ y \circ z \cup z \circ y \circ x) \cup D(z \circ x \circ y \cup y \circ x \circ z).$$

On the other hand

$$[x, x] \star ([y, y] \star [z, z]) = ([z, z] \star [y, y]) \star [x, x]$$
$$= D(z \circ y \circ x \cup x \circ z \circ y \cup y \circ z \circ x \cup x \circ y \circ z)$$
$$= D(x \circ y \circ z \cup z \circ y \circ x) \cup D(x \circ z \circ y \cup y \circ z \circ x).$$

Thus $([x, x] \star [y, y]) \star [z, z] \cap [x, x] \star ([y, y] \star [z, z]) \supseteq D(x \circ y \circ z) \cup D(z \circ y \circ x) \neq \emptyset$.

(3) Let $x \in H$ be an arbitrary element. Then $x \circ H = H = H \circ x$ and we have

$$[x, x] \star \triangle_H = \bigcup_{y \in H} ([x, x] \star [y, y]) = \bigcup_{y \in H} D(x \circ y \cup y \circ x)$$

$$= \left(\bigcup_{y \in H} D(x \circ y) \right) \cup \left(\bigcup_{y \in H} D(y \circ x) \right) = D\left(\bigcup_{y \in H} x \circ y \right) \cup D\left(\bigcup_{y \in H} y \circ x \right)$$

$$= D(x \circ H) \cup D(H \circ x) = D(H) = \triangle_H.$$

Since a semihypergroup is also weakly associative, the assertion (4) follows from (2). ∎

Let $(R, +, \cdot)$ be an H_v-ring. We define the hyperoperations $\oplus$ and $\odot$ on the diagonal $D(R) = \triangle_R$ by

$$[x, x] \oplus [y, y] = \{[u, u] \mid u \in (x + y) \cup (y + x)\},$$
$$[x, x] \odot [y, y] = \{[v, v] \mid v \in (x \cdot y) \cup (y \cdot x)\}$$

for all $x, y \in R$. Then we have

Proposition 4.4. *Let $(R, +, \cdot)$ be an H_v-ring. Then, $(D(R), \oplus, \odot)$ is a commutative H_v-ring.*

Proof. According to Lemma 4.1, we obtain that $(D(R), \oplus)$ is a commutative weakly associative hypergroupoid satisfying the reproduction axiom, thus it is a commutative H_v-group. Similarly, $(D(R), \odot)$ is a commutative H_v-semigroup. Thus, it remains to prove that

$$[x, x] \odot ([y, y] \oplus [z, z]) \cap ([x, x] \odot [y, y]) \oplus ([x, x] \odot [z, z]) \neq \emptyset$$

for arbitrary elements $x, y, z \in R$.

Indeed, we have $[y, y] \oplus [z, z] = \{[u, u] \mid u \in (y + z) \cup (z + y)\}$ and

$$[x, x] \odot ([y, y] \oplus [z, z]) = \bigcup_{u \in (y+z) \cup (z+y)} [x, x] \odot [u, u]$$

$$= \left(\bigcup_{u \in y+z} [x, x] \odot [u, u] \right) \cup \left(\bigcup_{u \in z+y} [x, x] \odot [u, u] \right)$$

$$= \left(\bigcup_{u \in y+z} \{[v, v] \mid v \in x \cdot u \cup u \cdot x\} \right) \cup \left(\bigcup_{u \in z+y} \{[v, v] \mid v \in x \cdot u \cup u \cdot x\} \right)$$

$$= \{[v, v] \mid v \in x \cdot (y + z)\} \cup M(x, y, z),$$

where $M(x, y, z) = \bigcup\limits_{u \in y+z} \{[v,v] \mid v \in u \cdot x\} \cup \bigcup\limits_{u \in z+y} \{[v,v] \mid v \in x \cdot u \cup u \cdot x\}.$

On the other hand,

$$[x,x] \odot [y,y] = \{[v,v] \mid v \in x \cdot y \cup y \cdot x\}$$
$$= \{[v,v] \mid v \in x \cdot y\} \cup \{[v,v] \mid v \in y \cdot x\},$$
$$[x,x] \odot [z,z] = \{[v,v] \mid v \in x \cdot z\} \cup \{[v,v] \mid v \in z \cdot x\}$$

and then

$$([x,x] \odot [y,y]) \oplus ([x,x] \odot [z,z])$$

$$= \Big(\{[v,v] \mid v \in x \cdot y\} \cup \{[v,v] \mid v \in x \cdot z\}\Big) \oplus \Big(\{[v,v] \mid v \in z \cdot x\} \cup \{[v,v] \mid v \in y \cdot x\}\Big)$$

$$= \Big(\{[v,v] \mid v \in x \cdot y\} \oplus \{[v,v] \mid v \in x \cdot z\}\Big) \cup \Big(\{[v,v] \mid v \in x \cdot y\} \oplus \{[v,v] \mid v \in z \cdot x\}\Big)$$

$$\cup \Big(\{[v,v] \mid v \in y \cdot x\} \oplus \{[v,v] \mid v \in x \cdot z\}\Big) \oplus \Big(\{[v,v] \mid v \in y \cdot x\} \cup \{[v,v] \mid v \in z \cdot x\}\Big)$$

$$= \Big(\bigcup\limits_{\substack{v \in x \cdot y \\ u \in x \cdot z}} [v,v] \oplus [u,u]\Big) \cup \Big(\bigcup\limits_{\substack{v \in x \cdot y \\ u \in z \cdot x}} [v,v] \oplus [u,u]\Big)$$

$$\cup \Big(\bigcup\limits_{\substack{v \in y \cdot x \\ u \in x \cdot z}} [v,v] \oplus [u,u]\Big) \cup \Big(\bigcup\limits_{\substack{v \in y \cdot x \\ u \in z \cdot x}} [v,v] \oplus [u,u]\Big)$$

$$= \bigcup\limits_{\substack{v \in y \cdot x \\ u \in x \cdot z}} \{[t,t] \mid t \in (v+u) \cup (u+v)\} \cup K(x,y,z),$$

where

$$K(x,y,z) =$$

$$\Big(\bigcup\limits_{\substack{v \in x \cdot y \\ u \in z \cdot x}} [v,v] \oplus [u,u]\Big) \cup \Big(\bigcup\limits_{\substack{v \in y \cdot x \\ u \in x \cdot z}} [v,v] \oplus [u,u]\Big) \cup \Big(\bigcup\limits_{\substack{v \in y \cdot x \\ u \in z \cdot x}} [v,v] \oplus [u,u]\Big)$$

Now, we have

$$([x,x] \odot [y,y]) \oplus ([x,x] \odot [z,z])$$

$$= \Big(\bigcup\limits_{\substack{v \in x \cdot y \\ u \in x \cdot z}} \{[t,t] \mid t \in u + v\}\Big) \cup \Big(\bigcup\limits_{\substack{v \in x \cdot y \\ u \in x \cdot z}} \{[t,t] \mid t \in u + v\}\Big) \cup K(x,y,z).$$

From $(x \cdot y + x \cdot z) \cap x \cdot (y + x) \neq \emptyset$, it follows that $[t_0, t_0] \in \{[v,v] \mid v \in x \cdot (y+z)\}$ for some $t_0 \in x \cdot y + x \cdot z$, thus

$$\{[v,v] \mid v \in x \cdot (y+z)\} \cap \{[t,t] \mid t \in x \cdot y + x \cdot z\} \neq \emptyset,$$

consequently the sets $[x,x] \odot ([y,y] \oplus [z,z])$ and $([x,x] \odot [y,y]) \oplus ([x,x] \odot [z,z])$ have a nonempty intersection. ∎

From the above proof, it follows that only one (right or left) of the weak distributivity laws for $(R, +, \cdot)$ ensures the weak distributivity of $(D(R), \oplus, \odot)$.

Definition 4.3. Let R_1 and R_2 be two H_v-rings. The map $f : R_1 \to R_2$ is called an H_v-*homomorphism* or a *weak homomorphism* if, for all $x, y \in R_1$ the following conditions hold:

$$f(x + y) \cap (f(x) + f(y)) \neq \emptyset \text{ and } f(x \cdot y) \cap f(x) \cdot f(y) \neq \emptyset.$$

f is called an *inclusion homomorphism* if, for all $x, y \in R$, the following relations hold:

$$f(x + y) \subseteq f(x) + f(y) \text{ and } f(x \cdot y) \subseteq f(x) \cdot f(y).$$

Finally, f is called a *strong homomorphism* if for all x, y in R_1 we have

$$f(x + y) = f(x) + f(y) \text{ and } f(x \cdot y) = f(x) \cdot f(y).$$

If R_1 and R_2 are H_v-rings and there exists a strong one to one and onto homomorphism from R_1 to R_2, then R_1 and R_2 are called *isomorphic*.

Corollary 4.1. *Let* $(R, +, \cdot)$ *be an* H_v-*ring and* $r_R(x) = [x, x] \in D(R)$ *for any* $x \in R$. *Then, the mapping* $r_R : (R, +, \cdot) \to (D(R), \oplus, \odot)$ *is an inclusion homomorphism of* H_v-*rings.*

Theorem 4.1. *For any pair of* H_v-*rings* $(R, +, \cdot)$, $(S, +, \cdot)$ *and for any inclusion* H_v-*ring homomorphism* $f : (R, +, \cdot) \to (S, +, \cdot)$ *there exists exactly one inclusion homomorphism* $\psi : (D(R), \oplus, \odot) \to (D(S), \oplus, \odot)$ *such that the diagram*

$$
\begin{array}{ccc}
(R, +, \cdot) & \xrightarrow{\ \ f\ \ } & (S, +, \cdot) \\
{\scriptstyle r_R}\downarrow & & \downarrow{\scriptstyle r_S} \\
(D(R), \oplus, \odot) & \xrightarrow{\ \ \psi\ \ } & (D(S), \oplus, \odot)
\end{array}
$$

is commutative.

Proof. Consider an arbitrary inclusion homomorphism $f : R \to S$ and define $\psi : D(R) \to D(S)$ as the restriction of the mapping $f \times f : R \times R \to S \times S$ onto $D(R) \subseteq R \times R$, i.e., $\psi = (f \times f)|_{D(R)}$, hence $\psi([x, x]) = [f(x), f(x)]$ for any $x \in R$. Now, we have

$$
\begin{aligned}
\psi([x, x] \oplus [y, y]) &= \psi(\{[u, u] \mid u \in (x + y) \cup (y + x)\}) \\
&= \{[f(u), f(u)] \mid u \in (x + y) \cup (y + x)\} \\
&= \{[v, v] \mid v \in f(x + y) \cup f(y + x)\} \\
&\subseteq \{[v, v] \mid v \in (f(x) + f(y)) \cup (f(y) + f(x))\} \\
&= [f(x) + f(y)] \oplus [f(y) + f(x)] \\
&= \psi([x, x]) \oplus \psi([y, y])
\end{aligned}
$$

for any elements $x, y \in R$ and similarly $\psi([x,x] \odot [y,y]) \subseteq \psi([x,x]) \cdot \psi([y,y])$, which is obtained as above. Now, we show that the above diagram commutes.

Let us suppose that $f : R \to S$ is an inclusion homomorphism. Then evidently $\psi : D(R) \to D(S)$ is an inclusion homomorphism as well. For an arbitrary x in R, we have

$$(r_S \circ f)(x) = r_S(f(x)) = [f(x), f(x)] = (f \times f)(x,x)$$
$$= \psi([x,x]) = \psi(r_R(x)) = (\psi \circ r_R)(x),$$

and so $r_S \circ f = \psi \circ r_R$. Now, let $g : D(R) \to D(S)$ be an inclusion homomorphism such that $r_S \circ f = g \circ r_R$. Since $r_R : R \to D(R)$ and $r_S : S \to D(S)$ are bijections, there exist the maps $r_R^{-1} : D(R) \to R$ and $r_S^{-1} : D(S) \to S$. Then, we obtain

$$\psi = \psi \circ id_{D(R)} = \psi \circ r_R \circ r_R^{-1} = r_S \circ f \circ r_R^{-1} = g \circ r_R \circ r_R^{-1} = g \circ id_{D(R)} = g.$$

∎

From the above results we obtain the following theorem.

Theorem 4.2. *Let $\mathcal{H}_v\mathcal{R}$ be the category of all H_v-rings and their inclusion homomorphisms and $\mathcal{AH}_v\mathcal{R}$ be its full subcategory of all commutative H_v-rings. Then there exists the functor $\phi : \mathcal{H}_v\mathcal{R} \to \mathcal{AH}_v\mathcal{R}$ defined by*

$$\phi(R, +, \cdot) = (D(R), \oplus, \odot), \quad \phi(f) = \psi \text{ for any } (R, +, \cdot) \in Ob(\mathcal{H}_v\mathcal{R}),$$

and any morphism $f \in Mor(\mathcal{H}_v\mathcal{R})$, $f : (R, +, \cdot) \to (S, +, \cdot)$ is a reflector; more precisely the pair $(r_R, (\triangle_R, \oplus, \odot))$ is an $\mathcal{AH}_v\mathcal{R}$-reflection for any $(R, +, \cdot) \in Ob(\mathcal{H}_v\mathcal{R})$. Thus $\mathcal{AH}_v\mathcal{R}$ is a reflective full subcategory of the category $\mathcal{H}_v\mathcal{R}$.

4.2 Fundamental relations on H_v-rings

In what follows, we focus our attention on the β^* and γ^* relations defined on H_v-rings. Notice that two kinds of β^* relations can be defined on H_v-rings. We denote them by β_+^* and $\beta_{\cdot}^*$. They are β^* relations with respect to addition and multiplication, respectively. If $(R, +, \cdot)$ is an H_v-ring, then the relations β_+ and $\beta_{\cdot}$ are defined as follows:

$x \beta_+ y \Leftrightarrow$ there exist $z_1, ..., z_n \in R$ such that $\{x, y\} \subseteq z_1 + ... + z_n$,

$x \beta_{\cdot} y \Leftrightarrow$ there exist $z_1, ..., z_n \in R$ such that $\{x, y\} \subseteq z_1 \cdot ... \cdot z_n$.

β_+^* and $\beta_{\cdot}^*$ are the transitive closures of the relations β_+ and $\beta_{\cdot}$. Note that the quotient hyperstructures with respect to β_+^* and $\beta_{\cdot}^*$ are H_v-rings. In this section, the fundamental relations defined on an H_v-ring are studied. Especially, some connections among different types of fundamental relations are obtained.

Definition 4.4. Let $(R, +, \cdot)$ be an H_v-ring. We define γ^* as the smallest equivalence relation such that the quotient R/γ^* is a ring. γ^* is called the *fundamental equivalence relation* and R/γ^* is called the *fundamental ring*. An H_v-ring is called an H_v-*field* if its fundamental ring is a field.

Let us denote the set of all finite polynomials of elements of R over $\mathbb{N}$ by $\mathcal{U}$. We define the relation γ as follows:

$$x\gamma y \text{ if and only if } \{x, y\} \subseteq u, \text{ where } u \in \mathcal{U}.$$

The following theorem is similar to Theorem 3.4, where it is proved by Vougiouklis [120].

Theorem 4.3. *The fundamental equivalence relation γ^* is the transitive closure of the relation γ.*

Proof. Let $\widehat{\gamma}$ be the transitive closure of the relation γ. We denote the equivalence class of a by $\widehat{\gamma}(a)$. First, we prove that the quotient set $R/\widehat{\gamma}$ is a ring. The sum $\oplus$ and the product $\odot$ are defined in $R/\widehat{\gamma}$ in the usual manner:

$$\widehat{\gamma}(a) \oplus \widehat{\gamma}(b) = \{\widehat{\gamma}(c) \mid c \in \widehat{\gamma}(a) + \widehat{\gamma}(b)\},$$
$$\widehat{\gamma}(a) \odot \widehat{\gamma}(b) = \{\widehat{\gamma}(d) \mid d \in \widehat{\gamma}(a) \cdot \widehat{\gamma}(b)\}.$$

Take $a' \in \widehat{\gamma}(a)$ and $b' \in \widehat{\gamma}(b)$. Then, we have $a'\widehat{\gamma}a$ if and only if there exist $x_1, ..., x_{m+1}$ with $x_1 = a'$, $x_{m+1} = a$ and $u_1, ..., u_m \in \mathcal{U}$ such that $\{x_i, x_{i+1}\} \subseteq u_i$ $(i = 1, ..., m)$, and $b'\widehat{\gamma}b$ if and only if there exist $y_1, ..., y_{n+1}$ with $y_1 = b'$, $y_{n+1} = b$ and $v_1, ..., v_n \in \mathcal{U}$ such that $\{y_j, y_{j+1}\} \subseteq v_j$ $(j = 1, ..., n)$. Now, we obtain

$$\{x_i, x_{i+1}\} + y_1 \subseteq u_i + v_1 \quad (i = 1, ..., m-1),$$
$$x_{m+1} + \{y_j, y_{j+1}\} \subseteq u_m + v_j \quad (j = 1, ..., n).$$

The sums $u_i + v_1 = t_i$ $(i = 1, ..., m-1)$ and $u_m + v_j = t_{m+j-1}$ $(j = 1, ..., n)$ are polynomials and so $t_k \in \mathcal{U}$ for all $k \in \{1, ..., m+n-1\}$. Now, pick up the elements $z_1, ..., z_{m+n}$ such that $z_i \in x_i + y_1$ $(i = 1, ..., m)$ and $z_{m+j} \in x_{m+1} + y_{j+1}$ $(j = 1, ..., n)$. Hence, we obtain $\{z_k, z_{k+1}\} \subseteq t_k$ $(k = 1, ..., m+n-1)$. Therefore, every element $z_1 \in x_1 + y_1 = a' + b'$

is $\widehat{\gamma}$ equivalent to every element $z_{m+n} \in x_{m+1} + y_{n+1} = a + b$. Thus, $\widehat{\gamma}(a) \oplus \widehat{\gamma}(b) = \widehat{\gamma}(c)$ for all $c \in \widehat{\gamma}(a) + \widehat{\gamma}(b)$. In a similar way, it is proved that $\widehat{\gamma}(a) \odot \widehat{\gamma}(b) = \widehat{\gamma}(d)$ for all $d \in \widehat{\gamma}(a) \cdot \widehat{\gamma}(b)$.

The weak associativity and the weak distributivity on R guarantee that the associativity and distributivity are valid in the quotient $R/\widehat{\gamma}$. Therefore, $R/\widehat{\gamma}$ is a ring.

Now, let σ be an equivalence relation on R such that R/σ is a ring. Denote the equivalence class of a by $\sigma(a)$. Then, $\sigma(a) \oplus \sigma(b)$ and $\sigma(a) \odot \sigma(b)$ are singletons for all $a, b \in R$, i.e., $\sigma(a) \oplus \sigma(b) = \sigma(c)$ for all $c \in \sigma(a) + \sigma(b)$ and $\sigma(a) \odot \sigma(b) = \sigma(d)$ for all $d \in \sigma(a) \cdot \sigma(b)$. Thus, for every $a, b \in R$ and $A \subseteq \sigma(a)$, $B \subseteq \sigma(b)$ we can write

$$\sigma(a) \oplus \sigma(b) = \sigma(a+b) = \sigma(A + B) \text{ and } \sigma(a) \odot \sigma(b) = \sigma(a \cdot b) = \sigma(A \cdot B).$$

By induction, we extend these equalities on finite sums and products. So, for every $u \in \mathcal{U}$ and for all $x \in u$ we have $\sigma(x) = \sigma(u)$. Therefore, for every $a \in R$,

$$x \in \gamma(a) \text{ implies } x \in \sigma(a).$$

Since σ is transitive, we obtain that

$$x \in \widehat{\gamma}(a) \text{ implies } x \in \sigma(a).$$

This means that the relation $\widehat{\gamma}$ is the smallest equivalence relation on R such that $R/\widehat{\gamma}$ is a ring, i.e., $\widehat{\gamma} = \gamma^*$.   ∎

Definition 4.5. We define the γ_1^*, γ_2^* relations as the transitive closures of the relations γ_1, γ_2 respectively, which are defined as follows:

$x\gamma_1 y$ if and only if there exist $a_i \in R$ and I_k, K finite sets of indices such that

$$\{x, y\} \subseteq \sum_{k \in K} \left(\prod_{i \in I_k} a_i \right)$$

and $x\gamma_2 y$ if and only if there exist $b_j \in R$ and J_s, S finite sets of indices such that

$$\{x, y\} \subseteq \prod_{s \in S} \left(\sum_{j \in J_s} b_j \right).$$

In a multiplicative H_v-ring, the addition is an operation, while in an additive H_v-ring, the multiplication is an operation.

Proposition 4.5.

(1) R/γ_1^* *is a multiplicative H_v-ring.*
(2) R/γ_2^* *is an additive H_v-ring.*

Proof. We prove only (1) and similarly (2) can be proved. The sum of the classes is

$$\gamma_1^*(x) \oplus \gamma_1^*(y) = \{\gamma_1^*(z) \mid z \in \gamma_1^*(x) + \gamma_1^*(y)\}.$$

In the definition of γ_1^*, expressions of the type $v = \sum \left(\prod \right)$ are used. In the definition of $\oplus$, the element z belongs to the sums v of the above type, which means that z belongs to a sum of products. In other words, all the elements z are in the same γ_1^* class. So, the sum of γ_1^*-classes is a singleton. Therefore, R/γ_1^* is a multiplicative H_v-ring. ∎

Note that the γ_1^* classes are greater than the β^* classes. Actually, the γ_1^* is not the smallest equivalence relation such that $(R/\gamma_1^*, \oplus)$ is a group. In order to see this, consider a multiplicative H_v-ring R. Then $R/\beta_+^* \cong R$, but R/γ_1^* is not isomorphic to R.

Proposition 4.6. *For all additive H_v-rings, we have $\gamma_1^* = \beta_+^*$. For all multiplicative H_v-rings, we have $\gamma_2^* = \beta_\cdot^*$.*

Proof. We present the proof for a multiplicative H_v-ring R. In this case, every sum of elements of R is singleton. Therefore,

$$\prod_{s \in S} \left(\sum_{j \in J_s} b_j \right) = \prod_{s \in S} d_s, \quad \text{where } d_s = \sum_{j \in J_s} b_j.$$

This means that $x\gamma_2^* y$ if and only if $x\beta_\cdot^* y$. ∎

Using the above propositions, it follows that $(R/\gamma_1^*)/\gamma_2^* = (R/\gamma_1^*)/\beta^*$ is a multiplication H_v-ring and $(R/\gamma_2^*)/\gamma_1^* = (R/\gamma_2^*)/\beta_+^*$ is an additive H_v-ring.

Theorem 4.4. *Let $(R, +, \cdot)$ be an H_v-ring. Then, $R/\gamma^* \cong (R/\beta^*)/\beta_\uplus^*$, where $\beta_\uplus^*$ is the fundamental relation defined in $(R/\beta_\cdot^*, \uplus)$ by setting $\beta^*(a) \uplus \beta_\cdot^*(b) = \{\beta^*(c) \mid c \in \beta^*(a) + \beta^*(b)\}$.*

Proof. The quotient of the additive H_v-ring $(R/\beta^*, \uplus, \otimes)$ with respect to $\beta_\uplus^*$ is a ring. Let us denote the equivalence relation associated to the projection $\psi : R \to (R/\beta^*)/\beta_\uplus^*$ by σ. Since ψ is a ring homomorphism, then we obtain $\gamma^*(a) \subseteq \sigma(a)$ for all $a \in R$. On the other hand, since $\beta_\cdot^*(x) \subseteq \gamma^*(x)$ for all $x \in R$, we have

$$\bigcup_{\beta_\cdot^*(z) \in \beta^*(x) \uplus \beta^*(y)} \beta^*(z) = \bigcup_{z \in \beta^*(x) + \beta^*(y)} \beta^*(z) \subseteq \bigcup_{z \in \gamma^*(x) + \gamma^*(y)} \gamma^*(z).$$

From the fundamental property in $(R/\gamma^*, \oplus, \odot)$, we know that $\gamma^*(x)\oplus\gamma^*(y)$ is a singleton, so $\gamma^*(x) \oplus \gamma^*(y) = \gamma^*(w)$, where $w \in x + y$. Therefore,

$$\bigcup_{\beta^*(z)\in\beta^*(x)\uplus\beta^*(y)} \beta^*(z) \subseteq \gamma^*(w), \text{ where } w \in x + y.$$

Consequently, for every finite sum of elements in R/β^*, we have

$$\bigcup_{z\in\uplus \sum_{i\in I}\beta^*(x_i)} \beta^*(z) \subseteq \gamma^*(w), \text{ where } w \in \sum_{i\in I} x_i.$$

Moreover, since γ^* is transitive, it follows that

$$\sigma(a) = \bigcup_{\{z|(\beta^*(z))\, \beta^*_\uplus\, (\beta^*(a))\}} \beta^*(z) \subseteq \gamma^*(a) \text{ for all } a \in R.$$

Therefore, $\sigma = \gamma^*$. ∎

Theorem 4.5. *If $(H, \Diamond)$ is an H_v-group, then for every hyperoperation ∇ such that $\{x,y\} \subseteq x\nabla y$ for all $x, y \in H$, the hyperstructures $(H, \Diamond, \nabla)$ and $(H, \nabla, \Diamond)$ are H_v-rings.*

Proof. Every hyperoperation ∇ that satisfies the condition of hypothesis is weak associative, weak commutative and H/γ^* is a singleton. Moreover, every element $x \in H$ is a unit element, i.e., $y \in x\nabla y \cap y\nabla x$ for all $y \in H$, and every element $x \in H$ is symmetric with respect to the unit x, i.e., $x \in x\nabla y \cap y\nabla x$.

In order to prove that $(H, \Diamond, \nabla)$ is an H_v-ring we need only to prove the weak distributivity on the left. For every x, y, z in H we have

$$x\nabla(y\Diamond z) \supseteq \{x\} \cup (y\Diamond z)$$

and

$$(x\nabla y)\Diamond(x\nabla z) \supseteq \{x,y\}\Diamond\{x,z\} = (x\Diamond x) \cup (x\Diamond z) \cup (y\Diamond x) \cup (y\Diamond z).$$

Therefore, $y\Diamond z \subseteq [x\nabla(y\Diamond z)] \cap [(x\nabla y)\Diamond(x\nabla z)] \neq \emptyset$. Thus, the left and similarly the right weak distributivity are valid.

Similarly, we need to prove the weak distributivity on the left for $(H, \nabla, \Diamond)$. For every x, y, x in H we have

$$x\Diamond(y\nabla z) \supseteq x\Diamond\{y,z\} = (x\Diamond y) \cup (x\Diamond z)$$

and

$$(x\Diamond y)\nabla(x\Diamond z) \supseteq (x\Diamond y) \cup (x\Diamond z).$$

So the left distributivity is valid, because

$$(x\Diamond y) \cup (x\Diamond z) \subseteq [x\Diamond(y\nabla z)] \cap [(x\Diamond y)\nabla(x\Diamond z)] \neq \emptyset.$$

H_v-rings $(H, \nabla, \Diamond)$ and $(H, \Diamond, \nabla)$ are called *associated H_v-rings*. ∎

In the theory of representations of the hypergroups in the sense of Marty, there are three types of associated hyperrings $(H, \oplus, \cdot)$ with the hypergroup $(H, \cdot)$. The hyperoperation $\oplus$ is defined, respectively, for all x, y in H, as follows:

$$\text{type a: } x \oplus y = \{x, y\},$$
$$\text{type b: } x \oplus y = \beta^*(x) \cup \beta^*(y),$$
$$\text{type c: } x \oplus y = H.$$

In all the above types, the strong associativity and strong or inclusion distributivity are valid. However, in H_v-structures there exists only one class of associated H_v-rings instead of three types.

Theorem 4.6. *Let $(H, +)$ be an H_v-group with a scalar zero element 0. Then, for every hyperoperation $\otimes$ such that*

$$\{x, y\} \subseteq x \otimes y, \text{ for all } x, y \in H \setminus \{0\},$$
$$x \otimes 0 = 0 \otimes x = 0, \text{ for all } x \in H,$$

the hyperstructure $(H, +, \otimes)$ is an H_v-ring.

Proof. For every nonzero elements x, y, z in H, we have

$$y + z \subseteq [x \otimes (y + z)] \cap [(x \otimes y) + (x \otimes z)] \neq \emptyset.$$

Moreover, if one of the elements x, y, z is zero, then the strong distributivity is valid. The rest of the weak axioms are also valid.   ∎

Theorem 4.7. *Let $(H, \cdot)$ be an H_v-group. Take an element $0 \notin H$ and denote $H' = H \cup \{0\}$. We define the hyperoperation $+$ as follows:*

$$0 + 0 = 0, \ 0 + x = H = x + 0, \ x + y = 0 \text{ for all } x, y \in H,$$

and we extend the hyperoperation $\cdot$ in H' by putting

$$0 \cdot 0 = 0 \cdot x = x \cdot 0 = 0 \text{ for all } x \in H.$$

Then, the hyperstructure $(H', +, \cdot)$ is an H_v-field with $H'/\gamma^* \cong \mathbb{Z}_2$, where 0 is an absorbing and $\gamma^*(0)$ is a singleton.

Proof. From the definition it is clear that 0 is an absorbing element. The hyperoperation $+$ is (strongly) associative because if in any triple (x, y, z) of elements of H' there are one or three non-zero elements, then their hypersum is 0; in the other cases, the result is H.

The $\cdot$ is weak associative because 0 is an absorbing and $(H, \cdot)$ is an H_v-group. The strong distributivity of $+$ with respect to $\cdot$ is valid, because the only one nonzero case is for $x, y \in H$ in which we have

$$x \cdot (0 + y) = (0 + y) \cdot x = x \cdot 0 + x \cdot y = 0 \cdot x + y \cdot x = H.$$

Finally, one can check that $\gamma^*(0)$ is a singleton and that there are only two fundamental classes in H'. Thus, $(H', +, \cdot)$ is an H_v-field and $H'/\gamma^* \cong \mathbb{Z}_2$. ∎

Notice that if the H_v-group $(H, \cdot)$ is strongly associative, then $(H', +, \cdot)$ is a hyperfield instead of an H_v-field; moreover, the strong distributivity is valid.

Since $\gamma^*(0)$ is a singleton, the H_v-fields of this type are very useful. This happens always in an H_v-group H, that we need to represent, for which the cardinality of the hyperproducts of the elements is equal to a power of $cardH$. On the other hand, the representations are normally of lower dimension and $cardH$ is a small number. The H_v-groups of constant length, such as the P-hypergroups, can be also represented on these H_v-fields.

Now, one can prove the following theorem. There is no need to check if the weak axioms are valid since they are obvious. Notice that non-degenerate fundamental rings or fields, which are desired actually, are obtained using this construction.

Theorem 4.8. *Let $(R, +, \cdot)$ be a ring and J be an ideal. Then, we can define two H_b-operations $\boxplus$ and $\boxdot$ greater than $+$ and $\cdot$, respectively, for all x, y in R as follows:*

$$x \boxplus y \subseteq x + y + J \quad and \quad x \boxdot y \subseteq xy + J.$$

Then, the hyperstructure $(R, \boxplus, \boxdot)$ is an H_v-ring for which the fundamental ring R/γ^* is a subring of R/J.

Notice that the maximum of the above hyperadditions, i.e., $x \boxplus y = x + y + J$, is a P-hyperoperation so that the H_v-ring $(R, \boxplus, \boxdot)$ can be a P-H_v-ring (see Section 6.4). Remark that for any maximal ideal J, one obtains $R/\gamma^* = R/J$. This construction leads to an enormous number of H_v-rings. Let us point out that if the products of the ring R are enlarged, then all hyperproducts with any cardinality can be represented and the main theorem of this theory is not trivial.

4.3   Uniting elements

The *uniting elements* method was introduced by Corsini and Vougiouklis [18] in 1989. With this method one puts in the same class, two or more elements. This leads, through hyperstructures, to structures satisfying additional properties.

The *uniting elements* method is described as follows: Let G be an algebraic structure and let d be a property, which is not valid. Suppose that d is described by a set of equations; then, consider the partition in G for which it is put together, in the same partition class, every pair of elements that causes the non-validity of the property d. The quotient by this partition G/d is an H_v-structure. Then, quotient out the H_v-structure G/d by the fundamental relation β^*, a stricter structure $(G/d)/\beta^*$ for which the property d is valid, is obtained.

An interesting application of the uniting elements is when more than one properties are desired. The reason for this is some of the properties lead straighter to the classes than others. Therefore, it is better to apply the straightforward classes followed by the more complicated ones. The commutativity is one of the easy applicable properties. Moreover it is clear that the reproductivity property is also easily applicable. One can do this because the following is valid.

Theorem 4.9. *Let $(G, \cdot)$ be a groupoid, and*

$$F = \{f_1, ..., f_m, f_{m+1}, ..., f_{m+n}\}$$

be a system of equations on G consisting of two subsystems

$$F_m = \{f_1, ..., f_m\} \text{ and } F_n = \{f_{m+1}, ..., f_{m+n}\}.$$

Let σ, σ_m be the equivalence relations defined by the uniting elements procedure using the systems F and F_m respectively, and let σ_n be the equivalence relation defined using the induced equations of F_n on the groupoid $G_m = (G/\sigma_m)/\beta^*$. Then,

$$(G/\sigma)/\beta^* \cong (G_m/\sigma_n)/\beta^*.$$

Analogous to the above theorem can be proved for rings.

Theorem 4.10. *Let $(R, +, \cdot)$ be a ring, and*

$$F = \{f_1, ..., f_m, f_{m+1}, ..., f_{m+n}\}$$

be a system of equations on R consisting of two subsystems

$$F_m = \{f_1, ..., f_m\} \text{ and } F_n = \{f_{m+1}, ..., f_{m+n}\}.$$

Let σ, σ_m be the equivalence relations defined by the uniting elements procedure using the systems F and F_m respectively, and let σ_n be the equivalence relation defined using the induced equations of F_n on the ring

$R_m = (R/\sigma_m)/\gamma^*$. Then

$$(R/\sigma)/\gamma^* \cong (R_m/\sigma_n)/\gamma^*$$

i.e., the following diagram is commutative

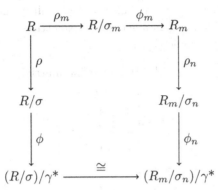

where all the maps ρ, ϕ, ρ_m, ϕ_m, ρ_n and ϕ_n are the canonicals.

From the above it is clear that the fundamental structure is very important, mainly if it is known from the beginning. This is the problem to construct hyperstructures with desired fundamental structures.

4.4   Multiplicative H_v-rings

Algebraic hyperstructures are a generalization of the classical algebraic structures which, among others, are appropriate in two directions: (a) to represent a lot of application in an algebraic model, (b) to overcome restrictions ordinary structures usually have. Concerning the second direction the restrictions of the ordinary matrix algebra can be overcome by the helix-operations. More precisely, the helix addition and the helix-multiplication can be defined on every type of matrices. In [47], Davvaz et al. studied properties and examples on special classes of matrices.

Let $A = (a_{ij}) \in M_{m \times n}$ be a matrix and $s, t \in \mathbb{N}$ be two natural numbers such that $1 \leq s \leq m$ and $1 \leq t \leq n$. Then, we define the *characteristic-like map* $\underline{cst}$ from $M_{m \times n}$ to $M_{s \times t}$ by corresponding to A the matrix $A\underline{cst} = (a_{ij})$, where $1 \leq i \leq s$ and $1 \leq j \leq t$. We call this the *cut-projection map* of type $\underline{st}$. In other words, $A\underline{cst}$ is a matrix obtained from A by cutting the lines and columns greater than s and t respectively.

Let $A = (a_{ij}) \in M_{m \times n}$ be a matrix and $s, t \in \mathbb{N}$ be two natural numbers such that $1 \leq s \leq m$ and $1 \leq t \leq n$. Then, we define the *mod-like* map $\underline{st}$

from $M_{m \times n}$ to $M_{s \times t}$ by corresponding to A the matrix $A\underline{st} = (A_{ij})$ which has as entries the sets

$$A_{ij} = \{a_{i+ks,j+\lambda t} \mid k, \lambda \in \mathbb{N}, \ i + ks \leq m, \ j + \lambda t \leq n\},$$

for $1 \leq i \leq s$ and $1 \leq j \leq t$. We call this multivalued map *helix-projection* of type $\underline{st}$. Therefore, $A\underline{st}$ is a set of $s \times t$-matrices $X = (x_{ij})$ such that $x_{ij} \in A_{ij}$ for all i, j. Obviously, $A\underline{mn} = A$.

Example 4.2. Let us consider the following matrix:

$$A = \begin{bmatrix} 2 & 1 & 3 & 4 & 2 \\ 3 & 2 & 0 & 1 & 2 \\ 2 & 4 & 5 & 1 & -1 \\ 1 & -1 & 0 & 0 & 8 \end{bmatrix}.$$

Suppose that $s = 3$ and $t = 2$. Then,

$$A\underline{c32} = \begin{bmatrix} 2 & 1 \\ 3 & 2 \\ 2 & 4 \end{bmatrix}$$

and $A\underline{32} = (A_{ij})$, where

$$A_{11} = \{a_{11}, a_{13}, a_{15}, a_{41}, a_{43}, a_{45}\} = \{2, 3, 2, 1, 0, 8\},$$
$$A_{12} = \{a_{12}, a_{14}, a_{42}, a_{44}\} = \{1, 4, -1, 0\},$$
$$A_{21} = \{a_{21}, a_{23}, a_{25}\} = \{3, 0, 2\},$$
$$A_{22} = \{a_{22}, a_{24}\} = \{2, 1\},$$
$$A_{31} = \{a_{31}, a_{33}, a_{35}\} = \{2, 5, -1\},$$
$$A_{32} = \{a_{32}, a_{34}\} = \{4, 1\}.$$

Therefore,

$$A\underline{32} = (A_{ij}) = \begin{bmatrix} \{2, 3, 1, 0, 8\} & \{1, 4, -1, 0\} \\ \{3, 0, 2\} & \{2, 1\} \\ \{2, 5, -1\} & \{4, 1\} \end{bmatrix}$$
$$= \{(x_{ij}) \mid x_{11} \in \{0, 1, 2, 3, 8\}, x_{12} \in \{-1, 0, 1, 4\}, x_{21} \in \{0, 2, 3\},$$
$$x_{22} \in \{1, 2\}, x_{31} \in \{-1, 2, 5\}, x_{32} \in \{1, 4\}\}.$$

Therefore $|A\underline{32}| = 720$.

Let $A = (a_{ij}) \in M_{m \times n}$ and $B = (a_{ij}) \in M_{u \times v}$ be two matrices and $s = \min(m, u)$, $t = \min(n, u)$. We define an addition, which we call *cut-addition*, as follows:

$$\oplus_c : M_{m \times n} \times M_{u \times v} \to M_{s \times t}$$
$$(A, B) \mapsto A \oplus_c B = A\underline{cst} + B\underline{cst}.$$

Let $A = (a_{ij}) \in M_{m \times n}$ and $B = (a_{ij}) \in M_{u \times v}$ be two matrices and $s = \min(n, u)$. Then, we define a multiplication, which we call *cut-multiplication*, as follows:

$$\otimes_c : M_{m \times n} \times M_{u \times v} \to M_{m \times v}$$
$$(A, B) \mapsto A \otimes_c B = A\underline{cms} \cdot B\underline{csv}.$$

The cut-addition is associative and commutative.

Let $A = (a_{ij}) \in M_{m \times n}$ and $B = (a_{ij}) \in M_{u \times v}$ be two matrices and $s = \min(m, u)$, $t = \min(n, v)$. We define a hyper-addition, which we call *helix-addition* or *helix-sum*, as follows:

$$\oplus : M_{m \times n} \times M_{u \times v} \to \mathcal{P}(M_{s \times t})$$
$$(A, B) \mapsto A \oplus B = A\underline{st} +_h B\underline{st},$$

where

$$A\underline{st} +_h B\underline{st} = \{(c_{ij}) = (a_{ij} + b_{ij}) \mid a_{ij} \in A_{ij}, b_{ij} \in B_{ij}\}.$$

Example 4.3. Suppose that

$$A = \begin{bmatrix} 2 & 1 \\ 0 & 1 \\ 2 & 3 \end{bmatrix} \quad \text{and} \quad B = \begin{bmatrix} 1 & 4 & 0 \\ 2 & 0 & 1 \end{bmatrix}.$$

Then,

$$A\underline{22} = \begin{bmatrix} A_{11} & A_{12} \\ A_{21} & A_{22} \end{bmatrix} \quad \text{and} \quad B\underline{22} = \begin{bmatrix} B_{11} & B_{12} \\ B_{21} & B_{22} \end{bmatrix},$$

where

$$A_{11} = \{a_{11}, a_{31}\} = \{2\}, \quad B_{11} = \{b_{11}, b_{13}\} = \{1, 0\},$$
$$A_{12} = \{a_{12}, a_{32}\} = \{1, 3\}, \quad B_{12} = \{b_{12}\} = \{4\},$$
$$A_{21} = \{a_{21}\} = \{0\}, \quad B_{21} = \{b_{21}, b_{23}\} = \{2, 1\},$$
$$A_{22} = \{a_{22}\} = \{1\}, \quad B_{22} = \{b_{22}\} = \{0\}.$$

So

$$A\underline{22} = \left\{ \begin{bmatrix} 2 & 1 \\ 0 & 1 \end{bmatrix}, \begin{bmatrix} 2 & 3 \\ 0 & 1 \end{bmatrix} \right\},$$

and

$$B\underline{22} = \left\{ \begin{bmatrix} 1 & 4 \\ 2 & 0 \end{bmatrix}, \begin{bmatrix} 1 & 4 \\ 1 & 0 \end{bmatrix}, \begin{bmatrix} 0 & 4 \\ 2 & 0 \end{bmatrix}, \begin{bmatrix} 0 & 4 \\ 1 & 0 \end{bmatrix} \right\}.$$

Therefore, we have

$$A\underline{22} +_h B\underline{22} = \left\{ \begin{bmatrix} 3 & 5 \\ 2 & 1 \end{bmatrix}, \begin{bmatrix} 3 & 5 \\ 1 & 1 \end{bmatrix}, \begin{bmatrix} 2 & 5 \\ 2 & 1 \end{bmatrix}, \begin{bmatrix} 2 & 5 \\ 1 & 1 \end{bmatrix}, \begin{bmatrix} 3 & 7 \\ 2 & 1 \end{bmatrix}, \right.$$
$$\left. \begin{bmatrix} 3 & 7 \\ 1 & 1 \end{bmatrix}, \begin{bmatrix} 2 & 7 \\ 2 & 1 \end{bmatrix}, \begin{bmatrix} 2 & 7 \\ 1 & 1 \end{bmatrix} \right\}.$$

Lemma 4.2. *The helix-addition is commutative.*

Let $A = (a_{ij}) \in M_{m \times n}$ and $B = (a_{ij}) \in M_{u \times v}$ be two matrices and $s = \min(n, u)$. Then, we define a hyper-multiplication, which we call *helix-hyperoperation*, as follows:

$$\otimes : M_{m \times n} \times M_{u \times v} \to \mathcal{P}(M_{m \times v})$$
$$(A, B) \mapsto A \otimes B = A\underline{ms} \cdot_h B\underline{sv},$$

where

$$A\underline{ms} \cdot_h B\underline{sv} = \left\{ (c_{ij}) = \left(\sum a_{it} b_{tj} \right) \mid a_{ij} \in A_{ij}, b_{ij} \in B_{ij} \right\}.$$

Example 4.4. We consider the matrices A and B as follows:

$$A = \begin{bmatrix} 1 & 0 & 2 & 0 \\ 3 & 1 & 3 & 2 \end{bmatrix} \quad \text{and} \quad B = \begin{bmatrix} -1 & 1 \\ 0 & 2 \end{bmatrix}.$$

Then,

$$A\underline{22} = \begin{bmatrix} \{1, 2\} & 0 \\ 3 & \{1, 2\} \end{bmatrix}.$$

Therefore,

$$A \otimes B = \begin{bmatrix} \{-1, -2\} & \{1, 2\} \\ -3 & \{5, 7\} \end{bmatrix}.$$

Proposition 4.7.

(1) *The cut-multiplication $\otimes_c$ is associative.*

(2) *The helix-multiplication $\otimes$ is weak associative.*

Proof. It is straightforward. ∎

Note that the helix-multiplication is not distributive (not even weak) with respect to the helix-addition. But if all matrices which are used in the distributivity are of the same type $M_{m \times n}$, Then, we have

$$A \otimes (B \oplus C) = A \otimes (B + C) \quad \text{and} \quad (A \otimes B) \oplus (A \otimes C) = (A \otimes B) + (A \otimes C).$$

Therefore, the weak distributivity is valid and more precisely the inclusion distributivity is valid.

One of the kinds of hyperrings was introduced and studied by Rota [90] in 1982, where the multiplication is a hyperoperation, while the addition is an ordinary operation. Many properties of multiplication hyperrings are investigated in [42, 90]. A multiplicative H_v-ring is a generalization of a multiplicative hyperring and a special case of H_v-rings, where the associative law and the distributive law are replaced by their corresponding weak axioms.

Definition 4.6. The hyperstructure $(R, +, \cdot)$ is a multiplicative H_v-ring if

(1) $(R, +)$ is an abelian group,

(2) $(R, \cdot)$ is an H_v-semigroup,

(3) $\cdot$ is weak distributive with respect to $+$, i.e.,

$$x \cdot (y + z) \cap (x \cdot y + x \cdot z) \neq \emptyset \text{ and } (x + y) \cdot z \cap (x \cdot z + y \cdot z) \neq \emptyset,$$

for all $x, y, z \in R$.

The *fundamental relation* γ^* is defined in multiplicative H_v-rings as the smallest equivalence relation, so that the quotient would be ordinary ring. The way to find the fundamental classes is given by Theorem 4.3 to the following.

Theorem 4.11. *Let* $(R, +, \cdot)$ *be a multiplicative H_v-ring and let us denote by $\mathcal{U}$ the set of all finite sum of finite product of elements of R. We define the relation γ in R as follows:*

$$x\gamma y \iff \{x, y\} \subseteq u, \text{ for some } u \in \mathcal{U}.$$

Then, the fundamental relation γ^* is the transitive closure of the relation γ.

Theorem 4.12. *Let* $R = M_{m \times n}$. *Then, R together with addition and helix-multiplication becomes a multiplicative H_v-ring.*

Remark that the helix-operations give a wide class of H_v-rings, therefore the research can be restricted in special classes, finite or infinite, obtained by several types of matrices. Following this remark we can see several examples presented in this paragraph.

In the following examples, we denote E_{ij} any type of matrices which have the ij-entry 1 and in all the others entries we have 0.

Example 4.5. Consider the 1×2 matrices, with entries in $\mathbb{Z}_4$, of the following form:

$$A_{\bar{k}} = \bar{k}E_{11} \text{ and } B_{\bar{\lambda}} = \bar{\lambda}E_{11} + \bar{2}E_{12},$$

for $\bar{k}, \bar{\lambda} \in \mathbb{Z}_4$. Suppose that $R = \{A_{\bar{k}}, B_{\bar{\lambda}} \mid \bar{k}, \bar{\lambda} \in \mathbb{Z}_4\}$. Clearly, $(R, \oplus)$ is an abelian group. Also, we have

$$
\begin{aligned}
A_{\bar{k}} \otimes A_{\bar{\lambda}} &= \begin{bmatrix} \bar{k} & \bar{0} \end{bmatrix} \otimes \begin{bmatrix} \bar{\lambda} & \bar{0} \end{bmatrix} \\
&= \left[\{ \bar{k}, \bar{0} \} \right] \cdot_h \begin{bmatrix} \bar{\lambda} & \bar{0} \end{bmatrix} \\
&= \left[\{ \bar{k}\bar{\lambda}, \bar{0} \} \ \ \bar{0} \right] \\
&= \left\{ \begin{bmatrix} \bar{k}\bar{\lambda} & \bar{0} \end{bmatrix}, \begin{bmatrix} \bar{0} & \bar{0} \end{bmatrix} \right\} \\
&= \{ A_{\bar{k}\bar{\lambda}}, A_{\bar{0}} \}, \\
B_{\bar{k}} \otimes B_{\bar{\lambda}} &= \begin{bmatrix} \bar{k} & \bar{2} \end{bmatrix} \otimes \begin{bmatrix} \bar{\lambda} & \bar{2} \end{bmatrix} \\
&= \left[\{ \bar{k}, \bar{2} \} \right] \cdot_h \begin{bmatrix} \bar{\lambda} & \bar{2} \end{bmatrix} \\
&= \left[\{ \bar{k}\bar{\lambda}, \bar{2}\bar{\lambda} \} \ \ \{ \bar{2}\bar{k}, \bar{0} \} \right] \\
&= \left\{ \begin{bmatrix} \bar{k}\bar{\lambda} & \bar{2}\bar{k} \end{bmatrix}, \begin{bmatrix} \bar{k}\bar{\lambda} & \bar{0} \end{bmatrix}, \begin{bmatrix} \bar{2}\bar{\lambda} & \bar{2}\bar{k} \end{bmatrix}, \begin{bmatrix} \bar{2}\bar{\lambda} & \bar{0} \end{bmatrix} \right\}.
\end{aligned}
$$

Since $\bar{2}\bar{k} = \bar{0}$ or $\bar{2}$, it follows that

$$
B_{\bar{k}} \otimes B_{\bar{\lambda}} = \begin{cases} \left\{ \begin{bmatrix} \bar{k}\bar{\lambda} & \bar{0} \end{bmatrix}, \begin{bmatrix} \bar{2}\bar{\lambda} & \bar{0} \end{bmatrix} \right\} = \{ A_{\bar{k}\bar{\lambda}}, A_{\bar{2}\bar{\lambda}} \} \\ \text{or} \\ \left\{ \begin{bmatrix} \bar{k}\bar{\lambda} & \bar{2} \end{bmatrix}, \begin{bmatrix} \bar{k}\bar{\lambda} & \bar{0} \end{bmatrix}, \begin{bmatrix} \bar{2}\bar{\lambda} & \bar{2} \end{bmatrix}, \begin{bmatrix} \bar{2}\bar{\lambda} & \bar{0} \end{bmatrix} \right\} = \{ B_{\bar{k}\bar{\lambda}}, A_{\bar{k}\bar{\lambda}}, B_{\bar{2}\bar{\lambda}}, A_{\bar{2}\bar{\lambda}} \}. \end{cases}
$$

Also, we have

$$
\begin{aligned}
A_{\bar{k}} \otimes B_{\bar{\lambda}} &= \begin{bmatrix} \bar{k} & \bar{0} \end{bmatrix} \otimes \begin{bmatrix} \bar{\lambda} & \bar{2} \end{bmatrix} \\
&= \left[\{ \bar{k}\bar{\lambda}, \bar{0} \} \ \ \{ \bar{2}\bar{k}, \bar{0} \} \right] \\
&= \left\{ \begin{bmatrix} \bar{k}\bar{\lambda} & \bar{2}\bar{k} \end{bmatrix}, \begin{bmatrix} \bar{k}\bar{\lambda} & \bar{0} \end{bmatrix}, \begin{bmatrix} \bar{0} & \bar{2}\bar{k} \end{bmatrix}, \begin{bmatrix} \bar{0} & \bar{0} \end{bmatrix} \right\}.
\end{aligned}
$$

Since $\bar{2}\bar{k} = \bar{0}$ or $\bar{2}$, it follows that

$$
A_{\bar{k}} \otimes B_{\bar{\lambda}} = \begin{cases} \left\{ \begin{bmatrix} \bar{k}\bar{\lambda} & \bar{0} \end{bmatrix}, \begin{bmatrix} \bar{0} & \bar{0} \end{bmatrix} \right\} = \{ A_{\bar{k}\bar{\lambda}}, A_{\bar{0}} \} \\ \text{or} \\ \left\{ \begin{bmatrix} \bar{k}\bar{\lambda} & \bar{2} \end{bmatrix}, \begin{bmatrix} \bar{k}\bar{\lambda} & \bar{0} \end{bmatrix}, \begin{bmatrix} \bar{0}\bar{\lambda} & \bar{2} \end{bmatrix}, \begin{bmatrix} \bar{0}\bar{\lambda} & \bar{0} \end{bmatrix} \right\} = \{ B_{\bar{k}\bar{\lambda}}, A_{\bar{k}\bar{\lambda}}, B_{\bar{0}}, A_{\bar{0}} \}. \end{cases}
$$

Finally, we have

$$
\begin{aligned}
B_{\bar{\lambda}} \otimes A_{\bar{k}} &= \begin{bmatrix} \bar{\lambda} & \bar{2} \end{bmatrix} \otimes \begin{bmatrix} \bar{k} & \bar{2} \end{bmatrix} \\
&= \left[\{ \bar{\lambda}, \bar{2} \} \right] \cdot_h \begin{bmatrix} \bar{k} & \bar{0} \end{bmatrix} \\
&= \left\{ \begin{bmatrix} \bar{k}\bar{\lambda} & \bar{0} \end{bmatrix}, \begin{bmatrix} \bar{2}\bar{k} & \bar{0} \end{bmatrix} \right\} \\
&= \{ A_{\bar{k}\bar{\lambda}}, A_{\bar{2}\bar{k}} \}.
\end{aligned}
$$

Therefore, $(R, \oplus, \otimes)$ is a multiplicative H_v-ring. The set $\{ A_{\bar{k}} \mid \bar{k} \in \mathbb{Z}_4 \}$ is a left H_v-ideal of R.

Example 4.6. Consider the 2×3 matrices of the form

$$A_{(k,\lambda)} = kE_{11} + \lambda E_{22},$$

where $k, \lambda \in \mathbb{Z}$. Let $R = \{A_{(k,\lambda)} \mid k, \lambda \in \mathbb{Z}\}$. Then, we have

$$A_{(k,\lambda)} \oplus A_{(k',\lambda')} = A_{(k+k',\lambda+\lambda')}.$$

Indeed, $(R, \oplus)$ is an abelian group. On the other hand, we have

$$
\begin{aligned}
A_{(k,\lambda)} \otimes A_{(k',\lambda')} &= \begin{bmatrix} k & 0 & 0 \\ 0 & \lambda & 0 \end{bmatrix} \otimes \begin{bmatrix} k' & 0 & 0 \\ 0 & \lambda' & 0 \end{bmatrix} \\
&= \begin{bmatrix} \{k,0\} & 0 \\ 0 & \lambda \end{bmatrix} \cdot_h \begin{bmatrix} k' & 0 & 0 \\ 0 & \lambda' & 0 \end{bmatrix} \\
&= \left\{ \begin{bmatrix} kk' & 0 & 0 \\ 0 & \lambda\lambda' & 0 \end{bmatrix}, \begin{bmatrix} 0 & 0 & 0 \\ 0 & \lambda\lambda' & 0 \end{bmatrix} \right\}. \\
&= \left\{ A_{(kk',\lambda\lambda')}, A_{(0,\lambda\lambda')} \right\}.
\end{aligned}
$$

Clearly, $\otimes$ is commutative and associative. Moreover, $\otimes$ is distributive respect to $\oplus$. Therefore, $(R, \oplus, \otimes)$ is a commutative multiplicative hyperring. The following sets

$$I = \left\{ \begin{bmatrix} k & 0 & 0 \\ 0 & 0 & 0 \end{bmatrix} \mid k \in \mathbb{Z} \right\} \text{ and } J = \left\{ \begin{bmatrix} 0 & 0 & 0 \\ 0 & \lambda & 0 \end{bmatrix} \mid \lambda \in \mathbb{Z} \right\}$$

are hyperideals of R.

Example 4.7. Consider the 2×3 matrices of the form

$$A_{(x,y,z)} = xE_{11} + yE_{21} + zE_{22},$$

where $x, y, z \in \mathbb{R}$. Let $R = \{A_{(x,y,x)} \mid x, y, z \in \mathbb{R}\}$. Then, $(R, \oplus)$ is an abelian group. On the other hand,

$$
\begin{aligned}
&A_{(x,y,z)} \otimes A_{(x',y',z')} \\
&= \begin{bmatrix} x & 0 & 0 \\ y & z & 0 \end{bmatrix} \otimes \begin{bmatrix} x' & 0 & 0 \\ y' & z' & 0 \end{bmatrix} \\
&= \begin{bmatrix} \{x,0\} & 0 \\ \{y,0\} & z \end{bmatrix} \cdot_h \begin{bmatrix} x' & 0 & 0 \\ y' & z' & 0 \end{bmatrix} \\
&= \left\{ \begin{bmatrix} 0 & 0 & 0 \\ zy' & zz' & 0 \end{bmatrix}, \begin{bmatrix} 0 & 0 & 0 \\ yx'+zy' & zz' & 0 \end{bmatrix}, \begin{bmatrix} xx' & 0 & 0 \\ zy' & zz' & 0 \end{bmatrix}, \begin{bmatrix} xx' & 0 & 0 \\ yx'+zy' & zz' & 0 \end{bmatrix} \right\} \\
&= \left\{ A_{(0,zy',zz')}, A_{(0,yx'+zy',zz')}, A_{(xx',zy',zz')}, A_{(xx',yx'+zy',zz')} \right\}.
\end{aligned}
$$

Therefore, $(R, \oplus, \otimes)$ is a multiplicative H_v-ring. Note that $\otimes$ is not (weak) commutative.

The unit matrix is $I_c = E_{11} + E_{22} = \begin{bmatrix} 1 & 0 & 0 \\ 0 & 1 & 0 \end{bmatrix}$. Let $\begin{bmatrix} x & 0 & 0 \\ y & z & 0 \end{bmatrix}$ is given.

This matrix is invertible if and only if $x \neq 0$ and $z \neq 0$. An element $\begin{bmatrix} a & 0 & 0 \\ b & c & 0 \end{bmatrix}$

is inverse of $\begin{bmatrix} x & 0 & 0 \\ y & z & 0 \end{bmatrix}$ if we have

$$\begin{bmatrix} 1 & 0 & 0 \\ 0 & 1 & 0 \end{bmatrix} \in \begin{bmatrix} a & 0 & 0 \\ b & c & 0 \end{bmatrix} \otimes \begin{bmatrix} x & 0 & 0 \\ y & z & 0 \end{bmatrix}$$

and

$$\begin{bmatrix} 1 & 0 & 0 \\ 0 & 1 & 0 \end{bmatrix} \in \begin{bmatrix} x & 0 & 0 \\ y & z & 0 \end{bmatrix} \otimes \begin{bmatrix} a & 0 & 0 \\ b & c & 0 \end{bmatrix}.$$

From the first one, we obtain

$$(ax = 1, \ cy = 0, \ cz = 1) \text{ or } (ax = 1, \ bx + cy = 0, \ cz = 1),$$

and from the second, we obtain

$$(xa = 1, \ zb = 0, \ zc = 1) \text{ or } (xa = 1, \ ya + zb = 0, \ zc = 1).$$

Therefore, we obtain solutions $a = \frac{1}{x}$, $b = \frac{-y}{xz}$, $c = \frac{1}{z}$. Thus, the inverse

of $\begin{bmatrix} x & 0 & 0 \\ y & z & 0 \end{bmatrix}$ is the matrix $\begin{bmatrix} \frac{1}{x} & 0 & 0 \\ \frac{-y}{xz} & \frac{1}{z} & 0 \end{bmatrix}$. Moreover the element $\begin{bmatrix} \frac{1}{x} & 0 & 0 \\ 0 & \frac{1}{z} & 0 \end{bmatrix}$ is also

an inverse element of $\begin{bmatrix} x & 0 & 0 \\ y & z & 0 \end{bmatrix}$ with respect to the unit matrix $\begin{bmatrix} 1 & 0 & 0 \\ 0 & 1 & 0 \end{bmatrix}$.

The set

$$I = \left\{ \begin{bmatrix} 0 & 0 & 0 \\ y & 0 & 0 \end{bmatrix} \mid y \in \mathbb{R} \right\}$$

is a hyperideal of R.

Example 4.8. (Finite case). Suppose that $R = \{O, A, B, C\}$, where

$$O = \begin{bmatrix} \bar{0} & \bar{0} & \bar{0} \\ \bar{0} & \bar{0} & \bar{0} \end{bmatrix}, \ A = \begin{bmatrix} \bar{1} & \bar{0} & \bar{0} \\ \bar{0} & \bar{0} & \bar{0} \end{bmatrix}, \ B = \begin{bmatrix} \bar{0} & \bar{0} & \bar{0} \\ \bar{1} & \bar{0} & \bar{0} \end{bmatrix}, \ C = \begin{bmatrix} \bar{1} & \bar{0} & \bar{0} \\ \bar{1} & \bar{0} & \bar{0} \end{bmatrix},$$

with entries in $\mathbb{Z}_2$. Clearly, $(R, \oplus)$ is an abelian group. For helix-multiplication we obtain the following Cayley's table:

$\otimes$	O	A	B	C
O	O	O	O	O
A	O	$\{O, A\}$	O	$\{O, A\}$
B	O	$\{O, B\}$	$\{O, B\}$	$\{O, B\}$
C	O	$\{O, A, B, C\}$	O	$\{O, A, B, C\}$

Then, $(R, \oplus, \otimes)$ is a multiplicative H_v-ring.

4.5 H_v-fields

More general structures can be defined by using the fundamental structures. An application in this direction is the general hyperfield.

Definition 4.7. An H_v-ring $(R, +, \cdot)$ is called H_v-field if R/γ^* is a field.

Definition 4.8. The H_v-semigroup $(H, \cdot)$ is called h/v-group if H/β^* is a group.

Definition 4.9. An H_v-structure is called *very thin* if all hyperoperations are operations except one, which has all hyperproducts singletons except one, which is a subset of cardinality more than one. Therefore, in a very thin H_v-structure in H there exists a hyperoperation $\cdot$ and a pair $(a, b) \in H^2$ for which $ab = A$, with $cardA > 1$, and all the other products, are singletons.

In this section, we denote by $[x]$ the fundamental class of the element $x \in H$. Therefore $\beta^*(x) = [x]$.

Analogous theorems are for H_v-rings, H_v-vector spaces and so on. An element is called *single* if its fundamental class is singleton, so $[x] = \{x\}$.

The class of h/v-groups is more general than the H_v-groups since in h/v-groups the reproductivity is not valid. The *reproductivity of classes* is valid, i.e., if H is partitioned into equivalence classes, then

$$x[y] = [xy] = [x]y, \text{ for all } x, y \in H,$$

because the quotient is reproductive. In a similar way the h/v-*rings, h/v-fields, h/v-modulus, h/v-vector spaces* etc. are defined. Note that h/v-structures were introduced by Vougiouklis for the first time.

Remark 4.1. From definition of the H_v-field, we remark that the reproduction axiom in the product, is not assumed, the same is also valid for the definition of the h/v-field. Therefore, an H_v-field is an h/v-field where the reproduction axiom for the sum is also valid.

We know that the reproductivity in the classical group theory is equivalent to the axioms of the existence of the unit element and the existence of an inverse element for any given element. From the definition of the h/v-group, since a generalization of the reproductivity is valid, we have to extend the above two axioms on the equivalent classes.

Definition 4.10. Let $(H, \cdot)$ be an H_v-semigroup, and denote $[x]$ the fundamental, or equivalent classe, of the element $x \in H$. We call $[e]$ the *unit* *class* if we have

$$([e] \cdot [x]) \cap [x] \neq \varnothing \text{ and } ([x] \cdot [e]) \cap [x] \neq \varnothing, \text{ for all } x \in H,$$

and for each element $x \in H$, we call *inverse class* of $[x]$, the class $[x']$, if we have

$$([x] \cdot [x']) \cap [e] \neq \varnothing \text{ and } ([x'] \cdot [x]) \cap [e] \neq \varnothing.$$

The "enlarged" hyperstructures were examined in the sense that a new element appears in one result. In enlargement or reduction, most useful are those H_v-structures or h/v-structures with the same fundamental structure.

Construction 4.13. (1) Let $(H, \cdot)$ be an H_v-semigroup and $v \notin H$. We extend the $\cdot$ into $\underline{H} = H \cup \{v\}$ as follows:

$$x \cdot v = v \cdot x = v, \text{ for all } x \in H, \text{ and } v \cdot v = H.$$

The $(\underline{H}, \cdot)$ is an h/v-group, called *attach*, where $(\underline{H}, \cdot)/\beta^* \cong \mathbb{Z}_2$ and v is a single element.

We have core $(\underline{H}, \cdot) = H$. The scalars and units of $(H, \cdot)$ are scalars and units, respectively, in $(\underline{H}, \cdot)$. If $(H, \cdot)$ is weak commutative (respectively, commutative), then $(\underline{H}, \cdot)$ is also weak commutative (respectively, commutative).

(2) Let $(H, \cdot)$ be an H_v-semigroup and $\{v_1, \ldots, v_n\} \cap H = \varnothing$, is an ordered set, where if $v_i < v_j$, when $i < j$. Extend $\cdot$ in $\underline{H}_n = H \cup \{v_1, \ldots, v_n\}$ as follows:

$$x \cdot v_i = v_i \cdot x = v_i, v_i \cdot v_j = v_j \cdot v_i = v_j, \text{ for all } i < j \text{ and}$$

$$v_i \cdot v_i = H \cup \{v_1, \ldots, v_{i-1}\}, \text{ for all } x \in H, i \in \{1, \ldots, n\}.$$

Then $(\underline{H}_n, \cdot)$ is h/v-group, called *attach elements*, where $(\underline{H}_n, \cdot)/\beta^* \cong \mathbb{Z}_2$ and v_n is single.

(3) Let $(H, \cdot)$ be an H_v-semigroup, $v \notin H$, and $(\underline{H}, \cdot)$ be its attached h/v-group. Take an element $0 \notin \underline{H}$ and define in $\underline{H}_o = H \cup \{v, 0\}$ two hyperoperations:

Hypersum $+$: $0 + 0 = x + v = v + x = 0$, $0 + v = v + 0 = x + y = v$, $0 + x = x + 0 = v + v = H$, for all $x, y \in H$.

Hyperproduct $\cdot$: remains the same as in $\underline{H}$. Moreover, $0 \cdot 0 = v \cdot x = x \cdot 0 = 0$, for all $x \in \underline{H}$.

Then, $(\underline{H}_o, +, \cdot)$ is h/v-field with $(\underline{H}_o, +, \cdot)/\gamma^* \cong \mathbb{Z}_3$. Moreover, $+$ is associative, $\cdot$ is weak associative and weak distributive with respect to $+$. Also, 0 is zero absorbing and single but not scalar in $+$. $(\underline{H}_o, +, \cdot)$ is called the *attached h/v-field* of the H_v-semigroup $(H, \cdot)$.

Let us denote by $\mathcal{U}$ the set of all finite products of elements of a hyper-groupoid $(H, \cdot)$. Consider the relation defined as follows:

$$xLy \Leftrightarrow \text{there exists } u \in \mathcal{U} \text{ such that } ux \cap uy \neq \varnothing.$$

Then the transitive closure L^* of L is called *left fundamental reproductivity relation*. Similarly, the *right fundamental reproductivity* relation R^* is defined.

Theorem 4.14. *If $(H, \cdot)$ is a commutative semihypergroup, i.e. the strong commutativity and the strong associativity is valid, then the strong expression of the above L relation: $ux = uy$, has the property: $L^* = L$.*

Proof. Suppose that two elements x and y of H are L^* equivalent. Therefore, there are $u_1, \ldots, u_{n+1}$ elements of $\mathcal{U}$ and $z_1, \ldots, z_n$ elements of H such that

$$u_1 x = u_1 z_1, \quad u_2 z_1 = u_2 z_2, \quad \ldots, \quad u_n z_{n-1} = u_n z_n, \quad u_{n+1} z_n = u_{n+1} y.$$

From these relations, using the strong commutativity, we obtain

$$u_{n+1} \ldots u_2 u_1 x = u_{n+1} \ldots u_2 u_1 z_1 = u_{n+1} \ldots u_1 u_2 z_1$$
$$= u_{n+1} \ldots u_2 u_1 z_2 = \cdots = u_{n+1} \ldots u_2 u_1 y.$$

Therefore, setting $u = u_{n+1} \ldots u_2 u_1 \in \mathcal{U}$, we have $ux = uy$. ∎

We present now the *small non-degenerate H_v-fields* on $(\mathbb{Z}_n, +, \cdot)$ which satisfy the following conditions, appropriate in Santilli's iso-theory:

(1) multiplicative very thin minimal,
(2) weak commutative (non-commutative),
(3) they have the elements 0 and 1, scalars,
(4) when an element has inverse element, then this is unique.

Note that last condition means than we cannot enlarge the result if it is 1 and we cannot put 1 in enlargement. Moreover we study only the upper triangular cases, in the multiplicative table, since the corresponding under, are isomorphic since the commutativity is valid for the underline rings. From the fact that the reproduction axiom in addition is valid, we have always H_v-fields.

Example 4.9. All multiplicative H_v-fields defined on $(\mathbb{Z}_4, +, \cdot)$, which have non-degenerate fundamental field, and satisfy the above 4 conditions, are the following isomorphic cases:
The only product which is set is $2 \otimes 3 = \{0, 2\}$ or $3 \otimes 2 = \{0, 2\}$.
The fundamental classes are $[0] = \{0, 2\}, [1] = \{1, 3\}$ and we have $(\mathbb{Z}_4, +, \otimes)/\gamma^* \cong (\mathbb{Z}_2, +, \cdot)$.

Example 4.10. Let us denote by E_{ij} the matrix with 1 in the ij-entry and zero in the rest entries. Take the following 2×2 upper triangular H_v-matrices on the above H_v-field $(\mathbb{Z}_4, +, \cdot)$ of the case that only $2 \otimes 3 = \{0, 2\}$ is a hyperproduct:

$$I = E_{11} + E_{22},$$
$$a = E_{11} + E_{12} + E_{22},$$
$$b = E_{11} + 2E_{12} + E_{22},$$
$$c = E_{11} + 3E_{12} + E_{22},$$
$$d = E_{11} + 3E_{22},$$
$$e = E_{11} + E_{12} + 3E_{22},$$
$$f = E_{11} + 2E_{12} + 3E_{22},$$
$$g = E_{11} + 3E_{12} + 3E_{22}.$$

Then, we obtain for $\mathbb{X} = \{I, a, b, c, d, e, f, g\}$, that $(\mathbb{X}, \otimes)$ is non-COW H_v-group and the fundamental classes are $\underline{a} = \{a, c\}, \underline{d} = \{d, f\}, \underline{e} = \{e, g\}$ and the fundamental group is isomorphic to $(\mathbb{Z}_2 \times \mathbb{Z}_2, +)$. In this H_v-group there is only one unit and every element has a unique double inverse.

Theorem 4.15. *All multiplicative H_v-fields defined on $(\mathbb{Z}_6, +, \cdot)$, which have non-degenerate fundamental field, and satisfy the above 4 conditions, are the following isomorphic cases:*
We have the only one hyperproduct,

- $2 \otimes 3 = \{0, 3\}$ *or* $2 \otimes 4 = \{2, 5\}$ *or*
  $3 \otimes 4 = \{0, 3\}$ *or* $3 \otimes 5 = \{0, 3\}$ *or* $4 \otimes 5 = \{2, 5\}$
  Fundamental classes: $[0] = \{0, 3\}, [1] = \{1, 4\}, [2] = \{2, 5\}$, *and*
  $(\mathbb{Z}_6, +, \cdot)/\gamma^* \cong (\mathbb{Z}_3, +, \cdot).$
- $2 \otimes 3 = \{0, 2\}$ *or* $2 \otimes 3 = \{0, 4\}$ *or* $2 \otimes 4 = \{0, 2\}$ *or* $2 \otimes 4 = \{2, 4\}$
  or
  $2 \otimes 5 = \{0, 4\}$ *or* $2 \otimes 5 = \{2, 4\}$ *or* $3 \otimes 4 = \{0, 2\}$ *or* $3 \otimes 4 = \{0, 4\}$
  or
  $3 \otimes 5 = \{3, 5\}$ *or* $4 \otimes 5 = \{0, 2\}$ *or* $4 \otimes 5 = \{2, 4\}$
  Fundamental classes: $[0] = \{0, 2, 4\}, [1] = \{1, 3, 5\}$, *and*
  $(\mathbb{Z}_6, +, \otimes)/\gamma^* \cong (\mathbb{Z}_2, +, \cdot).$

Example 4.11. All multiplicative H_v-fields defined on $(\mathbb{Z}_9, +, \cdot)$, which have non-degenerate fundamental field, and satisfy the above 4 conditions, are the following isomorphic cases:
We have the only one hyperproduct,

$2 \otimes 3 = \{0, 6\}$ or $\{3, 6\}$, $2 \otimes 4 = \{2, 8\}$ or $\{5, 8\}$, $2 \otimes 6 = \{0, 3\}$ or $\{3, 6\}$,

$2 \otimes 7 = \{2, 5\}$ or $\{5, 8\}$, $2 \otimes 8 = \{1, 7\}$ or $\{4, 7\}$, $3 \otimes 4 = \{0, 3\}$ or $\{3, 6\}$,
$3 \otimes 5 = \{0, 6\}$ or $\{3, 6\}$, $3 \otimes 6 = \{0, 3\}$ or $\{0, 6\}$, $3 \otimes 7 = \{0, 3\}$ or $\{3, 6\}$,
$3 \otimes 8 = \{0, 6\}$ or $\{3, 6\}$, $4 \otimes 5 = \{2, 5\}$ or $\{2, 8\}$, $4 \otimes 6 = \{0, 6\}$ or $\{3, 6\}$,
$4 \otimes 8 = \{2, 5\}$ or $\{5, 8\}$, $5 \otimes 6 = \{0, 3\}$ or $\{3, 6\}$, $5 \otimes 7 = \{2, 8\}$ or $\{5, 8\}$,
$5 \otimes 8 = \{1, 4\}$ or $\{4, 7\}$, $6 \otimes 7 = \{0, 6\}$ or $\{3, 6\}$, $6 \otimes 8 = \{0, 3\}$ or $\{3, 6\}$,
$$7 \otimes 8 = \{2, 5\} \text{ or } \{2, 8\},$$

Fundamental classes are $[0] = \{0, 3, 6\}$, $[1] = \{1, 4, 7\}$, $[2] = \{2, 5, 8\}$, and

$$(\mathbb{Z}_9, +, \otimes)/\gamma^* \cong (\mathbb{Z}_3, +, \cdot).$$

Example 4.12. All H_v-fields defined on $(\mathbb{Z}_{10}, +, \cdot)$, which have non-degenerate fundamental field, and satisfy the above 4 conditions, are the following isomorphic cases:

- We have the only one hyperproduct,
  $2 \otimes 4 = \{3, 8\}$, $2 \otimes 5 = \{2, 5\}$, $2 \otimes 6 = \{2, 7\}$, $2 \otimes 7 = \{4, 9\}$,
  $2 \otimes 9 = \{3, 8\}$,
  $3 \otimes 4 = \{2, 7\}$, $3 \otimes 5 = \{0, 5\}$, $3 \otimes 6 = \{3, 8\}$, $3 \otimes 8 = \{4, 9\}$,
  $3 \otimes 9 = \{2, 7\}$,
  $4 \otimes 5 = \{0, 5\}$, $4 \otimes 6 = \{4, 9\}$, $4 \otimes 7 = \{3, 8\}$, $4 \otimes 8 = \{2, 7\}$,
  $5 \otimes 6 = \{0, 5\}$,
  $5 \otimes 7 = \{0, 5\}$, $5 \otimes 8 = \{0, 5\}$, $5 \otimes 9 = \{0, 5\}$, $6 \otimes 7 = \{2, 7\}$,
  $6 \otimes 8 = \{3, 8\}$,
  $6 \otimes 9 = \{4, 9\}$, $7 \otimes 9 = \{3, 8\}$, $8 \otimes 9 = \{2, 7\}$.
  Fundamental classes: $[0] = \{0, 5\}$, $[1] = \{1, 6\}$, $[2] = \{2, 7\}$, $[3] = \{3, 8\}$, $[4] = \{4, 9\}$ and $(\mathbb{Z}_{10}, +, \otimes)/\gamma^* \cong (\mathbb{Z}_5, +, \cdot)$.
- The cases where we have two classes
  $[0] = \{0, 2, 4, 6, 8\}$ and $[1] = \{1, 3, 5, 7, 9\}$, thus we have fundamental field $(\mathbb{Z}_{10}, +, \otimes)/\gamma^* \cong (\mathbb{Z}_2, +, \cdot)$, can be described as follows:
  Taking in the multiplicative table only the results above the diagonal, we enlarge each of the products by putting one element of the same class of the results. We do not enlarge setting the element 1, and we cannot enlarge only the product $3 \otimes 7 = 1$. The number of those H_v-fields is 103.

Example 4.13. In order to see how hard is to realize the reproductivity of classes and the unit class and inverse class, we consider the above H_v-field $(\mathbb{Z}_{10}, +, \otimes)$ where we have $2 \otimes 4 = \{3, 8\}$. Then, the multiplicative table of the hyperproduct is the following:

⊗	0	1	2	3	4	5	6	7	8	9
0	0	0	0	0	0	0	0	0	0	0
1	0	1	2	3	4	5	6	7	8	9
2	0	2	4	6	3,8	0	2	4	6	8
3	0	3	6	9	2	5	8	1	4	7
4	0	4	8	2	6	0	4	8	2	6
5	0	5	0	5	0	5	0	5	0	5
6	0	6	2	8	4	0	6	2	8	4
7	0	7	4	1	8	5	2	9	6	3
8	0	8	6	4	2	0	8	6	4	2
9	0	9	8	7	6	5	4	3	2	1

On this table it is easy to see that the reproductivity is not valid but it is very hard to see that the reproductivity of classes is valid. We can see the reproductivity of classes easier if we reformulate the multiplicative table according to the fundamental classes, $[0] = \{0,5\}, [1] = \{1,6\}, [2] = \{2,7\}, [3] = \{3,8\}, [4] = \{4,9\}$. Then, we obtain

⊗	0	5	1	6	2	7	3	8	4	9
0	0	0	0	0	0	0	0	0	0	0
5	0	5	5	0	0	5	5	0	0	5
1	0	5	1	6	2	7	3	8	4	9
6	0	0	6	6	2	2	8	8	4	4
2	0	0	2	2	4	4	6	6	3,8	8
7	0	5	7	2	4	9	1	6	8	3
3	0	5	3	8	6	1	9	4	2	7
8	0	0	8	8	6	6	4	4	2	2
4	0	0	4	4	8	8	2	2	6	6
9	0	5	9	4	8	3	7	2	6	1

From this it is easy to see the unit class and the inverse class of each class.

4.6 H_v-rings endowed with P-hyperoperations

In this section, we study a wide class of H_v-rings obtained from an arbitrary ring by using P-hyperoperations. The notion of P-hyperoperations and their generalizations are introduced in [112, 114]. We use the results obtained by S. Spartalis [103, 105].

Let $(R, +, \cdot)$ be a ring and P_1, P_2 be nonempty subsets of R. We shall use of the following right P-hyperoperations:

$$xP_1^* y = x + y + P_1, \quad xP_2^* y = xyP_2 \quad \text{for all } x, y \in R.$$

We denote the center of the semigroup $(R, \cdot)$ by $Z(R)$.

Theorem 4.16. *If* $0 \in P_1$ *and* $P_2 \cap Z(R) \neq \emptyset$, *then* (R, P_1^*, P_2^*) *is an* H_v-*ring called a* P-H_v-*ring or an* H_v-*ring with* P-*hyperoperations.*

Proof.   The proof is straightforward.   ∎

Let J be an H_v-ideal of (R, P_1^*, P_2^*). Since $0 \in RP_2^* J \cap JP_2^* R \subseteq J$ and $P_1 = 0P_1^* 0 \subseteq J$, we have $JP_1^* x = J + x = xP_1^* J$ for all $x \in R$. Moreover, the addition $\oplus$ and the multiplication $\odot$ between classes are defined in a usual manner:

$$(JP_1^* x) \oplus (JP_1^* y) = \{JP_1^* z \mid z \in (JP_1^* x)P_1^*(JP_1^* y)\} = \{J + x + y\},$$

$$(JP_1^* x) \odot (JP_1^* y) = \{J + w \mid w \in (JP_1^* x)P_2^*(JP_1^* y)\} = \{J + w \mid w \in xyP_2\}.$$

Theorem 4.17. *If* (R, P_1^*, P_2^*) *is a* P-H_v-*ring and* J *is an* H_v-*ideal, then* $(R/J, \oplus, \odot)$ *is a multiplicative* H_v-*ring.*

Proof.   Obviously, $(R/J, \oplus)$ is an abelian group. Moreover, $(R/J, \odot)$ is an H_v-semigroup, because $\odot$ is well defined and for all $x, y, z \in R$, we have

$$(JP_1^* x) \odot [(JP_1^* y) \odot (JP_1^* z)] = \{J + v \mid v \in xyzP_2P_2\},$$
$$[(JP_1^* x) \odot (JP_1^* y)] \odot (JP_1^* z) = \{J + u \mid u \in xyP_2zP_2\}.$$

But, since $zP_2P_2 \cap P_2zP_2 \neq \emptyset$, it follows that the multiplication is weak associative. Finally,

$$\begin{aligned}
(JP_1^* x) \odot [(JP_1^* y) \oplus (JP_1^* z)] &= \{J + v \mid v \in x(y + z)P_2\} \\
&\subseteq \{J + u \mid u \in xyP_2 + xzP_2\} \\
&= [(JP_1^* x) \odot (JP_1^* y)] \oplus [(JP_1^* x) \odot (JP_1^* z)].
\end{aligned}$$

In the same way, the right distributivity is proved and so $(R/J, \oplus, \odot)$ is a multiplicative H_v-ring.   ∎

Theorem 4.18. *Let* (R, P_1^*, P_2^*) *be a* P-H_v-*ring over the ring* $(R, +, \cdot)$ *and* J *be an ideal of the ring* R, *containing* P_1. *Then* $(R/J, \oplus, \odot)$ *is a multiplicative* H_v-*ring, which is a* P-H_v-*ring.*

Proof. It is easy to see that J is an H_v-ideal of (R, P_1^*, P_2^*). So, according to the previous theorem, $(R/J, \oplus, \odot)$ is a multiplicative H_v-ring. On the other hand, consider the quotient ring $(R/J, +, \cdot)$ and take $L_1 = \{J\}$, $L_2 = \{J + a \mid a \in P_2\}$. Then, for all $x \in R$, we have

$$(J + x)L_2L_2 = \{J + v \mid v \in xP_2P_2\},$$
$$L_2(J + x)L_2 = \{J + w \mid w \in P_2xP_2\}.$$

Since for all $x \in R$, $xP_2P_2 \cap P_2xP_2 \neq \emptyset$, it follows that

$$(J + x)L_2L_2 \cap L_2(J + x)L_2 \neq \emptyset.$$

Consequently, since J is the zero element of R/J, from Theorem 4.16 it follows that $(R/J, L_1^*, L_2^*)$ is a P-H_v-ring over the ring $(R/J, +, \cdot)$. Therefore, the hyperoperation $\odot$ is the P-hyperoperation L_2^* and $\oplus$ is the degenerate P-hyperoperation L_1^*.   ■

Theorem 4.19. *Let (R, P_1^*, P_2^*) be a P-H_v-ring and J be an H_v-ideal of R. If H is an H_v-subring of R containing P_1, then $HP_1^*J/J \cong H/H \cap J$.*

Proof. It is easy to see that H, J are subgroups of $(R, +)$ and so $HP_1^*J = H + J$ and $H \cap J$ are two groups containing P_1. Since

$$(HP_1^*J)P_2^*(HP_1^*J) = (H + J)(H + J)P_2 \subseteq HHP_2 + J$$

we have that (HP_1^*J, P_1^*, P_2^*) is an H_v-subring of (R, P_1^*, P_2^*). Moreover, $J, J \cap H$ are H_v-ideals of the H_v-rings HP_1^*J and H respectively. On the other hand, the quotients

$$HP_1^*J/J = \{J + x \mid x \in H\} \text{ and } H/H \cap J = \{(H \cap J) + y \mid y \in H\}$$

are multiplicative H_v-rings. We consider the bijection map $f : HP_1^*J/J \to H/H \cap J$ such that $J + x \longmapsto (H \cap J) + x$. The map f is a homomorphism, since for all $x, y \in R$, we have

$$f(J+x \oplus J+y) = f(J + x + y) = (H \cap J) + x + y = f(J + x) \oplus f(J + y),$$
$$f(J+x \odot J+y) = f(J + s \mid s \in xyP_2\}) = \{(H \cap J) + s \mid s \in xyP_2\}$$
$$= f(J + x) \odot f(J + y).$$

Consequently, f is an isomorphism.   ■

Theorem 4.20. *Let J and K be two H_v-ideals of the $P - H_v$-ring (R, P_1^*, P_2^*). If $J \subseteq K$, then $(R/J)/(K/J) \cong R/K$.*

Proof. The quotients R/J, K/J and R/K are multiplicative H_v-rings and K/J is an H_v-ideal of $(R/J, \oplus, \odot)$. Indeed, $K/J \subseteq R/J$ and for all $x \in K$, $y \in R$, we have

$$(J + x) \odot (J + y) = \{J + z \mid z \in xP_2^*y\} \subseteq K/J.$$

Therefore, $K/J \odot R/J \subseteq K/J$. Similarly, $R/J \odot K/J \subseteq K/J$. On the other hand, $((R/J)/(K/J), \diamond, *)$ is a multiplicative H_v-ring, where $\diamond$ and $*$ are the usual addition and multiplication of classes. Now, we consider the map $f : (R/J)/(K/J) \to R/K$ such that $(K/J) \oplus (J + x) \longmapsto K + x$. Since, for all $x, y \in R$,

$$(K/J) \oplus (J + x) = (K/J) \oplus (J + y)$$
$$\Leftrightarrow \{J + z \mid z \in K + x\} = \{J + w \mid w \in K + y\}$$
$$\Leftrightarrow y - x \in K$$
$$\Leftrightarrow K + x = K + y$$

it follows that f is well defined and one to one. Obviously, it is onto, so it remains to prove that f is a homomorphism. Indeed, for all $x, y \in R$ we have

$$f[(K/J) \oplus (J + x) \diamond (K/J) \oplus (J + y)]$$
$$= f(\{(K/J) \oplus (J + z) \mid J + z \in (K/J) \oplus (J + x) \oplus (K/J) \oplus (J + y)\})$$
$$= f(K/J \oplus (J + x + y)) = K + x + y = (K + x) \oplus (K + y)$$
$$= f((K/J) \oplus (J + x)) \oplus f((K/J) \oplus (J + y))$$

and

$$f[(K/J) \oplus (J + x) * (K/J) \oplus (J + y)]$$
$$= f[\{(K/J) \oplus (J + w) \mid J + w \in (J + x) \odot (J + y)\}]$$
$$= f(\{(K/J) \oplus (J + w) \mid w \in xyP_2\}] = \{K + w \mid w \in xP_2^*y\}$$
$$= (K + x) \odot (K + y) = f((K/J) \oplus (J + x)) \odot f((K/J) \oplus (J + y)).$$

Hence f is an isomorphism. ■

Theorem 4.21. *Let R and A be two rings, $f \in Hom(R, A)$ and (R, P_1^*, P_2^*) be an H_v-ring with P-hyperoperations. Then, the following assertions hold:*

(1) *If (A, L_1^*, L_2^*) is an H_v-ring such that $f(P_2) \cap L_2 \neq \emptyset$, then $f : (R, P_1^*, P_2^*) \to (A, L_1^*, L_2^*)$ is an H_v-homomorphism.*

(2) *If $f(P_2) \cap Z(A) \neq \emptyset$, then $f : (R, P_1^*, P_2^*) \to (A, f(P_1)^*, f(P_2)^*)$ is a strong homomorphism.*
A particular case: $f : (R, P_1^*, P_2^*) \to (Imf, f(P_1)^*, f(P_2)^*)$.

Proof.   (1) For all $x, y \in R$ we have

$$f(xP_1^*y) = f(x) + f(y) + f(P_1) \text{ and } f(x)L_1^*f(y) = f(x) + f(y) + L_1.$$

From the hypothesis it follows that $0 \in P_1$ and so

$$0 = f(0) \in f(P_1) \cap L_1 \text{ and } f(xP_1^*y) \cap f(x)L_1^*f(y) \neq \emptyset.$$

Moreover, the condition $f(xP_2^*y) \cap f(x)L_2^*f(y) \neq \emptyset$ holds obviously. Hence f is an H_v-homomorphism.

(2) The structure $(A, f(P_1)^*, f(P_2)^*)$ is an H_v-ring, because $0 \in f(P_1)$ and $f(P_2) \cap Z(A) \neq \emptyset$. The H_v-homomorphism f is strong, since for all $x, y \in R$, we have

$$f(xP_1^*y) = f(x) + f(y) + f(P_1) = f(x)f(P_1)^*f(y),$$
$$f(xP_2^*y) = f(x)f(y)f(P_2) = f(x)f(P_2)^*f(y).$$

In the particular case when f is the H_v-homomorphism from (R, P_1^*, P_2^*) to $(Imf, f(P_1)^*, f(P_2)^*)$, from $P_2 \cap Z(R) \neq \emptyset$, we can deduce easily that $f(P_2) \cap Z(Imf) \neq \emptyset$. Hence, (2) is valid.   ∎

Proposition 4.8. *Let (R, P_1^*, P_2^*) be an H_v-ring with P-hyperoperations. If $\alpha \in Z(R)$, then the translation of the semigroup $(R, \cdot)$ by α:*

$$f_\alpha : x \to \alpha x$$

is a multiplicatively strong homomorphism from $(R, L_1^*, (\alpha P_2)^*)$ to (R, P_1^*, P_2^*), where $0 \in L_1 \subseteq R$.

Proof.   First, we can observe that the structure $(R, L_1^*, (\alpha P_2)^*)$ is an H_v-ring because $0 \in L_1$ and $\alpha P_2 \cap Z(R) \neq \emptyset$. Moreover, for all $x, y, z \in R$, we have

$$f_\alpha(xL_1^*y) = \alpha(x+y+L_1) = \alpha x + \alpha y + \alpha L_1 \text{ and } f_\alpha(x)P_1^*f_\alpha(y) = \alpha x + \alpha y + P_1$$

and since $0 \in \alpha L_1 \cap P_1$, it follows that

$$f_\alpha(xL_1^*y) \cap f_\alpha(x)P_1^*f_\alpha(y) \neq \emptyset.$$

Moreover,

$$f_\alpha(x(\alpha P_2)^*y) = \alpha(xy\alpha P_2) = (\alpha x)(\alpha y)P_2 = f_\alpha(x)P_2^*f_\alpha(y).$$

Hence, f_α is a multiplicatively strong homomorphism.   ∎

Proposition 4.9. *Let (R, P_1^*, P_2^*) be an H_v-ring with P-hyperoperations and $\alpha \in Z(R)$. If the element α is simplificable and reproductive in $(R, \cdot)$, then for each subset L_1 of R such that $\alpha L_1 = P_1$, we have*

$$(R, L_1^*, (\alpha P_2)^*) \cong (R, P_1^*, P_2^*).$$

Proof. Let us consider the translation of the semigroup $(R, \cdot)$ by α

$$f_\alpha : (R, L_1^*, (\alpha P_2)^*) \to (R, P_1^*, P_2^*), \quad f_\alpha(x) = \alpha x,$$

which is a multiplicatively strong homomorphism of H_v-rings. Moreover, f_α is additively strong because, for all $x, y \in R$,

$$f_\alpha(xL_1^*y) = \alpha x + \alpha y + \alpha L_1 = \alpha x + \alpha y + P_1 = f_\alpha(x)P_1^*f_\alpha(y).$$

Finally, according to the hypothesis, the map f_α is one to one and onto, hence f_α is an isomorphism of H_v-rings. ∎

Suppose that the conditions of Proposition 4.9 hold and $\alpha^2 P_1 = P_1$. Therefore, every translation of the sets P_1, P_2 by $\alpha \in Z(R)$ gives an isomorphism of H_v-rings, i.e., $(R, (\alpha P_1)^*, (\alpha P_2)^*) \cong (R, P_1^*, P_2^*)$. In case the H_v-ring (R, P_1^*, P_2^*) is derived from a ring with unit 1, we obtain the following isomorphism: $(R, (-P_1)^*, (-P_2)^*) \cong (R, P_1^*, P_2^*)$, since -1 satisfies the hypothesis of Proposition 4.9 and $(-1)^2 P_1 = P_1$.

Now, we calculate the number of H_v-rings with P-hyperoperations, which can be constructed starting from a finite ring $(R, +, \cdot)$. By Proposition 4.8, it follows that this number can be substantially reduced, because some of these H_v-rings are isomorphic.

Proposition 4.10. *Let $(R, +, \cdot)$ be a ring with $cardR = n$, $n > 1$ and $cardZ(R)=m$. The number of H_v-rings (R, P_1^*, P_2^*) is at most $2^{n-1}(2^n - 2^{n-m})$.*

Proof. The number of the subsets P_1 of R, which satisfy the condition $0 \in P_1$, is 2^{n-1}. The number of the subsets P_2 of R, which satisfy the condition $P_2 \cap Z(R) \neq \emptyset$, is $2^{n-1} + 2^{n-2} + ... + 2^{n-m} = 2^n - 2^{n-m}$. Hence, the number of H_v-rings is at most $2^{n-1}(2^n - 2^{n-m})$. ∎

If $\cdot$ is commutative, the above number is $2^{2n-1} - 2^{n-1}$.

Proposition 4.11. *If $(R, +, \cdot)$ is a commutative ring without non-zero divisors and $cardR = n$, $n > 1$, then the number of hyperrings (R, P_1^*, P_2^*) which are not rings is at most $5 \cdot 2^{n-1} - n - 4$.*

Proof. First of all it is easy to check that the structure (R, P_1^*) is a hypergroup. Moreover, because of the commutativity of the multiplication, we have

$$xP_2^*(yP_2^*z) = xyzP_2P_2 = (xP_2^*y)P_2^*z,$$

for all $x, y, z \in R$. The necessary and sufficient condition for the validity of the inclusion distributivity is $RP_2P_1 \subseteq P_1$. We suppose that there exist $P_1, P_2 \subseteq R$, $P_1 \neq R$, satisfying the previous condition. Then, for any $p_1 \in P_1$, $p_2 \in P_2$, $p_1 \neq 0 \neq p_2$, the condition $Rp_2p_1 \subseteq P_1$ is valid and so, there exist $a, b \in R$, $a \neq b$ such that $ap_2p_1 = bp_2p_1$. Therefore, $(a - b)p_2p_1 = 0$, i.e., p_1 or p_2 is a zero divisor, which is a contradiction. Hence, the only cases, in which the condition of distributivity is satisfied, are $P_1 = R$, $P_1 = \{0\}$ and $P_2 = \{0\}$. Therefore, the number of hyperrings (R, P_1^*, P_2^*) is at most $(2^n - 1 + 2^n - 1 + 2^{n-1}) - 2 = 5 \cdot 2^{n-1} - 4$, because the hyperrings $(R, \{0\}^*, \{0\}^*)$ and $(R, R^*, \{0\}^*)$ are calculated twice. Finally, the number of hyperrings which are not rings is at most $5 \cdot 2^{n-1} - n - 4$. $\blacksquare$

We remark that the following facts are also valid.

- There are $2^n - 1$ hyperrings of the form (R, R^*, P_2^*), where $P_2 \subseteq R$, because $xR^*y = R$ holds for all $x, y \in R$.
- There are $2^n - n - 1$ multiplicative hyperrings $(R, \{0\}^*, P_2^*)$ in the sense of Rota, where $P_2 \subseteq R$. Indeed, the hyperoperation $\{0\}^*$ of each hyperring of the above form is the addition $+$ of the ring $(R, +, \cdot)$. Moreover, we have $(-x)P_2^*y = -(xP_2^*y) = xP_2^*(-y)$.
- If $\cdot \equiv \circ$, where $x \circ y = 0$, for all $x, y \in R$, then the number of H_v-rings is at most 2^{n-1}, because $(R, P_1^*, \circ) \cong (R, P_1^*, \{0\}^*)$.

Example 4.14. In the case of the ring $(\mathbb{Z}_p, +, \cdot)$, where p is a prime number, there are at most $2^{2p-1} - 2^{p-1}$ H_v-rings, from which $5 \cdot 2^{p-1} - p - 4$ are hyperrings, that are not rings. In the particular case $p = 3$ we have 28 H_v-rings. Observe that 13 H_v-rings are hyperrings which are not rings. We have the following isomorphisms:

$$(\mathbb{Z}_3, \{0\}^*, \{1\}^*) \cong (\mathbb{Z}_3, \{0\}^*, \{2\}^*),$$
$$(\mathbb{Z}_3, \{0\}^*, \{0,1\}^*) \cong (\mathbb{Z}_3, \{0\}^*, \{0,2\}^*),$$
$$(\mathbb{Z}_3, \{0,1\}^*, \{0\}^*) \cong (\mathbb{Z}_3, \{0,2\}^*, \{0\}^*),$$
$$(\mathbb{Z}_3, \{0,1\}^*, \{1\}^*) \cong (\mathbb{Z}_3, \{0,2\}^*, \{2\}^*),$$
$$(\mathbb{Z}_3, \{0,1\}^*, \{2\}^*) \cong (\mathbb{Z}_3, \{0,2\}^*, \{1\}^*),$$
$$(\mathbb{Z}_3, \{0,1\}^*, \{0,1\}^*) \cong (\mathbb{Z}_3, \{0,2\}^*, \{0,2\}^*),$$
$$(\mathbb{Z}_3, \{0,1\}^*, \{0,2\}^*) \cong (\mathbb{Z}_3, \{0,2\}^*, \{0,1\}^*),$$
$$(\mathbb{Z}_3, \{0,1\}^*, \{1,2\}^*) \cong (\mathbb{Z}_3, \{0,2\}^*, \{1,2\}^*),$$
$$(\mathbb{Z}_3, \{0,1\}^*, \mathbb{Z}_3^*) \cong (\mathbb{Z}_3, \{0,2\}^*, \mathbb{Z}_3^*),$$
$$(\mathbb{Z}_3, \mathbb{Z}_3^*, \{1\}^*) \cong (\mathbb{Z}_3, \mathbb{Z}_3^*, \{2\}^*),$$
$$(\mathbb{Z}_3, \mathbb{Z}_3^*, \{0,1\}^*) \cong (\mathbb{Z}_3, \mathbb{Z}_3^*, \{0,2\}^*).$$

So, the number of H_v-rings is reduced to 17 and observe that 9 of them are hyperrings which are not rings.

Example 4.15. We consider the finite field $(F, +, \cdot)$ and suppose that $cardF = p^n$, where p is a prime number, $p > 2$, $n \geq 1$. Consequently, if we consider the translation of F

$$f_{a^{-1}b} : x \longmapsto a^{-1}bx, \text{ for all } a, b \in R \setminus \{0\},$$

then we have $f_{a^{-1}b}(a) = b$. Now, we obtain the following isomorphisms between H_v-rings with P-hyperoperations:

$$(F, \{0\}^*, \{a\}^*) \cong (F, \{0\}^*, \{b\}^*),$$
$$(F, F^*, \{a\}^*) \cong (F, F^*, \{b\}^*),$$
$$(F, \{0\}^*, \{0, a\}^*) \cong (F, \{0\}^*, \{0, b\}^*),$$
$$(F, F^*, \{0, a\}^*) \cong (F, F^*, \{0, b\}^*),$$

So, the number of H_v-rings (R, P_1^*, P_2^*) is at most $2^{2p^n-1} - 2^{p^n-1} - 4(p^n - 2)$.

Theorem 4.22. *Let (R, P_1^*, P_2^*) be an H_v-ring with the P-hyperoperations P_1^*, P_2^* over the ring $(R, +, \cdot)$. Consider the subgroup $< P_1 >$ of $(R, +)$ generated by P_1. Then for all a in R, we have $\beta_+^*(a) = < P_1 > +a$ and R/β_+^* is a multiplicative H_v-ring with the inclusion distributivity.*

Proof. We denote the fundamental class of $a \in R$ by $\beta_+^*(a)$ and any hypersum with respect to the P_1^* by $\overset{*}{\sum}$. Let $a \in R$ and $x \in \beta_+^*(a)$. Then, there exist $z_1, ..., z_{n+1}$ and there are $y_j \in R$ and the finite sets of indices I_i, $i = 1, ..., n$ such that

$$\{z_i, z_{i+1}\} \subseteq \overset{*}{\sum_{j \in I_i}} y_j \text{ for } i = 1, ..., n.$$

Set $u_i = \overset{*}{\underset{j \in I_i}{\sum}} y_j$ and $s_i = cardI_i$. Then, $\{z_i, z_{i+1}\} \subseteq u_i + (s_i - 1)P_1$ for $i = 1, ..., n$. Therefore, for $i = 1, ..., n - 1$ we have

$$z_{i+1} \in (u_i + (s_i - 1)P_1) \cap (u_{i+1} + (s_{i+1} - 1)P_1)$$

and so $u_i \in z_{i+1} - (s_i - 1)P_1$.

Consequently, $u_i \in u_{i+1} + (s_{i+1} - 1)P_1 - (s_i - 1)P_1$. We obtain

$$u_1 \in u_n + (s_2 + s_3 + ... + s_n - n + 1)P_1 - (s_1 + s_2 + ... + s_{n-1} - n + 1)P_1.$$

But $z_{n+1} \in u_n + (s_n - 1)P_1$, so $u_n \in z_{n+1} - (s_n - 1)P_1$.

Moreover, $z_1 \in u_1 + (s_1 - 1)P_1$. Thus, we have

$$z_1 \in z_{n+1} + (s_1 + s_2 + ... + s_n - n)P_1 - (s_1 + s_2 + ... + s_n - n)P_1.$$

Finally, $x \in (s_1 + s_2 + ... + s_n - n)(P_1 - P_1) + a$. This means that $x \in < P_1 > +a$, so $\beta_+^* \subseteq < P_1 > +a$.

Now, let $a \in R$ and take $x \in < P_1 > +a$. Then, there exists $s \in \mathbb{N}$ such that $x \in s(P_1 - P_1) + a$. So $\{x, a\} \subseteq s(P_1 - P_1) + a = aP_1^*(-P_1)...P_1^*(-P_1)$ which means that $x\beta_+^* a$. Therefore, we proved that $\beta_+^*(a) = < P_1 > +a$. The sum $\uplus$ and the product $\otimes$ of the elements of R/β_+^* are defined in the usual manner, and $(R/\beta_+^*, \uplus)$ is a group. Moreover, the weak associativity of $\otimes$ is valid. Finally, for all $x, y, z \in R$, we have

$$\beta_+^*(x) \otimes (\beta_+^*(y) \uplus \beta_+^*(z)) = \beta_+^*(x) \otimes \beta_+^*(y + z)$$
$$= \{\beta_+^*(u) \mid u \in (< P_1 > +x)(< P_1 > +y + z)P_2\}.$$

On the other hand,

$$(\beta_+^*(x) \otimes \beta_+^*(y)) \uplus (\beta_+^*(x) \otimes \beta_+^*(z))$$
$$= \{\beta_+^*(v) \mid v \in (< P_1 > +x)(< P_1 > +y)P_2\}$$
$$\uplus \{\beta_+^*(w) \mid w \in (< P_1 > +x)(< P_1 > +z)P_2\}$$
$$= \{\beta_+^*(v + w) \mid v + w \in (< P_1 > +x)(< P_1 > +y)P_2$$
$$+ (< P_1 > +x)(< P_1 > +z)P_2\}.$$

Consequently, the inclusion distributivity is valid. ∎

Let (R, P_1^*, P_2^*) be an H_v-ring with the P-hyperoperations P_1^*, P_2^* over the ring $(R, +, \cdot)$ such that $RP_2 \subseteq P_2$. Denote the set of all finite polynomials of elements of R by A. Then, for every $a_i \in A$, $i \in \mathbb{N}$, there exist $r_i \in R$, I_i finite set of indices, $P_{2j} \in \mathcal{P}(P_2)$, $j \in I_i$ and $s_i \in \mathbb{N}$ such that

$$a_i = r_i + \sum_{j \in I_i} P_{2j} + s_i P_i,$$

where $s_i P_1 = \underbrace{P_1 + ... + P_1}_{s_i}$.

Theorem 4.23. *Let (R, P_1^*, P_2^*) be an H_v-ring with the P-hyperoperations P_1^*, P_2^* over the ring $(R, +, \cdot)$. If $RP_2 \subseteq P_2$ and $< P_1, P_2 >$ is the subgroup of $(R, +)$ generated by $P_1 \cup P_2$, then for all $x \in R$, $\gamma^*(x) \subseteq < P_1, P_2 > +x$.*

Proof. Suppose that $x \in R$ and $y \in \gamma^*(x)$. Then, there exist $z_1, ..., z_{m+1} \in R$ with $z_1 = y$, $z_{m+1} = x$ and $a_1, ..., a_m \in A$ such that $\{z_i, z_{i+1}\} \subseteq a_i$, $(i = 1, ..., m)$. Then, for $i = m$, we have $x = z_{m+1} \in r_m + \sum_{j \in I_m} P_{2j} + s_m P_1$. So $r_m \in x + t_m(-P_2) + s_m(-P_1)$, where $t_m = card I_m$. Moreover, for all $i = 1, ..., m$, we have

$$z_{i+1} \in (r_i + \sum_{j \in I_i} P_{2j} + s_i P_1) \cap (r_{i+1} + \sum_{j \in I_{i+1}} P_{2j} + s_{i+1} P_1)$$

and hence $r_i \in r_{i+1} + t_{i+1}P_2 + t_i(-P_2) + s_{i+1}P_1 + s_i(-P_1)$, where $t_{i+1} = cardI_{i+1}$, $t_i = cardI_i$. We obtain $y = z_1 \in x + t(P_2 - P_2) + s(P_1 - P_1)$, where $t = t_1 + ... + t_m$, $t_i = cardI_i$, $i \in \{1, ..., m\}$, $s = s_1, ..., s_m$. This means that $y \in x+ < P_2 > + < P_1 >$. Hence, $\gamma^*(x) \subseteq < P_1, P_2 > +x$. ∎

Theorem 4.24. *Let (R, P_1^*, P_2^*) be an H_v-ring with the P-hyperoperations P_1^*, P_2^* over the unitary ring $(R, +, \cdot)$. If P_2 is a left ideal and $< P_1 >$ is the subgroup of $(R, +)$ generated by P_1, then $R/\gamma^* = R/(< P_1 > +P_2)$.*

Proof. Suppose that $x \in R$. From the previous theorem, we have that $\gamma^*(x) \subseteq < P_1, P_2 > +x$. Since P_2 is a subgroup of $(R, +)$, it follows that $\gamma^*(x) \subseteq < P_1 > +P_2 + x$.

Conversely, for all $z \in < P_1 > +P_2 + x$ there exist $p_2 \in P_2$ and $n \in \mathbb{N}$ such that $z \in x + p_2 + n(P_1 - P_1)$. Moreover, $p_2 \in P_2 = 1P_2^*1$ and so

$$\{z, x\} \subseteq xP_1^*(1P_2^*1)P_1^*(-P_1)P_1^*...P_1^*(-P_1)$$

where $P_1^*(-P_1)$ appears n times. Hence $\gamma^*(x) = < P_1 > +P_2 + x$. ∎

Corollary 4.2. *If $(R, +, \cdot)$ is a ring and P_2 is a left ideal, then for all multiplicative $P - H_v$-rings over R, we have $R/\gamma^* = R/P_2$.*

Proof. Suppose that (R, P_1^*, P_2^*) is a multiplicative P-H_v-ring over the ring $(R, +, \cdot)$. Then, $cardP_1 = 1$ and since the necessary and sufficient condition for the weak distributivity is $0 \in P_1$, we have $P_1 = \{0\}$. From the previous theorem, it follows that $R/\gamma^* = R/P_2$. ∎

4.7 ∂-hyperoperations and H_v-rings

In this section, we consider a hyperoperation, denoted by ∂_f. This hyperoperation studied by Vougiouklis in [129, 132]. The motivation for this hyperoperation is the property which the "derivative" has on the product of functions. Since there is no confusion, we can simply write ∂. To more precise, we introduce the following definition.

Definition 4.11. Let H be a set with n operations (or hperoperations) $\otimes_1, \otimes_2, ..., \otimes_n$ and one map (or multivalued map) $f : H \to H$, then n hyperoperations $\partial_1, \partial_2, ..., \partial_n$ on H are defined, called ∂-*hyperoperations* by putting

$$x\partial_i y = \{f(x) \otimes_i y, x \otimes_i f(y)\}, \text{ for all } x, y \in H, i \in \{1, 2, ..., n\}$$

or in case where $\otimes_i$ is hyperoperation or f is multivalued map we have

$$x\partial_i y = (f(x) \otimes_i y) \cup (x \otimes_i f(y)), \text{ for all } x, y \in H, i \in \{1, 2, ..., n\}.$$

One can see that if $\otimes_i$ is associative then ∂_i is weak associative.

Remark that one can use several maps f, instead of only one, in an analogous way. We can define ∂-*hyperoperations* on the union of maps:

Definition 4.12. Let $(G, \cdot)$ groupoid and $f_i : G \to G, i \in I$, set of maps on G. Take the map $f_\cup : G \to \mathcal{P}^*(G)$ such that $f_\cup(x) = \{f_i(x) | i \in I\}$ and we call it the union of the $f_i(x)$. We call the *union ∂-hyperoperation (∂)*, on G if we consider the map $f_\cup(x)$. An important case for a map f, is to take the union of this with the identity id. Thus, we consider the map $\underline{f} \equiv f_\cup(id)$, so $\underline{f}(x) = \{x, f(x)\}$, for all $x \in G$, which is called *b-∂-hyperoperation*, we denote it by $(\underline{\partial})$, so we have

$$x\underline{\partial}y = \{xy, f(x) \cdot y, x \cdot f(y)\}, \text{ for all } x, y \in G.$$

Remark that $\underline{\partial}$ contains the operation $(\cdot)$, so it is b-operation. Moreover, if $f : G \to \mathcal{P}^*(G)$ is multivalued then the b-∂-hyperoperations is defined by using the

$$\underline{f}(x) = \{x\} \cup f(x), \text{ for all } x \in G.$$

Motivation for the definition of ∂-hyperoperation is the derivative where only multiplication of functions can be used. Therefore, for functions $s(x)$, $t(x)$, we have $s\partial t = \{s't, st'\}$, $(')$ is the derivative.

Example 4.16. Application on derivative: consider all polynomials of first degree $g_i(x) = a_i x + b_i$. We have

$$g_1 \partial g_2 = \{a_1 a_2 x + a_1 b_2, a_1 a_2 x + b_1 a_2\},$$

so it is a hyperoperation in the set of first degree polynomials. Moreover, all polynomials $x + c$, where c be a constant, are units.

Example 4.17. If $\mathbb{R}^+$ is the set of positive reals, and $a \in \mathbb{R}^+$, then we take the map $f : x \mapsto x^a$. The theta-operation is $x\partial y = \{x^a y, xy^a\}$, for all $x, y \in \mathbb{R}^+$. The only unit is 1, and any element $x \in \mathbb{R}^+$ has two inverses, x^{-a} and $x^{-1/a}$.

Lemma 4.3. *Let $(G, \cdot)$ semigroup. Then,*

(1) For every $f : G \to G$, the hyperoperation ∂ is weak associative.
(2) For every $f : G \to G$, the b-∂-hyperoperation $\underline{\partial}$ is weak associative.

(3) If f is homomorphism and projection, i.e., $f^2 = f$, then (∂) is associative.

Proof. (1) For all $x, y, z \in G$, we have

$$(x\partial y)\partial z = \{f(f(x) \cdot y) \cdot z, \ f(x) \cdot y \cdot f(z), \ f(x \cdot f(y)) \cdot z, \ x \cdot f(y) \cdot f(z)\}.$$
$$x\partial(y\partial z) = \{f(x) \cdot f(y) \cdot z, \ x \cdot f(f(y)) \cdot z), \ f(x) \cdot y \cdot f(z), \ x \cdot f(y \cdot f(z))\}.$$

Therefore, $(x\partial y)\partial z \cap x\partial(y\partial z) = \{f(x) \cdot y \cdot f(z)\} \neq \emptyset$, so ∂ is weak associative.

(2) Since $\underline{\partial}$ is greater than ∂, we obtain that $\underline{\partial}$ is weak associative. Moreover, for every $x, y, z \in G$, we have

$$(x\underline{\partial}y)\underline{\partial}z = \{f(x \cdot y) \cdot z, \ x \cdot y \cdot z, \ x \cdot y \cdot f(z), \ f(f(x) \cdot y) \cdot z,$$
$$f(x) \cdot y \cdot z, \ f(x) \cdot y \cdot f(z), \ f(x \cdot f(y)) \cdot z, \ x \cdot f(y) \cdot z, \ x \cdot f(y) \cdot f(z)\},$$

$$x\underline{\partial}(y\underline{\partial}z) = \{f(x) \cdot y \cdot z, \ x \cdot y \cdot z, \ x \cdot f(y \cdot z), \ f(x) \cdot f(y) \cdot z,$$
$$x \cdot f(y) \cdot z, \ x \cdot f(f(y) \cdot z), \ f(x) \cdot y \cdot f(z), \ x \cdot y \cdot f(z), \ x \cdot f(y \cdot f(z))\}.$$

Thus, we obtain

$$(x\underline{\partial}y)\underline{\partial}z \cap x\underline{\partial}(y\underline{\partial}z) = \{x \cdot y \cdot z, \ f(x) \cdot y \cdot z, \ x \cdot f(y) \cdot z, \ x \cdot y \cdot f(z), \ f(x) \cdot y \cdot f(z)\} \neq \emptyset.$$

(3) If f is a homomorphism and projection, then we obtain

$$(x\partial y)\partial z = \{f(x) \cdot f(y) \cdot z, \ f(x) \cdot y \cdot f(z), \ x \cdot f(y) \cdot f(z)\} = x\partial(y\partial z).$$

Therefore, ∂ is an associative hyperoperation. ∎

Note that projection without homomorphism does not give the associativity. Furtheremore, commutativity does not improve the result.

Properties 4.25. *Reproductivity.* For the reproductivity we must have

$$x\partial G = \bigcup_{g \in G} \{f(x) \cdot g, \ x \cdot f(g)\} = G$$

and

$$G\partial x = \bigcup_{g \in G} \{f(g) \cdot x, \ g \cdot f(x)\} = G.$$

So, if $(\cdot)$ is reproductive, then (∂) is reproductive, since

$$\bigcup_{g \in G} \{f(x) \cdot g\} = G \text{ and } \bigcup_{g \in G} \{g(x) \cdot f\} = G.$$

Commutativity. If $\cdot$ is commutative, then ∂ is commutative. If f is into the centre of G, then ∂ is commutative. If $\cdot$ is weak commutative, then ∂ is weak commutative.

Unit elements. u is a right unit element if $x\partial u = \{f(x)\cdot u, x\cdot f(u)\} \ni x$. So, $f(u) = e$, where e is a unit in $(G,\cdot)$. The elements of the kernel of f, are the units of (G,∂).

Inverse elements. Let $(G,\cdot)$ is a monoid with unit e and u be a unit in (G,∂). Then, $f(u) = e$. For given x, the element x' is an inverse with respect to u, if

$$x\partial x' = \{f(x)\cdot x', x\cdot f(x')\} \ni u \text{ and } x'\partial x = \{f(x')\cdot x, x'\cdot f(x)\} \ni u.$$

So, $x' = (f(x))^{-1}u$ and $x' = u(f(x))^{-1}$, are the right and left inverses, respectively. We have two-sided inverses if and only if $f(x)u = uf(x)$.

Similar properties for multivalued maps, are obtained.

Proposition 4.12. *If $(G,\cdot)$ is a group, then (G,∂) is an H_v-group, for all $f : G \to G$.*

Proof. It is straightforward. ∎

Proposition 4.13. *Let $(G,\cdot)$ be a group and $f(x) = a$, constant map on G. Then, $(G,\partial)/\beta^*$ is singleton. If $f(x) = e$, then we obtain $x\partial y = \{x,y\}$, the smallest incidence hyperoperation.*

Proof. For every $x \in G$, we have

$$a^{-1}\partial(a^{-1}\cdot x) = \{f(a^{-1}\cdot a^{-1}\cdot x), \ a^{-1}\cdot f(a^{-1}\cdot x)\} = \{x,e\}.$$

Thus, $x\beta e$, for all $x \in G$. So, $\beta^*(x) = \beta^*(e)$ and $(G,\partial)/\beta^*$ is singleton. ∎

Every $f : G \to G$ defines a partition of G, by setting two elements x, y in the same class, if and only if $f(x) = f(y)$. We call this partition f-*partition* (this is the associated map), and we denote the class of x, by $f[x]$. So, in Proposition 4.13, for constant maps, we have that $f[x] = G = \beta^*(x)$, for all $x \in G$, where $\beta^*(x)$ is referred to (G,∂).

Proposition 4.14. *Let $(G,\cdot)$ be a group and f be a homomorphism. Then, $f[x] \subseteq \beta^*(x)$, for all $x \in G$.*

Proof. It is straightforward. ∎

Example 4.18. Let $(G,\cdot)$ is a commutative semigroup and $P \subseteq G$. Consider the multivalued map f such that $f(x) = P\cdot x$, for all $x \in G$. Then, we have

$$x\partial y = x\cdot y\cdot P, \text{ for all } x,y \in G.$$

So, the ∂-operation coincides with the well known class of P-hyperoperations.

Example 4.19. Let $(G, \cdot)$ be a group and $f : x \mapsto x^{-1}$, for all $x \in G$, be the inverse map. Then, the ∂-hyperoperation is defined by

$$x\partial y = \{x^{-1} \cdot y, \ x \cdot y^{-1}\}, \ \text{for all } x, y \in G.$$

The hyperstructure (G, ∂) has a unique unit, the unit e, of $(G, \cdot)$. All elements in (G, ∂) are self inverses and unique.

Now, we define ∂-hyperoperation on rings and analogously to other more complicate structures, where more than one θ-operations can be defined. Moreover, one can replace structures by hyperstructures or by H_v-structures as well.

Definition 4.13. Let $(R, +, \cdot)$ be ring and $f : R \to R$, $g : R \to R$ be two maps. We define two hyperoperations ∂_+ and $(\partial.)$, called both ∂-*hyperoperations*, on R as follows:

$$x\partial_+ y = \{f(x) + y, \ x + f(y)\} \ \text{and} \ x\partial. y = \{g(x) \cdot y, \ x \cdot g(y)\}, \ \text{for all } x, y \in G.$$

A hyperstructure $(R, +, \cdot)$, where $+, \cdot$ are two hyperoperations which satisfy all H_v-ring axioms, except the weak distributivity, will be called an H_v-*semi-near-ring*.

Proposition 4.15. *Let $(R, +, \cdot)$ ring and $f : R \to R$, $g : R \to R$ maps. The hyperstructure $(R, \partial_+, \partial.)$, called theta, is an H_v-semi-near-ring. Moreover, $+$ is commutative.*

Proof. One can see that all properties of H_v-rings, expect the distributivity, are valid. ∎

More properties are valid if we replace ∂ by the corresponding b-∂-*hyperoperations* $\underline{\partial}$.

Proposition 4.16. *Let $(R, +, \cdot)$ ring and $f : R \to R, g : R \to R$ maps. The $(R, \partial_+, \partial.)$, is an H_v-ring.*

Proof. The only axiom we check is the one of distributivity. Therefore, for all $x, y, z \in R$, we have

$$x\partial.(y\partial_+ z) = \{g(x)(y + z), \ g(x)(f(y) + z), \ g(x)(y + f(z)),$$
$$xg(y + z), \ xg(f(y) + z), \ xg(y + f(z))\}$$

and

$$(x\partial.y)\underline{\partial}_+(x\partial.z) = \{g(x)(y+z)f(g(x)y) + g(x)z, \ g(x)y + f(g(x)z),$$
$$g(x)y + xg(z), f(g(x)y) + xg(z), \ g(x)y + f(xg(z)),$$
$$xg(y) + g(x)z, \ f(xg(y)) + g(x)z, \ xg(y) + f(g(x)z),$$
$$x(g(y) + g(z)), \ f(xg(y)) + xg(z), \ xg(y) + f(xg(z))\}.$$

Therefore, we obtain $x\partial.(y\partial_+z) \cap (x\partial.y)\underline{\partial}_+(x\partial.z) = \{g(x)(y+z)\} \neq \emptyset.$ ∎

Properties 4.26. (*Special classes*). The theta hyperstructure $(R, \partial_+, \partial.)$ takes a new form and have some properties in several forms as the following ones.

(1) If $f(x) \equiv g(x), \forall x \in R$, i.e., the two maps coincide, $(R, \partial_+, \partial.)$ is an H_v-ring.

(2) If $g(x) = x, \forall x \in R$, i.e., only the f in addition is used, then we have

$$x(y\partial_+z) = \{xf(y) + xz, xy + xf(z)\}, (xy)\partial_+(xz)$$
$$= \{f(xy) + xz, xy + f(xz)\}$$

Therefore, $x(y\partial_+z) \cap (xy)\partial_+(xz) = \emptyset.$

(3) If $f(x) = x$, for all $x \in R$, then $(R, +, \partial.)$ becomes a multiplicative H_v-ring.

Example 4.20. Consider the ring $(\mathbb{Z}_6, +, \cdot)$, the map $f : \overline{0} \mapsto \overline{2}$ and $f(x) = x$, for all $x \in \mathbb{Z}_6 \setminus \{\overline{0}\}$. Then the operations ∂_+ and $\partial.$ are given in the following tables:

∂_+	$\overline{0}$	$\overline{1}$	$\overline{2}$	$\overline{3}$	$\overline{4}$	$\overline{5}$
$\overline{0}$	$\overline{2}$	$\{\overline{1},\overline{3}\}$	$\{\overline{2},\overline{4}\}$	$\{\overline{3},\overline{5}\}$	$\{\overline{4},\overline{0}\}$	$\{\overline{5},\overline{1}\}$
$\overline{1}$	$\{\overline{1},\overline{3}\}$	$\overline{2}$	$\overline{3}$	$\overline{4}$	$\overline{5}$	$\overline{0}$
$\overline{2}$	$\{\overline{2},\overline{4}\}$	$\overline{3}$	$\overline{4}$	$\overline{5}$	$\overline{0}$	$\overline{1}$
$\overline{3}$	$\{\overline{3},\overline{5}\}$	$\overline{4}$	$\overline{5}$	$\overline{0}$	$\overline{1}$	$\overline{2}$
$\overline{4}$	$\{\overline{4},\overline{0}\}$	$\overline{5}$	$\overline{0}$	$\overline{1}$	$\overline{2}$	$\overline{3}$
$\overline{5}$	$\{\overline{5},\overline{1}\}$	$\overline{0}$	$\overline{1}$	$\overline{2}$	$\overline{3}$	$\overline{4}$

and

$\partial.$	$\overline{0}$	$\overline{1}$	$\overline{2}$	$\overline{3}$	$\overline{4}$	$\overline{5}$
$\overline{0}$	$\overline{0}$	$\{\overline{0},\overline{2}\}$	$\{\overline{0},\overline{4}\}$	$\overline{0}$	$\{\overline{0},\overline{2}\}$	$\{\overline{0},\overline{4}\}$
$\overline{1}$	$\{\overline{0},\overline{2}\}$	$\overline{1}$	$\overline{2}$	$\overline{3}$	$\overline{4}$	$\overline{5}$
$\overline{2}$	$\{\overline{0},\overline{4}\}$	$\overline{2}$	$\overline{4}$	$\overline{0}$	$\overline{2}$	$\overline{4}$
$\overline{3}$	$\overline{0}$	$\overline{3}$	$\overline{0}$	$\overline{3}$	$\overline{0}$	$\overline{3}$
$\overline{4}$	$\{\overline{0},\overline{2}\}$	$\overline{4}$	$\overline{2}$	$\overline{0}$	$\overline{4}$	$\overline{2}$
$\overline{5}$	$\{\overline{0},\overline{4}\}$	$\overline{5}$	$\overline{4}$	$\overline{3}$	$\overline{2}$	$\overline{1}$

We obtain

$$(\mathbb{Z}_6, \partial_+, \partial.)/\gamma^* = \{\{\overline{0},\overline{2},\overline{4}\}, \{\overline{1},\overline{3},\overline{5}\} \cong \mathbb{Z}_2.$$

Generalization of Example 4.20, on rings and groups, can be the following.

Theorem 4.27. *Consider the group of integers $(\mathbb{Z}, +)$ and let $n \neq 0$ be a natural number. Take the map f such that $f(0) = n$ and $f(x) = x$, for all $x \in \mathbb{Z} \setminus \{0\}$. Then,*

$$(\mathbb{Z}, \partial)/\beta^* \cong (\mathbb{Z}_n, +).$$

Proof. The proof is clear from the following:

$$0\partial 0 = \{n\}, \ 0\partial(0\partial 0) = \{n, 2n\}, \ 0\partial\{n, 2n\} = \{n, 2n, 3n\},$$
$$0\partial(-n) = \{0, -n\}, \ 0\partial x = \{x, x + n\}, \ \text{for all } x \in \mathbb{Z} \setminus \{0\}$$

and

$$x\partial\{y, y + n\} = \{x + y, x + y + n\}, \ \text{for all } x, y \in \mathbb{Z} \setminus \{0\}. \quad \blacksquare$$

4.8 (H, R)-H_v-rings

In this section, we present the following structures. The main reference is [104].

Let $(H, *, \circ)$ be an H_v-ring, $(R, +, \cdot)$ be a ring with the zero element denoted by 0 and $\{A_i\}_{i \in R}$ be a family of nonempty sets indexed in R such that $A_0 = H$ and for all $i, j \in R$, $i \neq j$, $A_i \cap A_j = \emptyset$. Moreover, for all A_i, $i \in R^*$ there exists a set of indices I_i and a unique family $\{B_k\}_{k \in I_i}$

such that $B_k \subseteq A_i$ and $\bigcup_{k \in I_i} B_k \neq \emptyset$. Set $K = \bigcup_{i \in R} A_i$ and consider the hyperoperations $\oplus$, $\odot$ defined in K as follows:

$$\forall (x,y) \in H^2, \; x \oplus y = x * y, \; x \odot y = x \circ y,$$

$$\forall (x,y) \in A_i \times A_j \neq H^2, \; x \oplus y = A_{i+j}, \; x \odot y = \begin{cases} H & \text{if } ij = 0 \\ \bigcup_{k \in I_{ij}} B_k & \text{if } ij \neq 0. \end{cases}$$

It is clear that $(K, \oplus)$ is an H_v-group. Moreover, $(K, \odot)$ is an H_v-semigroup since the hyperoperation $\circ$ is weak associative and for all $(x,y,z) \in A_i \times A_j \times A_r \neq H^3$, we have

(1) if $i = 0 = j$ (similarly, if $i = 0, j \neq 0 \neq r$), then $x \odot (y \odot z) \subseteq (x \odot y) \odot z = H$;

(2) if $j = 0 = r$ (similarly, if $r = 0, i \neq 0 \neq j$), then $H = x \odot (y \odot z) \supseteq (x \odot y) \odot z$;

(3) if $i = 0 = r$, then $(x \circ H = x \odot (y \odot z)) \cap ((x \odot y) \odot z = H \circ z) \neq \emptyset$;

(4) if $j = 0, i \neq 0 \neq r$, then $x \odot (y \odot z) = H = (x \odot y) \odot z$;

(5) if $i \neq 0 \neq j \neq 0 \neq r$, then $x \odot (y \odot z) \cap (x \odot y) \odot z \neq \emptyset$.

Finally, the weak distributive law is verified and so $(K, \oplus, \odot)$ is an H_v-ring.

Definition 4.14. *The previous H_v-ring $(K, \oplus, \odot)$ is called $(H, R) - H_v$-ring with the support $K = \bigcup_{i \in R} A_i$.*

Theorem 4.28. *If γ^* is the fundamental equivalence relation in K, then $K/\gamma^* \cong R$.*

Proof. Let $a \in K$. Then there exists $r \in R$ such that $a \in A_r$. In order to determine $\gamma^*(a)$, we consider $x \in \gamma^*(a)$. Then, there exist $z_1, ..., z_{n+1} \in K$ such that $z_1 = x$, $z_{n+1} = a$ and $u_i \in \mathcal{U}$, $i \in \{1, ..., n\}$ such that $\{z_i, z_{i+1}\} \subseteq u_i$, $(i = 1, ..., n)$. Moreover, it is clear that for all $u_i \in \mathcal{U}$, $i = 1, ..., n$ there exists an appropriate $r_i \in R$, such that $u_i \subseteq A_{r_i}$. Consequently, $A_{r_i} = A_{r_{i+1}}$, $i = 1, ..., n-1$, because $z_{i+1} \in A_{r_i} \cap A_{r_{i+1}}$. Hence, $\{x, a\} \subseteq A_{r_n} = A_r$ and so $\gamma^*(a) \subseteq A_r$.

For the converse, let $y \in A_r$. If $r \in R^*$, then we consider $u \in H$, $w \in A_r$, and we have $\{y, a\} \subseteq u \odot w = A_r$. Hence $y \gamma^* a$, i.e., $y \in \gamma^*(a)$. If $r = 0$, then we have $\{y, a\} \subseteq u \odot w = H$, i.e., $y \in \gamma^*(a)$. Therefore, $A_r \subseteq \gamma^*(a)$ and consequently, $\gamma^*(a) = A_r$.

Finally, the map $f : K/\gamma^* \to R$ such that $f(A_i) = i$ is an isomorphism and so $K/\gamma^* \cong R$. ∎

We denote the kernel of the canonical map $\phi_K : K \to K/\gamma^*$ such that $\phi_K(x) = \gamma^*(x)$ by ω_k. According to the previous theorem, for all $i \in R$, $x \in A_i$, we have $\gamma^*(x) = A_i$ and hence, we can write $K/\gamma^* = \{\phi_K(A_i) \mid i \in R\}$. Consequently, $\omega_K = H$.

Now, we consider the (H_1, R_1)-H_v-ring $(K_1, \oplus, \odot)$ with the support $K_1 = \bigcup\limits_{i \in R_1} A_i$ and the (H_2, R_2)-H_v-ring $(K_2, \boxplus, \boxdot)$ with support $K_2 = \bigcup\limits_{j \in R_2} G_j$. We prove the following theorems.

Theorem 4.29. *If $f : K_1 \to K_2$ is an inclusion homomorphism, then*

(1) $f(\gamma^*(x)) \subseteq \gamma^*(f(x))$ *for all $x \in K_1$.*
(2) *We define the induced homomorphism $f^* : K_1/\gamma^* \to K_2/\gamma^*$ of f by $f^*(\phi_{K_1}(x)) = \phi_{K_2}(f(x))$.*
(3) $f(H_1) \subseteq H_2$.

Proof. (1) Let $x \in \gamma^*(x) = A_i$, $i \in R$. Then, for $y \in A_j$, $z \in A_{i-j}$, we have

$$f(x) \in f(\gamma^*(x)) = f(y \oplus z) \subseteq f(y) \boxplus f(z) = \gamma^*(f(x)).$$

Therefore, $f(\gamma^*(x)) \subseteq \gamma^*(f(x))$.

(2) The map f^* is well defined. In fact, if $\phi_{K_1}(x) = \phi_{K_1}(y)$, then $x\gamma^*y$ and so $f(x)\gamma^*f(y)$, i.e., $\phi_{K_2}(f(x)) = \phi_{K_2}(f(y))$ and hence $f^*(\phi_{K_1}(x)) = f^*(\phi_{K_1}(y))$. Moreover, f^* is a homomorphism, because for all $x, y \in K_1$, $z \in x \oplus y$, $w \in x \odot y$, we obtain

$$f^*(\phi_{K_1}(x) + \phi_{K_2}(y)) = f^*(\phi_{K_1}(z)) = \phi_{K_2}(f(z)) = \phi_{K_2}(f(x) \boxplus f(y))$$
$$= \phi_{K_2}(f(x)) + \phi_{K_2}(f(y)) = f^*(\phi_{K_1}(x)) + f^*(\phi_{K_2}(y))$$

and

$$f^*(\phi_{K_1}(x) \cdot \phi_{K_2}(y)) = f^*(\phi_{K_1}(w)) = \phi_{K_2}(f(w)) = \phi_{K_2}(f(x) \boxdot f(y))$$
$$= \phi_{K_2}(f(x)) \cdot \phi_{K_2}(f(y)) = f^*(\phi_{K_1}(x)) \cdot f^*(\phi_{K_2}(y)).$$

(3) From (1) and (2) it follows that $f(x) = \phi_{K_2}^{-1}(f^*(\phi_{K_1}(x)))$. Therefore, if $x \in H$ and $\bar{0}_{K_1/\gamma^*}$, $\bar{0}_{K_2/\gamma^*}$ are the zero elements of the rings K_1/γ^*, K_2/γ^*, respectively, then

$$f(x) \in \phi_{K_2}^{-1}(f^*(\bar{0}_{K_1/\gamma^*})) = \phi_{K_2}^{-1}(\bar{0}_{K_2/\gamma^*}) = H_2,$$

which implies that $f(H_1) \subseteq H_2$.   ∎

Theorem 4.30. *If $K_1 \cong K_2$, then $H_1 \cong H_2$ and $R_1 \cong R_2$.*

Proof. From the previous theorem it follows that for all $x \in K_1$, $f(\gamma^*(x)) = \gamma^*(f(x))$ and f^* is a homomorphism. Therefore, $K_1/\gamma^* \cong K_2/\gamma^*$. Consequently, from Theorem 4.28, we obtain $R_1 \cong R_2$. Next we consider the map $g : H_1 \to H_2$ defined by $g(x) = f(x)$. The map g is well defined, one to one and onto. We show that it is a strong homomorphism. In fact, for all $(x,y) \in H^2$, we obtain

$$g(x *_1 y) = f(x \oplus y) = f(x) \boxplus f(y) = g(x) *_2 g(y),$$
$$g(x \circ_1 y) = f(x \odot y) = f(x) \boxdot f(y) = g(x) \circ_2 g(y).$$

Theorem 4.31. *If $g : H_1 \to H_2$, $f : R_1 \to R_2$ are homomorphisms and for all $i \in R_1^*$, $cardA_i \leq cardG_{f(i)}$, then there is an additively strong and one to one homomorphism from K_1 to K_2.*

Proof. Let $i \in R_1^*$ and let $F_i = F_i(A_i, G_{f(i)})$ be the set of all the one to one maps from A_i to $G_{f(i)}$. If $\{B_k\}_{k \in I_i}$, $\{B_s\}_{s \in I_{f(i)}}$ are the families of subsets of A_i and $G_{f(i)}$, respectively, then we denote

$$F_i^* = \left\{ h_{f(i)} \in F_i \mid h_{f(i)}\left(\bigcup_{k \in I_i} B_k\right) \cap \left(\bigcup_{s \in I_{f(i)}} B_s\right) \neq \emptyset \right\}$$

and we consider the map

$$t : K_1 \to K_2 : \quad x \mapsto t(x) = \begin{cases} g(x) & \text{if } x \in H_1 \\ h_{f(i)}(x) \text{ where } h_{f(i)} \in F_i^* & \text{if } x \in A_i \neq H_1. \end{cases}$$

The map t is well defined and one to one. We will show that it is an additively strong homomorphism.

If $(x,y) \in H_1^2$, then

$$t(x \oplus y) = g(x *_1 y) = g(x) *_2 g(y) = t(x) \boxplus t(y)$$
$$t(x \odot y) = g(x \circ_1 y) = g(x) \circ_2 g(y) = t(x) \boxdot t(y).$$

If $(x,y) \in A_i \times A_j \neq H_1^2$, then

$$t(x \oplus y) = t(A_{i+j}) = G_{f(i+j)} \text{ and } t(x) \boxplus t(y) = G_{f(i)+f(j)} = G_{f(i+j)}.$$

In order to check the multiplications, we notice that:

If $ij = 0$, then $t(x \odot y) = g(H_1) = H_2$ and $t(x) \boxdot t(y) = G_{f(ij)} = H_2$.

If $ij \neq 0$, then $t(x \odot y) = h_{f(ij)}(\bigcup_{r \in I_{ij}} B_r)$, where $B_r \subseteq A_{ij}$,

$$t(x) \boxdot t(y) = \bigcup_{m \in I_{f(ij)}} B_m, \text{ where } B_m \subseteq G_{f(ij)}.$$

Since $h_{f(ij)}\left(\bigcup_{r \in I_{ij}} B_r\right) \cap \left(\bigcup_{m \in I_{f(ij)}} B_m\right) \neq \emptyset$, it follows that

$$t(x \odot y) \cap t(x) \boxdot t(y) \neq \emptyset.$$

4.9   The H_v-ring of fractions

It is well-known that if S is a multiplicatively closed subset of a commutative
ring R, then there is a natural way to define the ring of fractions of R with
respect to S. This ring is denoted by $S^{-1}R$. A natural question that arises,
is the following one: how the H_v-ring of fractions can be defined? In this
section, our aim is to answer the above question and obtain some properties
of the H_v-ring of fractions. We use the results obtained by Darafsheh and
Davvaz [19].

Throughout this section, R is a commutative (general) hyperring with
a unit denoted by 1. Recall that a hyperstructure $(R, +, \cdot)$ is a hyperring
if $(R, +)$ is a hypergroup, $(\cdot)$ is an associative hyperoperation and the distributivity is valid, that is $x \cdot (y + z) = x \cdot y + x \cdot z$, $(x + y) \cdot z = x \cdot z + y \cdot z$
for all $x, y, z \in R$. Under the above conditions, we define the H_v-ring of
fractions of R.

Definition 4.15. *A non-empty subset S of R is called a strong multiplicatively closed subset (s.m.c.s.) if the following axioms hold:*

(1) $1 \in S$;
(2) $a \cdot S = S \cdot a = S$, *for all* $a \in S$.

Now, as we have indicated earlier, suppose that R is a commutative
hyperring with scalar unit. Furthermore we assume that S is a s.m.c.s. of
R. Let $\mathcal{M}$ be the set of all the ordered pairs (r, s) where $r \in R$, $s \in S$. For
$A \subseteq R$ and $B \subseteq S$, we denote the set $\{(a, b) | a \in A, \ b \in B\}$ by (A, B). We
define the following relation $\sim$.

$(A, B) \sim (C, D)$ if and only if there exists a subset X of S such that
$X \cdot (A \cdot D) = X \cdot (B \cdot C)$.

Lemma 4.4. $\sim$ *is an equivalence relation.*

Proof.   Obviously $\sim$ is reflexive and symmetric. To verify that $\sim$ is transitive, we assume $(A_1, B_1) \sim (A_2, B_2)$ and $(A_2, B_2) \sim (A_3, B_3)$, where
$(A_i, B_i) \in \mathcal{P}(\mathcal{M})$, $1 \leq i \leq 3$. By definition of $\sim$ there exist the subsets X_1
and X_2 of S such that

$$X_1 \cdot (A_1 \cdot B_2) = X_1 \cdot (A_2 \cdot B_1), \tag{4.1}$$

$$X_2 \cdot (A_2 \cdot B_3) = X_2 \cdot (A_3 \cdot B_2). \tag{4.2}$$

Multiplying both sides of (4.2) by $X_1.B_1$ we get $X_1 \cdot X_2 \cdot A_2 \cdot B_3 \cdot B_1 = X_1 \cdot$
$X_2 \cdot A_3 \cdot B_2 \cdot B_1$ which implies that $X_2 \cdot (X_1 \cdot A_2 \cdot B_1) \cdot B_3 = X_1 \cdot X_2 \cdot A_3 \cdot B_2 \cdot B_1$.

Using (4.1), we obtain $X_2 \cdot (X_1 \cdot A_1 \cdot B_2) \cdot B_3 = X_1 \cdot X_2 \cdot A_3 \cdot B_2 \cdot B_1$ which implies that $(X_1 \cdot X_2 \cdot B_2) \cdot (A_1 \cdot B_3) = (X_1 \cdot X_2 \cdot B_2) \cdot (A_3 \cdot B_1)$. If we take $X = X_1 \cdot X_2 \cdot B_2$, then $X \cdot (A_1 \cdot B_3) = X \cdot (A_3 \cdot B_1)$ which implies that $(A_1, B_1) \sim (A_3, B_3)$. ∎

The equivalence class containing (A, B) is denoted by $\| A, B \|$.

We consider the restriction of the relation $\sim$ on $\mathcal{M}$. We obtain the following two corollaries.

Corollary 4.3. *For* $(r, s), (r_1, s_1) \in \mathcal{M}$*, we have* $(r, s) \sim (r_1, s_1)$ *if and only if there exists* $A \subseteq S$ *such that* $A \cdot (r \cdot s_1) = A \cdot (r_1 \cdot s)$*.*

Corollary 4.4. $\sim$ *is an equivalence relation on* $\mathcal{M}$*.*

In $\mathcal{M}$, the equivalence class containing (r, s) is denoted by $[r, s]$ and we denote the set of all the equivalence classes by $S^{-1}R$.

We define:

$$\ll A, B \gg = \bigcup_{(A_1, B_1) \in \| A, B \|} \{[a_1, b_1] | a_1 \in A_1, b_1 \in B_1\}.$$

Now, we define the following hyperoperations on $S^{-1}R$,

$$[r_1, s_1] \uplus [r_2, s_2] = \bigcup_{(A, B) \in \| r_1 \cdot s_2 + r_2 \cdot s_1, s_1 \cdot s_2 \|} \{[r, s] | r \in A, s \in B\}$$

$$= \ll r_1 \cdot s_2 + r_2 \cdot s_1, s_1 \cdot s_2 \gg,$$

$$[r_1, s_1] \otimes [r_2, s_2] = \bigcup_{(A, B) \in \| r_1 \cdot r_2, s_1 \cdot s_2 \|} \{[r, s] | r \in A, s \in B\}$$

$$= \ll r_1 \cdot r_2, s_1 \cdot s_2 \gg.$$

Lemma 4.5. $\uplus$ *and* $\otimes$ *defined above are well-defined hyperoperations.*

Proof. Suppose that $[r_1, s_1] = [a_1, t_1]$ and $[r_2, s_2] = [a_2, t_2]$. Then there exist subsets A and B of S such that

$$A \cdot r_1 \cdot t_1 = A \cdot a_1 \cdot s_1, \tag{4.3}$$

$$B \cdot r_2 \cdot t_2 = B \cdot a_2 \cdot s_2. \tag{4.4}$$

Multiplying (4.3) by $B \cdot s_2 \cdot t_2$ and (4.4) by $A \cdot t_1 \cdot s_1$ we obtain $A \cdot B \cdot s_1 \cdot s_2 \cdot t_2 \cdot a_1 = A \cdot B \cdot s_2 \cdot t_2 \cdot t_1 \cdot r_1$ and $A \cdot B \cdot s_1 \cdot s_2 \cdot t_1 \cdot a_2 = A \cdot B \cdot s_1 \cdot t_1 \cdot t_2 \cdot r_2$. Adding the above equalities, we obtain

$$A \cdot B \cdot (s_1 \cdot s_2 \cdot (t_2 \cdot a_1 + t_1 \cdot a_2)) = A \cdot B \cdot (t_1 \cdot t_2 \cdot (s_2 \cdot r_1 + s_1 \cdot r_2)).$$

Therefore, $\| r_1 \cdot s_2 + r_2 \cdot s_1, s_1 \cdot s_2 \| = \| a_1 \cdot t_2 + a_2 \cdot t_1, t_1 \cdot t_2 \|$ which implies that

$$\ll r_1 \cdot s_2 + r_2 \cdot s_1, s_1 \cdot s_2 \gg = \ll a_1 \cdot t_2 + a_2 \cdot t_1, t_1 \cdot t_2 \gg,$$

hence ⊎ is well defined.

Now, multiplying (4.3) with (4.4) we obtain $A \cdot B \cdot (r_1 \cdot r_2) \cdot (t_1 \cdot t_2) = A \cdot B \cdot (a_1 \cdot a_2) \cdot (s_1 \cdot s_2)$ and so $\| r_1 \cdot r_2, s_1 \cdot s_2 \| = \| a_1 \cdot a_2, t_1 \cdot t_2 \|$ which implies that

$$\ll r_1 \cdot r_2, s_1 \cdot s_2 \gg = \ll a_1 \cdot a_2, t_1 \cdot t_2 \gg.$$

Therefore, ⊗ is well defined. ∎

Corollary 4.5. *For all $r \in R$, $s \in S$, we have $\ll r, s \gg = \ll r \cdot s, s \cdot s \gg$.*

Theorem 4.32. $(S^{-1}R, ⊎, ⊗)$ *is an H_v-ring, that we shall call the H_v-ring of fractions.*

Proof. If $[r_1, s_1], [r_2, s_2], [r_3, s_3] \in S^{-1}R$, then we have:

$$\{[r, s] \mid r \in r_1 \cdot (s_2 \cdot s_3) + (r_2 \cdot s_3 + r_3 \cdot s_2) \cdot s_1, s \in s_1 \cdot (s_2 \cdot s_3)\}$$
$$\subseteq [r_1, s_1] ⊎ ([r_2, s_2] ⊎ [r_3, s_3]),$$
$$\{[r, s] \mid r \in (r_1 \cdot s_2 + r_2 \cdot s_1)s_3 + r_3(s_1 \cdot s_2), s \in (s_1 \cdot s_2) \cdot s_3\}$$
$$\subseteq ([r_1, s_1] ⊎ [r_2, s_2]) ⊎ [r_3, s_3].$$

Since R is associative and distributive, we obtain that $(S^{-1}R, ⊎)$ is weak associative. The weak distributivity of $(S^{-1}R, ⊗)$ can be proved in a similar way.

Now, we prove the reproduction axioms for $(S^{-1}R, ⊎)$.

For every $[r, s], [r_1, s_1] \in S^{-1}R$, we have $s \in S, s_1 \in S$ and then by the definition of S there exists $s_2 \in S$ such that $s \in s_1 \cdot s_2$. On the other hand, since reproduction axioms hold for the additive law in R, we obtain $r_1 \cdot s_2 + (s_1 + 1)R = R$. Therefore, there exists $r_2 \in R$ such that $r \in r_1 \cdot s_2 + s_1 \cdot r_2 + r_2$ which implies that $r \in r_1 \cdot s_2 + (r_2 + r_2 \cdot s_3) \cdot s_1$ where $1 \in s_3 \cdot s_1$. Therefore, there exists $a \in r_2 + r_2 \cdot s_3$ such that $r \in r_1 \cdot s_2 + a \cdot s_1$. Hence

$$[r, s] \in [r_1, s_1] ⊎ [a, s_2] = \ll r_1 \cdot s_2 + a \cdot s_1, s_1 \cdot s_2 \gg$$

which implies that $S^{-1}R \subseteq [r_1, s_1] ⊎ S^{-1}R$, therefore $S^{-1}R = [r_1, s_1] ⊎ S^{-1}R$.

Finally, we prove the weak distributivity of $\otimes$ with respect to $\uplus$. We have $\{[r,s] | r \in s_1 \cdot r_1 \cdot (r_2 \cdot s_3 + s_2 \cdot r_3), s \in s_1 \cdot s_1 \cdot (s_2 \cdot s_3)\} \subseteq [r_1, s_1] \otimes ([r_2, s_2] \uplus [r_3, s_3])$, by Corollary 4.5, and hence

$$\{[r,s] \mid r \in (r_1 \cdot r_2) \cdot (s_1 \cdot s_3) + (r_1 \cdot r_3) \cdot (s_1 \cdot s_2), s \in (s_1 \cdot s_2) \cdot (s_1 \cdot s_3)\}$$
$$\subseteq ([r_1, s_1] \otimes [r_2, s_2]) \uplus ([r_1, s_1] \otimes [r_3, s_3]).$$

Therefore,

$$[r_1, s_1] \otimes ([r_2, s_2] \uplus [r_3, s_3]) \cap ([r_1, s_1] \otimes [r_2, s_2]) \uplus ([r_1, s_1] \otimes [r_3, s_3]) \neq \emptyset$$

and in the similar way we obtain

$$(([r_1, s_1] \uplus [r_2, s_2]) \otimes [r_3, s_3]) \cap (([r_1, s_1] \otimes [r_3, s_3]) \uplus ([r_2, s_2] \otimes [r_3, s_3])) \neq \emptyset$$

thus, $(S^{-1}R, \uplus, \otimes)$ is an H_v-ring. ∎

Theorem 4.33. *Let R_1 and R_2 be two commutative hyperrings with scalar unit and S be a s.m.c.s. of R_1 and let $g : R_1 \to R_2$ be a strong homomorphism of H_v-rings such that $g(1) = 1$. Then, g induces an H_v-homomorphism $\overline{g} : S^{-1}R_1 \to g(S)^{-1}R_2$ by setting*

$$\overline{g}([r,s]) = [g(r), g(s)].$$

Proof. It is clear that $g(S)$ is a s.m.c.s. of R_2. First, we prove that $\overline{g}$ is well defined. If $[r,s] = [r_1, s_1]$ then there exists $A \subseteq S$ such that $A \cdot r \cdot s_1 = A \cdot r_1 \cdot s$ which implies that $g(A \cdot r \cdot s_1) = g(A \cdot r_1 \cdot s)$ or $g(A) \cdot g(r) \cdot g(s_1) = g(A) \cdot g(r_1) \cdot g(s)$. Since $g(A) \subseteq g(S)$, it follows that $[g(r), g(s)] = [g(r_1), g(s_1)]$ or $\overline{g}([r,s]) = \overline{g}([r_1, s_1])$. Thus, $\overline{g}$ is well defined.

Moreover, $\overline{g}$ is an H_v-homomorphism because we have

$$\{[a,b] \mid a \in g(r_1 \cdot s_2 + r_2 \cdot s_1), b \in g(s_1 \cdot s_2)\} \subseteq \overline{g}([r_1, s_1] \uplus [r_2, s_2])$$

and

$$\{[a,b] | a \in g(r_1) \cdot g(s_2) + g(r_2) \cdot g(s_1), b \in g(s_1) \cdot g(s_2)\} \subseteq \overline{g}([r_1, s_1]) \uplus \overline{g}([r_2, s_2]).$$

Therefore,

$$\overline{g}([r_1, s_1] \uplus [r_2, s_2]) \cap (\overline{g}([r_1, s_1]) \uplus \overline{g}([r_2, s_2])) \neq \emptyset$$

and similarly we obtain

$$\overline{g}([r_1, s_1] \otimes [r_2, s_2]) \cap (\overline{g}([r_1, s_1]) \otimes \overline{g}([r_2, s_2])) \neq \emptyset,$$

which proves that $\overline{g}$ is an H_v-homomorphism. ∎

Definition 4.16. Let R be an H_v-ring. A non-empty subset I of R is called an H_v-*ideal* if the following conditions hold:

(1) $(I, +)$ is an H_v-subgroup of $(R, +)$;

(2) $I \cdot R \subseteq R$ and $R \cdot I \subseteq I$.

An H_v-ideal I is called an H_v-*isolated ideal* if it satisfies the following axiom,

- For all $X \subseteq I, Y \subseteq S$ if $(M, N) \in \| X, Y \|$, then $M \subseteq I$.

Lemma 4.6. *If I is an H_v-isolated ideal of R, then the set $S^{-1}I = \{[a, s] \mid a \in I, s \in S\}$ is an H_v-ideal of $S^{-1}R$.*

Proof. First, we prove that $(S^{-1}I, \uplus)$ is an H_v-subgroup of $(S^{-1}R, \uplus)$. For every $[a_1, s_1], [a_2, s_2] \in S^{-1}I$, we have

$$[a_1, s_1] \uplus [a_2, s_2] = \bigcup_{(A,B) \in \| a_1 \cdot s_2 + a_2 \cdot s_1, s_1 \cdot s_2 \|} \{[a, s] \mid a \in A, s \in B\}.$$

From $a_1, a_2 \in I$ we obtain $a_1 \cdot s_2 + a_2 \cdot s_1 \subseteq I$ and since I is an H_v-isolated ideal of R, it follows that $A \subseteq I$. Therefore, $[a_1, s_1] \uplus [a_2, s_2] \subseteq S^{-1}I$.

Now, we prove the equality $S^{-1}I = [a_1, s_1] \uplus S^{-1}I$, for all $[a_1, s_1] \in S^{-1}I$. Suppose that $[a, s] \in S^{-1}I$, $a \in I$. Since $s, s_1 \in S$, there exists $s_2 \in S$ such that $s \in s_1 \cdot s_2$. Moreover, since I is an H_v-ideal, we have $a_1 \cdot s_2 + (s_1 + 1)I = I$. Hence there exists $a_2 \in I$, such that $a \in a_1 \cdot s_2 + s_1 \cdot a_2 + a_2$ and so $a \in a_1 \cdot s_2 + (a_2 + a_2 \cdot s_3) \cdot s_1$, whence $1 \in s_3 \cdot s_1$. So there exists $b \in a_2 + a_2 \cdot s_3$ such that $a \in a_1 \cdot s_2 + b \cdot s_1$, therefore $[a, s] \in [a_1, s_1] \uplus [b, s_2]$ implying $S^{-1}I \subseteq [a_1, s_1] \uplus S^{-1}I$.

It remains to prove the second condition of the definition of an H_v-ideal. In order to do this, suppose that $[a, t] \in S^{-1}I$ and $[r, s] \in S^{-1}R$. Then

$$[a, t] \otimes [r, s] = \bigcup_{(A,B) \in \| a \cdot r, t \cdot s \|} \{[x, y] \mid x \in A, y \in B\}.$$

Since $a \in I$ and I is an H_v-isolated ideal of R, we have $a \cdot r \subseteq I$ and so $A \subseteq I$. Consequently $[a, t] \otimes [r, s] \subseteq S^{-1}I$. Therefore, $S^{-1}I$ is an H_v-ideal of $S^{-1}R$. ∎

Lemma 4.7. *If I, J are two H_v-isolated ideals of R, then*

(1) $S^{-1}(I \cap J) = S^{-1}I \cap S^{-1}J$,

(2) $S^{-1}(I \cdot J) = S^{-1}I \otimes S^{-1}J$,

(3) $S^{-1}(I + J) \subseteq S^{-1}I \uplus S^{-1}J$.

Proof. The proof is straightforward and is omitted. ∎

The *natural mapping* $\psi : R \to S^{-1}R$, where $\psi(r) = [r, 1]$, is an inclusion homomorphism.

Theorem 4.34. *Let I be an H_v-isolated ideal of R. Then, $S \cap I \neq \emptyset$ if and only if $S^{-1}I = S^{-1}R$.*

Proof. If $t \in S \cap I$, then $[t, t] = [1, 1] \in S^{-1}I$. Therefore, for every $[r, s] \in S^{-1}R$, we have $[1, 1] \otimes [r, s] \subseteq S^{-1}I$. From $[r, s] \in [1, 1] \otimes [r, s]$ we obtain $[r, s] \in S^{-1}I$ and this prove that $S^{-1}R \subseteq S^{-1}I$.

Conversely, assume that $S^{-1}I = S^{-1}R$. If we consider the natural inclusion homomorphism $\psi : R \to S^{-1}R$, then $\psi(1) = [1, 1]$. On the other hand, $\psi(1) \in S^{-1}R$, consequently $\psi(1) \in S^{-1}I$ and so $\psi(1) = [a, s]$ for some $a \in I, s \in S$. Now, we have $[1, 1] = [a, s]$, therefore, there exists $A \subseteq S$ such that $A \cdot s = A \cdot a$. Since $A \cdot s \subseteq S$ and $A \cdot a \subseteq I$, we get $I \cap S \neq \emptyset$. ∎

Theorem 4.35. *Let I be an H_v-isolated ideal of R. Then the following assertions hold:*

(1) $I \subseteq \psi^{-1}(S^{-1}I)$,
(2) *If $I = \psi^{-1}(J)$ for some H_v-ideal J of $S^{-1}R$, then $S^{-1}I = J$.*

Proof. The proof of (1) is obvious. In order to prove (2), let $I = \psi^{-1}(J)$ where J is an H_v-ideal of $S^{-1}R$. Then $[r, s] \in S^{-1}I$ implies $r \in I$ and so $\psi(r) = [r, 1] \in J$. Therefore, $[1, s] \otimes [r, 1] \subseteq J$. Since $[r, s] \in [1, s] \otimes [r, 1]$, we obtain $[r, s] \in J$ which implies that $S^{-1}I \subseteq J$. Now, let $[r, s] \in J$. Then $\psi(r) = [r, 1] \in [r, 1] \otimes [s, s] = [r, s] \otimes [s, 1] \subseteq J$. Therefore $r \in \psi^{-1}(J) = I$, hence $[r, s] \in S^{-1}I$, and this proves that $J \subseteq S^{-1}I$. ∎

Definition 4.17. *Let A be a commutative H_v-ring. An H_v-ideal P is called an H_v-prime ideal of A, if $a \cdot b \subseteq P$ implies $a \in P$ or $b \in P$.*

Theorem 4.36. *If P is an H_v-isolated prime ideal of R such that $S \cap P = \emptyset$, then $S^{-1}P$ is an H_v-prime ideal of $S^{-1}R$ and $\psi^{-1}(S^{-1}P) = P$.*

Proof. By Lemma 4.6, $S^{-1}P$ is an H_v-ideal of $S^{-1}R$. Now, we check that $S^{-1}P$ is prime. If $[r, s] \otimes [r_1, s_1] \subseteq S^{-1}P$, then $\{[b, s_2] \mid b \in r \cdot r_1, s_2 \in s \cdot s_1\} \subseteq \ll r \cdot r_1, s \cdot s_1 \gg \subseteq S^{-1}P$. It follows that for every $b \in r \cdot r_1$, $s_2 \in s \cdot s_1$ there exists $a \in P$ and $t \in S$ such that $[b, s_2] = [a, t]$. Therefore, there exists a subset A of S such that $A \cdot b \cdot t = A \cdot a \cdot s_2$. Since $A \cdot a \cdot s_2 \subseteq P$, we have $A \cdot b \cdot t \subseteq P$. Now, for every $x \in A \cdot t$ we obtain $x \cdot b \subseteq P$. Since $A \cdot t \subseteq S$ and $S \cap P = \emptyset$, it follows that $x \notin P$ and so $b \in P$. Consequently

$r \cdot r_1 \subseteq P$ which implies that $r \in P$ or $r_1 \in P$. Therefore, $[r, s] \in S^{-1}P$ or $[r_1, s_1] \in S^{-1}P$.

On the other hand, by Theorem 4.35, we have $P \subseteq \psi^{-1}(S^{-1}P)$.

Conversely, assume that $r \in \psi^{-1}(S^{-1}P)$. Then, $\psi(r) \in S^{-1}P$ and since $\psi(r) = [r, 1]$, there exists $a \in P, t \in S$ such that $[r, 1] = [a, t]$. Therefore, there is a subset A of S such that $A \cdot r \cdot t = A \cdot a$. Since $A \cdot a \subseteq P$, we have $A \cdot r \cdot t \subseteq P$. Now, for every $x \in A \cdot t$, we obtain $x \cdot r \subseteq P$. Since $A \cdot t \subseteq S$, it follows that $x \notin P$ and so, $r \in P$. Therefore, $\psi^{-1}(S^{-1}P) = P$. ∎

Lemma 4.8. *For every $a \in S$, $\gamma^*(a)$ is invertible in R/γ^*.*

Proof.   Since $1 \in R$, it follows that $\gamma^*(1) \in R/\gamma^*$. Now, for every $\gamma^*(x) \in R/\gamma^*$, we have $\gamma^*(x) \odot \gamma^*(1) = \gamma^*(1) \odot \gamma^*(x) = \gamma^*(x)$, i.e., $\gamma^*(1)$ is the identity of the ring R/γ^*. On the other hand, by the definition of S, for every $a \in S$ there exists $b \in S$ such that $1 \in a \cdot b = b \cdot a$. Therefore, $\gamma^*(1) = \gamma^*(a \cdot b) = \gamma^*(b \cdot a)$ and so $\gamma^*(1) = \gamma^*(a) \odot \gamma^*(b) = \gamma^*(b) \odot \gamma^*(a)$ which implies that $\gamma^*(b)$ is the inverse of $\gamma^*(a)$. ∎

Theorem 4.37. *If all subsets A of S are finite polynomials of elements of R over $\mathbb{N}$, then there exists an H_v-homomorphism $f : S^{-1}R \to R/\gamma^*$ such that $f\psi = \varphi$, i.e., the following diagram is commutative.*

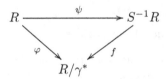

Proof.   We define $f : S^{-1}R \to R/\gamma^*$ by setting $f([r, s]) = \gamma^*(r) \odot \gamma^*(s)^{-1}$. First, we prove that f is well defined. If $[r, s] = [r_1, s_1]$, then there exists $A \subseteq S$ such that $A \cdot r \cdot s_1 = A \cdot r_1 \cdot s$ and so $\varphi(A \cdot r \cdot s_1) = \varphi(A \cdot r_1 \cdot s)$ which implies that $\gamma^*(A) \odot \gamma^*(r) \odot \gamma^*(s_1) = \gamma^*(A) \odot \gamma^*(r_1) \odot \gamma^*(s)$. By hypothesis $\gamma^*(A) = \gamma^*(a)$ for every $a \in A$, so we obtain $\gamma^*(a) \odot \gamma^*(r) \odot \gamma^*(s_1) = \gamma^*(a) \odot \gamma^*(r_1) \odot \gamma^*(s)$.

Multiplying the above relation by $\gamma^*(a)^{-1} \odot \gamma^*(s)^{-1} \odot \gamma^*(s_1)^{-1}$, we have $\gamma^*(r) \odot \gamma^*(s)^{-1} = \gamma^*(r_1) \odot \gamma^*(s_1)^{-1}$. Therefore, $f([r, s]) = f([r_1, s_1])$. Thus, f is well defined. Moreover, f is an H_v-homomorphism, because we have

$$\gamma^*(r_1 \cdot s_2 + r_2 \cdot s_1) \odot \gamma^*(s_1 \cdot s_2)^{-1} \in f([r_1, s_1)] \uplus [r_2, s_2]),$$
$$(\gamma^*(r_1) \odot \gamma^*(s_1)^{-1}) \oplus (\gamma^*(r_2) \odot \gamma^*(s_2)^{-1}) = f([r_1, s_1]) \oplus f([r_2, s_2]),$$
$$\gamma^*(r_1 \cdot r_2) \odot \gamma^*(s_1 \cdot s_2)^{-1} \in f([r_1, s_1]) \otimes [r_2, s_2]),$$
$$(\gamma^*(r_1) \odot \gamma^*(s_1)^{-1}) \odot (\gamma^*(r_2) \odot \gamma^*(s_2)^{-1}) = f([r_1, s_1]) \odot f([r_2, s_2]).$$

Finally, it is clear that $f\psi = \varphi$.   ∎

Let γ_s^* be the fundamental equivalence relation on $S^{-1}R$ and let $\mathcal{U}_s$ denotes the set of finite polynomials of elements of $S^{-1}R$ over $\mathbb{N}$.

Theorem 4.38. *There exists a homomorphism* $h : R/\gamma^* \to S^{-1}R/\gamma_s^*$.

Proof.   We define $h(\gamma^*(r)) = \gamma_s^*([r,1])$. First, we prove that h is well defined. Suppose that $\gamma^*(r_1) = \gamma^*(r_2)$, so $r_1\gamma^* r_2$. Hence there exist $x_1,...,x_{m+1} \in R$; $u_1,...,u_m \in \mathcal{U}$ with $x_1 = r_1$, $x_{m+1} = r_2$ such that $\{x_i, x_{i+1}\} \subseteq u_i$, $i = 1,...,m$ which implies that

$$\{[x_i,1],[x_{i+1},1]\} \subseteq \ll u_i, 1 \gg \in \mathcal{U}_s.$$

Therefore, $[r_1,1]\gamma_s^*[r_2,1]$ and so $\gamma_s^*([r_1,1]) = \gamma_s^*([r_2,1])$. Thus, h is well defined. h is a homomorphism, because

$$h(\gamma^*(a) \oplus \gamma^*(b)) = h(\gamma^*(c)) = \gamma_s^*([c,1]) \text{ for all } c \in \gamma^*(a) + \gamma^*(b)$$

and

$$h(\gamma^*(a)) \oplus h(\gamma^*(b)) = \gamma_s^*([a,1]) \oplus \gamma_s^*([b,1])$$
$$= \gamma_s^*([d,s]) \text{ for all } [d,s] \in \gamma_s^*([a,1]) \uplus \gamma_s^*([b,1]).$$

Thus, setting $d = c \in a + b$, $s = 1$, we obtain

$$h(\gamma^*(a) \oplus \gamma^*(b)) = h(\gamma^*(a)) \oplus h(\gamma^*(b)).$$

Similarly, we obtain

$$h(\gamma^*(a) \odot \gamma^*(b)) = h(\gamma^*(a)) \odot h(\gamma^*(b)).$$

Therefore, h is a homomorphism of rings.   ∎

Corollary 4.6. *The following diagram is commutative, i.e.,* $\varphi_s\psi = h\varphi$ *where* φ *and* φ_s *are the canonical projections.*

$$
\begin{array}{ccc}
R & \xrightarrow{\psi} & S^{-1}R \\
\varphi \downarrow & & \downarrow \varphi_s \\
R/\gamma^* & \xrightarrow{h} & S^{-1}R/\gamma_s^*
\end{array}
$$

Corollary 4.7. *If* $\varphi : R \to R/\gamma^*$ *is the canonical projection, then the map* $\theta : S^{-1}R \to \varphi(S)^{-1}(R/\gamma^*)$ *defined by* $\theta([r,s]) = [\gamma^*(r), \gamma^*(s)]$ *is an* H_v-*homomorphism.*

Corollary 4.8. *The following diagram is commutative, i.e.,* $\theta\psi = \psi_1\varphi$.

$$
\begin{array}{ccc}
R & \xrightarrow{\psi} & S^{-1}R \\
\varphi \downarrow & & \downarrow \theta \\
R/\gamma^* & \xrightarrow{\psi_1} & \varphi(S)^{-1}(R/\gamma^*)
\end{array}
$$

4.10 H_v-group rings

In an H_v-group, several convolutions can be defined. In this paragraph, we present a convolution and obtain an H_v-group ring. The results are mainly from [106, 120]. Examples and applications in known classes of hyperstructures are also investigated.

Definition 4.18. Let $(H, \cdot)$ be a hypergroupoid. The following set is called a *set of fundamental maps on H with respect to $\cdot$* :

$$\Theta = \{\theta : H \times H \to^{onto} H \mid \theta(x, y) \in x \cdot y\}.$$

Any subset $\Theta_\mu \subseteq \Theta$ define a hyperoperation $\circ_\mu$ on H as follows:

$$x \circ_\mu y = \{z \mid z = \theta(x, y) \text{ for some } \theta \in \Theta_\mu\}.$$

Obviously, $\circ_\mu \leq \cdot$ and $\Theta_\mu \subseteq \Theta_{\circ_\mu}$, where $\Theta_{\circ_\mu}$ denotes of the fundamental maps on H with respect to $\circ_\mu$. A set $\Theta_\alpha \subseteq \Theta$ is called *associative* (respectively weak associative) if and only if for every subset $\Theta_\mu \subseteq \Theta_\alpha$ the hyperoperation $\circ_\mu$ is associative (respectively weak associative). A hypergroupoid $(H, \cdot)$ will be called Θ-weak associative if there exists an element $\theta_\circ \in \Theta$ which defines an associative operation $\circ$ in H.

- For every $\theta \in \Theta$ we have $1 \leq |\theta^{-1}(g)| \leq n^2 - n + 1$ for all $g \in H$ where $\theta^{-1}(g)$ is the inverse image of g. However, for every $\theta \in \Theta$ we have $\sum_{g \in H} |\theta^{-1}(g)| = n^2$.
- If $(H, \cdot)$ is Θ-weak associative, then every greater hypergroupoid is Θ-weak associative.
- All H_b-groups are Θ-weak associative.
- Any H_b-semigroup, which has a b-structure with the property $H^2 = H$ is Θ-weak associative.
- If $(H, \cdot)$ is an H_v-group containing a scalar, then all the maps $f : H \times H \to H$ with $f(x, y) \in xy$ for all $x, y \in H$, are onto, i.e., $f \in \Theta$. Indeed, if s is a scalar, then $f(s, H) = H$.

Example 4.21. An example of a type of b-semigroups is defined as follows: Fix an element $s \in H$. We define the product $\circ$ by setting:

$$x \circ y = \begin{cases} s \text{ if } x \neq y \\ x \text{ if } x = y. \end{cases}$$

Then $H \circ H = H$ and $\circ$ is associative, since

$$(x \circ y) \circ z = x \circ (y \circ z) = \begin{cases} x \text{ if } x = y = z \\ s \text{ otherwise.} \end{cases}$$

Definition 4.19. Let $(H, \cdot)$ be Θ-weak associative with $|G| = n$ and $\theta_\circ \in \Theta$ be associative. Let F be a field and $F[H]$ be the set of formal linear combinations of elements of H with coefficients from F. In $F[H]$ the ordinary addition $+$ can be defined by setting

$$(f_1 + f_2)(g) = f_1(g) + f_2(g) \text{ for all } g \in H \text{ and } f_1, f_2 \in F[H].$$

Furthermore, consider the hyperproduct $\star$, called *convolution*, defined for every f_1, f_2 of $F[H]$ as follows:

$$f_1 \star f_2 = \left\{ f_\theta \mid f_\theta(g) = \sum_{\theta(x,y)=g} f_1(x)f_2(y); \ \theta \in \Theta \right\}.$$

Theorem 4.39. *The structure $(F[H], +, \star)$ is a multiplicative H_v-ring where the inclusion distributivity is valid.*

Proof.   Obviously, $(F[H], +)$ is a group. Let $f_1, f_2, f_3 \in F[H]$. Then

$$(f_1 \star f_2) \star f_3 = \left\{ f_\theta \mid f_\theta(u) = \sum_{\theta(x,y)=u} f_1(x)f_2(y); \theta \in \Theta \right\} \star f_3$$

$$= \left\{ f_\varphi \mid f_\varphi(g) = \sum_{\varphi(u,z)=g} \sum_{\theta(x,y)=u} f_1(x)f_2(y)f_3(z); \theta, \varphi \in \Theta \right\} \ni f_{\theta_\circ}$$

where $f_{\theta_\circ}(g) = \sum_{\theta_\circ(u,z)=g} \sum_{\theta_\circ(x,y)=u} f_1(x)f_2(y)f_3(z)$. On the other hand,

$$f_1 \star (f_2 \star f_3) = \left\{ f_{\underline{\varphi}} \mid f_{\underline{\varphi}}(g) = \sum_{\varphi(x,v)=g} \sum_{\underline{\theta}(y,z)=v} f_1(x)f_2(y)f_3(z); \underline{\theta}, \varphi \in \Theta \right\} \ni f_{\theta_\circ},$$

where $f_{\theta_\circ}(g) = \sum_{\theta_\circ(x,v)=g} \sum_{\theta_\circ(y,z)=v} f_1(x)f_2(y)f_3(z)$. Since $\theta_\circ$ is associative.

Thus, $(f_1 \star f_2) \star f_3 \cap f_1 \star (f_2 \star f_3) \neq \emptyset$. Now, for the left distributivity, we have

$$f_1 \star (f_2 + f_3) = \left\{ f_\theta \mid f_\theta(g) = \sum_{\theta(x,y)=g} f_1(x)[(f_2 + f_3)(y)]; \ \theta \in \Theta \right\}$$

$$= \left\{ f_\theta \mid f_\theta(g) = \sum_{\theta(x,y)=g} [f_1(x)f_2(y) + f_1(x)f_3(y)]; \ \theta \in \Theta \right\}$$

and

$$f_1 \star f_2 + f_1 \star f_3 = \left\{ f_\rho \mid f_\rho(g) = \sum_{\rho(x,y)=g} f_1(x)f_2(y); \ \rho \in \Theta \right\}$$

$$+ \left\{ f_\varphi \mid f_\varphi(g) = \sum_{\varphi(x,y)=g} f_1(x)f_3(y); \ \varphi \in \Theta \right\}$$

$$= \left\{ f_{\rho\varphi} = f_\rho + f_\varphi \mid f_{\rho\varphi}(g) = \sum_{\rho(x,y)=g} f_1(x)f_2(y) \right.$$

$$\left. + \sum_{\varphi(r,s)=g} f_1(r)f_3(s); \ \rho, \varphi \in \Theta \right\}.$$

Thus, $f_1 \star (f_2 + f_3) \subseteq f_1 \star f_2 + f_1 \star f_3$, which is the left inclusion distributively. Similarly, the right inclusion distributivity is valid. ∎

Definition 4.20. The above H_v-ring is called a *hypergroupoid H_v-algebra* or an *H_v-group ring*.

Given a hypergroupoid $(H, \cdot)$ one can define an H_v-group ring by "enlarging" the hyperoperation $\cdot$ as follows: Take any $\star$ on H such that $(H, \star)$ is Θ-weak associative hypergroupoid then take the union $\Diamond = \cdot \cup \star$, i.e., $x \Diamond y = (xy) \cup (x \star y)$, for all $x, y \in H$. An H_v-group ring is defined on $(H, \Diamond)$.

The most important condition in order to define the H_v-group ring is the Θ-weak associative condition. Therefore, in what follows we focus on our attention on classes which satisfy the Θ-weak associative condition.

Now, we prove the following theorem.

Theorem 4.40. *In every H_v-group ring $(F[H], +, \star)$ we have*

$$(-f_1) \star f_2 = -(f_1 \star f_2) = f_1 \star (-f_2), \text{ for all } f_1, f_2 \in F[H],$$

and there exists an absorbing element.

Proof. For every $f_1, f_2 \in F[H]$, we have

$$-f_1 \star f_2 = \left\{ f_\theta \mid f_\theta(h) = \sum_{\theta(x,y)=h} (-f_1)(x) f_2(y); \ \theta \in \Theta \right\}$$

$$= \left\{ -f_\theta \mid f_\theta(h) = \sum_{\theta(x,y)=h} f_1(x) f_2(y); \ \theta \in \Theta \right\}$$

$$= -(f_1 \star f_2).$$

Take the element $f_0 \in F[H]$ such that $f_0(h) = 0$, for all $h \in H$. Then, for every $f \in F[H]$ we have

$$f_0 \star f = \left\{ f_\theta \mid f_\theta(h) = \sum_{\theta(x,y)=h} f_0(x) f(y); \ \theta \in \Theta \right\}$$

$$= \left\{ f_\theta \mid f_\theta(h) = 0; \ h \in H, \ \theta \in \Theta \right\}$$

$$= f_0.$$

Therefore, f_0 is the absorbing element. ∎

Example 4.22. Consider the H_b-group $(\mathbb{Z}_3, \oplus)$ which has the b-group $(\mathbb{Z}_3, +)$ and the non-singleton products are: $\overline{1} \oplus \overline{1} = \{\overline{0}, \overline{1}\}$, $\overline{2} \oplus \overline{2} = \{\overline{0}, \overline{1}, \overline{2}\}$. The $\overline{0}$ is scalar. Therefore, every map $\theta : \mathbb{Z}_3^2 \to \mathbb{Z}_3$ with $\theta(x, y) \in x \oplus y$, is an element of Θ. Thus, $|\Theta| = 2 \cdot 3 = 6$ and we see $\Theta = \{\theta_1, \theta_2, \theta_3, \theta_4, \theta_5, \theta_6\}$. Then, for all $\theta \in \Theta$, we have

$$\theta(\bar{0},\bar{0})=\bar{0}, \quad \theta(\bar{0},\bar{1})=\theta(\bar{1},\bar{0})=\bar{1}, \quad \theta(\bar{0},\bar{2})=\theta(\bar{2},\bar{0})=\bar{2}, \quad \theta(\bar{1},\bar{2})=\theta(\bar{2},\bar{1})=\bar{0},$$

and

$$\theta_1(\bar{1},\bar{1})=\bar{0}, \quad \theta_1(\bar{2},\bar{2})=\bar{0}; \theta_2(\bar{1},\bar{1})=\bar{0}, \quad \theta_2(\bar{2},\bar{2})=\bar{1}; \theta_3(\bar{1},\bar{1})=\bar{0}, \quad \theta_3(\bar{2},\bar{2})=\bar{2};$$

$$\theta_4(\bar{1},\bar{1})=\bar{1}, \quad \theta_4(\bar{2},\bar{2})=\bar{0}; \theta_5(\bar{1},\bar{1})=\bar{1}, \quad \theta_5(\bar{2},\bar{2})=\bar{1}; \theta_6(\bar{1},\bar{1})=\bar{1}, \quad \theta_6(\bar{2},\bar{2})=\bar{2}.$$

Every hyperproduct of elements of $F[\mathbb{Z}_3]$ has at most 6 elements. Let $r, s \in F[\mathbb{Z}_3]$, then

$$r \star s = \left\{ t_\theta \mid t_\theta(g) = \sum_{\theta(x,y)=g} r(x)s(y); \ \theta \in \Theta \right\}.$$

For every $\theta \in \Theta$, we have to calculate 9 products of the form $r(x)s(y)$ in order to obtain t_θ.

$$t_{\theta_1}(\bar{0}) = r(\bar{0})s(\bar{0}) + r(\bar{1})s(\bar{2}) + r(\bar{2})s(\bar{1}) + r(\bar{1})s(\bar{1}) + r(\bar{2})s(\bar{2}),$$
$$t_{\theta_1}(\bar{1}) = r(\bar{0})s(\bar{1}) + r(\bar{1})s(\bar{0}),$$
$$t_{\theta_1}(\bar{2}) = r(\bar{0})s(\bar{2}) + r(\bar{2})s(\bar{0}),$$
$$t_{\theta_2}(\bar{0}) = r(\bar{0})s(\bar{0}) + r(\bar{1})s(\bar{2}) + r(\bar{2})s(\bar{1}) + r(\bar{1})s(\bar{1}),$$
$$t_{\theta_2}(\bar{1}) = r(\bar{0})s(\bar{1}) + r(\bar{1})s(\bar{0}) + r(\bar{2})s(\bar{2}),$$
$$t_{\theta_2}(\bar{2}) = r(\bar{0})s(\bar{2}) + r(\bar{2})s(\bar{0}),$$
$$t_{\theta_3}(\bar{0}) = r(\bar{0})s(\bar{0}) + r(\bar{1})s(\bar{2}) + r(\bar{2})s(\bar{1}) + r(\bar{1})s(\bar{1}),$$
$$t_{\theta_3}(\bar{1}) = r(\bar{0})s(\bar{1}) + r(\bar{1})s(\bar{0}),$$
$$t_{\theta_3}(\bar{2}) = r(\bar{0})s(\bar{2}) + r(\bar{2})s(\bar{0}) + r(\bar{2})s(\bar{2})$$

and similar for the $t_{\theta_4}, t_{\theta_5} t_{\theta_6}$.

Example 4.23. Consider the H_b-group $(\mathbb{Z}_{mn}, \oplus)$ defined in Example 3.2. Then, Θ has only two elements $\Theta = \{\theta_1, \theta_2\}$. For all $\theta \in \Theta$, we have $\theta(x,y) = x + y$ if $(x,y) \neq (\bar{0}, \overline{m})$ and $\theta_1(\bar{0}, \overline{m}) = \bar{0}$, $\theta_2(\bar{0}, \overline{m}) = \overline{m}$. The map θ_2 leads to the known convolution on $(\mathbb{Z}_{mn}, +)$. For every element $g \neq \bar{0}$ and m, the sum $t_{\theta_1}(g) = \sum_{\theta_1(x,y)=g} r(x)s(y)$ has mn elements. Moreover,

$$t_{\theta_1}(\bar{0}) = \sum_{\theta_1(x,y)=\bar{0}} r(x)s(y)$$

is a sum of $mn + 1$ terms of the form $r(x)s(y)$ and

$$t_{\theta_1}(\overline{m}) = \sum_{\theta_1(x,y)=\overline{m}} r(x)s(y)$$

is a sum of $mn - 1$ terms of $r(x)s(y)$.

Let $(H, \circ)$ be an H_v-group, $(G, +)$ be a group with the zero element 0, $\{A_i\}_{i \in G}$ be a family of non empty sets with $A_0 = H$ and $A_i \cap A_j = \emptyset$, for all $i, j \in G, i \neq j$. Set $K = \bigcup_{i \in G} A_i$ and consider the hyperoperation $\odot$ defined in K as follows:

$$x \odot y = \begin{cases} x \circ y \text{ if } (x,y) \in H^2 \\ A_{i+j} \text{ if } (x,y) \in A_i \times A_j \neq H^2. \end{cases}$$

Then, $(K, \odot)$ becomes an H_v-group. which is called an (H,G)-H_v-group. It is easy to see that $K/\beta^* \cong G$.

Theorem 4.41. *If* $card A_i = n$ *for all* $i \in G$ *and* $(H, \circ)$ *is* Θ-*weak associative, then* $(K, \odot)$ *is* Θ-*weak associative. Moreover,*

$$card\Theta_K \geq (n!)^{m-1} card\Theta_H,$$

where $m = cardG$.

Proof. We consider a family of one to one maps $\{p_i\}_{i \in G}$ such that $p_i : H \to A_i$, $i \neq 0$ and p_0 is the identity map. Notice that all these maps are also onto. Take $\theta \in \Theta_H$ which defines an associative operation $\cdot$ and consider the mapping $\theta' : K \times K \to K$ which defines the operation $\Diamond$ in K, as follows:

$$x \Diamond y = p_{i+j}(p_i^{-1}(x) \cdot p_j^{-1}(y)) \text{ for all } x \in A_i, y \in A_j.$$

This mapping is, obviously, onto, so it remain to prove that $\Diamond$ is associative. Suppose that $(x, y, z) \in A_i \times A_j \times A_r$, we have

$$x\Diamond(y\Diamond z) = x \Diamond p_{j+r}(p_j^{-1}(y) \cdot p_r^{-1}(z))$$
$$= p_{i+(j+r)}[p_i^{-1}(x) \cdot (p_j^{-1}(y) \cdot p_r^{-1}(z))] = (x\Diamond y)\Diamond z.$$

Therefore, $(K, \oplus)$ is Θ-weak associative. Now, we remark that the number of the families p_i, $i \in G$, $i \neq 0$ of bijective maps is $(n!)^{m-1}$. Therefore,

$$card\Theta_K \geq (n!)^{m-1} card\Theta_H.$$ ∎

Theorem 4.42. *Let* $(K, \cdot)$ *be* Θ-*weak associative such that* $card A_i = 1$, *for all* $i \neq 0$. *Then, for all* $\theta \in \Theta_K$ *which define an associative operation* $\Diamond$ *in* K, *there exists an element* $x \in H$ *such that*

$y\Diamond x = x = x\Diamond y$, *for all* $y \in H$,
$z\Diamond w = x$, *for all* $(z,w) \in A_r \times A_s \neq H^2$ *for which* $r + s = 0$.

Proof.   Take $(z, w) \in A_r \times A_s \neq H^2$ for which $r + s = 0$ and set $z \Diamond w = x \in H$. Then, for all $y \in H$,

$$y \Diamond x = y \Diamond (z \Diamond w) = (y \Diamond z) \Diamond w = z \Diamond w = x.$$

Similarly, $x \Diamond y = x$. This element x is unique, because if there exists another element x' such that $(u, v) \in A_p \times A_q \neq H^2$ with $p + q = 0$ and $u \Diamond v = x'$, then $x = x \Diamond x' = x'$.   ∎

Note that in the above theorem, the operation induced by the restriction of θ to $H \times H$ is weak associative. This means that $(H, \circ)$ is Θ-weak associative.

From the above theorem we obtain the following construction.

Theorem 4.43. *Let $(H, \circ)$ be such that there exists an associative operation $\Diamond$ on H and a special element $x \in H$ such that $x \Diamond y = x = y \Diamond x$ for all $y \in H$. Then there exists a Θ-weak associative (H, G)-H_v-group $(K, \odot)$ with $card A_i = 1$, for all $i \in G$, $i \neq 0$.*

Proof.   Consider the extension of $\Diamond$ to K for which $z \Diamond w = x$ for all $(z, w) \in A_i \times A_j \neq H$ with $i + j = 0$. It is easy to check that this operation is associative on K.   ∎

In the case of the (H, G)-H_v-groups with $card A_i = 1$, for all $i \in G \setminus \{0\}$, the cardinality of the set of onto maps

$$K \times K \to K : (x, y) \mapsto z \in x \odot y$$

is less or equal to $n^{m + n^2 - 1}$, where $n = card H$ and $m = card G$.

Definition 4.21. Let $\{S_i\}_{i \in I}$ be a pairwise disjoint family of H_v-semigroups, where $|I| > 1$. We define a hyperoperation $\otimes$, called an S-*hyperoperation* on the set $S = \bigcup_{i \in I} S_i$ as follows:

$$x_i \otimes y_i = x_i y_i \text{ for all } (x_i, y_i) \in S_i^2,$$
$$x_i \otimes x_j = S_i \cup S_j \text{ for all } (x_i, x_j) \in S_i \times S_j, \ i \neq j.$$

Then, the hyperstructure $(S, \otimes)$, called an S-*construction*, is an H_v-group.

Let $\{S_i\}_{i \in I}$ be a family of pairwise disjoint sets, where $card I > 1$. On S_i we consider the *total hyperoperation* $ab = S_i$, for all $a, b \in S_i$ or the *least incidence hyperoperation* $ab = \{a, b\}$, for all $a, b \in S_i$. In each case, we obtain the S-construction $(S, \otimes)$.

In what follows, we consider the finite case. Let $card I = n$ and

$cardS_i = s_i$, $i \in I$ and suppose that for each $i \in I$, S_i is a group or a groupoid with the associated set of fundamental maps $\Theta_i \neq \emptyset$. Let Θ_i be the set of fundamental maps on S with respect to $\otimes$. We obtain

$$card\Theta = 2 \cdot \prod_{\substack{i,j \in I}}^{i<j} (s_i + s_j).$$

In the particular case when $S_i = \{x_i\}$, $i \in I$, the S-construction coincides with the incidence operation and we obtain $cardI = n$, we have

$$card\Theta = 2 \cdot 2^{n(n-1)}.$$

Theorem 4.44. *If each one of the following conditions is valid:*

(1) S_i *is a group for all* $i \in I$;
(2) S_i *is a semigroup such that* $S_i^2 = S_i$ *for all* $i \in I$;
(3) *Every* S_i *has a scalar element;*

then every map $S \times S \to S : (x,y) \mapsto z \in xy$ *is onto.*

Proof.   It is clear.   ■

Theorem 4.45. *For each one of the cases of Theorem 4.44 we have*

$$card\Theta = \left(\prod_{\substack{i,j \in I}}^{i<j} (n_i + n_j)^{n_i n_j} \right)^2.$$

Proof.   It is clear.   ■

In the particular case when $S_i = \{x_i\}$, the S-construction coincides with the incidence operation and we obtain $card\Theta = 2^{n(n-1)}$.

Theorem 4.46. *Every S-construction, where $(S, \cdot)$ is an H_b-semigroup with a b-semigroup S_i, is Θ-weak associative.*

Proof.   Suppose that the b-operation of $(S_i, \cdot)$ is $\circ$. Consider the operation $\otimes$ on S, where $\otimes$ coincides with $\circ$ on S_i and

$$s_i \otimes s_j = s_j \circ s_i = s_j \text{ for } s_i \in S_i, \ s_j \in S_j \text{ and } i \neq j.$$

The operation $\otimes$ guarantees that $\Theta \neq \emptyset$ and it is easy to check that it is associative.   ■

4.11 H_v-near rings

In this section we will introduce the concept of an H_v-near ring and we will study several properties of this class of hyperstructures. The main reference for this section is [24].

Definition 4.22. An H_v-*near ring* is an algebraic structure $(R, +, \cdot)$ which satisfies the following axioms:

(1) $(R, +)$ is a weak canonical hypergroup, i.e.,

   (i) for every $x, y, z \in R$, $x + (y + z) \cap (x + y) + z \neq \emptyset$,
   (ii) there exists $0 \in R$ such that $x + 0 = 0 + x = x$, for all $x \in R$,
   (iii) for every $x \in R$ there exists one and only one $x' \in R$ such that
        $0 \in x + x' \cap x' + x$,
        (We shall write $-x$ for x' and we call it the opposite of x),
   (iv) $z \in x + y$ implies $y \in -x + z$ and $x \in z - y$;

(2) Relating to the multiplication, $(R, \cdot)$ is a semigroup;

(3) The multiplication is weak distributive with respect to the hyperoperation $+$ on the left side or on the right side, i.e.,

$$x \cdot (y + z) \cap x \cdot y + x \cdot z \neq \emptyset \text{ or } (x + y) \cdot z \cap x \cdot z + y \cdot z \neq \emptyset,$$

for all $x, y, z \in R$.

Of course, there are various modifiers for the various H_v-near rings. Note that for all $x, y \in R$ we have

$$-(-x) = x, \quad 0 = -0, \quad 0 \text{ is unique and } -(x + y) = -y - x.$$

Definition 4.23. For an H_v-near ring, if the right weak distributive law is valid, then $(R, +, \cdot)$ is called a *right H_v-near ring*. You guessed it, if the left weak distributive is valid, $(R, +, \cdot)$ is called a *left H_v-near-ring*. If both the right and left weak distributive are valid, $(R, +, \cdot)$ is a *distributive H_v-near ring*.

Here, when considering H_v-near rings in general, we shall assume the H_v-near ring to be a left H_v-near ring.

An H_v-near ring $(R, +, \cdot)$ for which $a \cdot 0 = 0 \cdot a = 0$ for each $a \in R$ is called a *zero-symmetric H_v-near ring*. In the following we give an example of an H_v-near ring.

Example 4.24. Let $R := \{0, a, b, c\}$ be a set, and consider addition and multiplication tables below.

+	0	a	b	c
0	0	a	b	c
a	a	$\{0,a\}$	c	b
b	b	c	$\{0,b\}$	a
c	c	b	a	$\{0,c\}$

$\cdot$	0	a	b	c
0	0	0	0	0
a	0	a	a	a
b	0	b	b	b
c	0	c	c	c

Then, we can easily see that $(R, +, \cdot)$ is a zero-symmetric distributive H_v-near ring.

Let $(A, +, \cdot)$ be a zero-symmetric near-ring and I an ideal of A; for $a, b \in A$ we say a is congruent to b $mod(I)$, written $a \equiv b$ $mod(I)$ if $a - b \in I$. The relation $a \equiv b$ $mod(I)$ is an equivalence relation and it is denoted by $a\sigma b$ if and only if $a \equiv b$ $mod(I)$. Let $\sigma^*(a)$ be the equivalence class of the element $a \in A$. Suppose that

$$A/\sigma = \{\sigma(x) \mid x \in A\}.$$

On A/σ we consider the hyperoperation $\oplus$ and multiplication $\odot$ defined as follow:

$$\sigma(a) \oplus \sigma(b) = \{\sigma(c) \mid c \in \sigma(a) + \sigma(b)\}, \ \sigma(a) \odot \sigma(b) = \sigma(a \cdot b).$$

Our aim in the following proposition is to show that A/σ is an H_v-near ring.

Proposition 4.17. $(A/\sigma, \oplus, \odot)$ *is an H_v-near ring.*

Proof. For all $a, b, c \in R$, we have

$$(a + b) + c \in (\sigma(a) \oplus \sigma(b)) \oplus \sigma(c), a + (b + c) \in \sigma(a) \oplus (\sigma(b) \oplus \sigma(c)).$$

Therefore, $\oplus$ is weak associative. It is easy to see that I is the zero element in A/σ and $\sigma(-x)$ is the opposite of $\sigma(x)$ in A/σ. Now, we show that $\sigma(c) \in \sigma(a) \oplus \sigma(b)$ implies $\sigma(a) \in \sigma(c) \oplus \sigma(-b)$ and $\sigma(b) \in \sigma(-a) \oplus \sigma(c)$.

We have $\sigma(c) \in \sigma(a) \oplus \sigma(b)$, and hence $\sigma(c) = \sigma(x)$ for some $x \in \sigma(a) + \sigma(b)$. Therefore, there exist $y \in \sigma(a)$ and $z \in \sigma(b)$ such that $x = y + z$, so $y = x - z$. This implies $\sigma(y) = \sigma(x - z) \in \sigma(x) \oplus \sigma(-z)$, and so $\sigma(a) \in \sigma(c) \oplus \sigma(-b)$. Similarly, we get $\sigma(b) \in \sigma(-a) \oplus \sigma(c)$. Therefore, $(A/\sigma, \oplus)$ is a weak canonical hypergroup. Also, we have

$$(a \cdot b) \cdot c \in \sigma((a \cdot b) \cdot c), \ a \cdot (b \cdot c) \in \sigma(a \cdot (b \cdot c)),$$
$$a \cdot (b + c) \in \sigma(a) \odot (\sigma(b) \oplus \sigma(c)), \ a \cdot b + a \cdot c \in \sigma(a \cdot b) \oplus \sigma(a \cdot c).$$

Therefore, $(A/\sigma, \oplus, \odot)$ is an H_v-near ring. ∎

Definition 4.24. A weak subhypergroup K of R is called *normal* if for all $x \in R$ holds:

$$x + K = K + x.$$

Definition 4.25. A normal weak subhypergroup K of the weak canonical hypergroup $(R, +)$ is

(1) a *left H_v-ideal* of R if $x \cdot a \in K$, for all $x \in R$ and $a \in K$,
(2) a *right H_v-ideal* of R if $(x + K) \cdot y - x \cdot y \subseteq K$, for all $x, y \in R$,
(3) a *bilaterally H_v-ideal* of R if $(x + K) \cdot y - x \cdot y \cup z \cdot K \subseteq K$, for all $x, y, z \in R$.

Definition 4.26. Let R_1 and R_2 be two H_v-near rings. The mapping $\varphi : R_1 \to R_2$ is an *H_v-homomorphism* of H_v-near rings, if

$$\varphi(x + y) \cap \varphi(x) + \varphi(y) \neq \emptyset, \text{ for all } x, y \in R,$$
$$\varphi(x \cdot y) = \varphi(x) \cdot \varphi(y), \text{ for all } x, y \in R,$$
$$\varphi(0) = 0.$$

φ is called a *strong H_v-homomorphism*, if

$$\varphi(x + y) = \varphi(x) + \varphi(y) \text{ for all } x, y \in R,$$
$$\varphi(x \cdot y) = \varphi(x) \cdot \varphi(y) \text{ for all } x, y \in R,$$
$$\varphi(0) = 0.$$

If φ is one to one, onto and a strong H_v-homomorphism, then it is called an *isomorphism*.

Definition 4.27. If I is a bilaterally H_v-ideal of R, then we define the relation $x \equiv y \, mod(I)$ if and only if there exists a set $\{z_0, z_1, \ldots, z_{k+1}\} \subseteq R$, where $z_0 = x$, $z_{k+1} = y$ such that

$$(x - z_1) \cap I \neq \emptyset, \ (z_1 - z_2) \cap I \neq \emptyset, \ \ldots, \ (z_k - y) \cap I \neq \emptyset.$$

This relation is called the *chain relation* and it is denoted by $x\sigma^*y$ if and only if $x \equiv y \, mod(I)$.

Lemma 4.9. *The chain relation σ^* is an equivalence relation.*

Proof. Since $0 \in (x - x) \cap I$ for all $x \in R$, it follows that $x\sigma^*x$, i.e., σ^* is reflexive.

Suppose that $x\sigma^*y$. Then, there exists $\{z_0, z_1, \ldots, z_{k+1}\} \subseteq R$, where $z_0 = x$, $z_{k+1} = y$ such that

$$(x - z_1) \cap I \neq \emptyset, \ (z_1 - z_2) \cap I \neq \emptyset, \ \ldots, \ (z_k - y) \cap I \neq \emptyset.$$

Therefore, there exist $a_i \in (z_i - z_{i+1}) \cap I$ $(i = 1, \ldots, k)$ which implies $-a_i \in z_{i+1} - z_i$ and $-a_i \in I$. This means that $y\sigma^*x$, and so σ^* is symmetric.

(3) Let $x\sigma^*y$ and $y\sigma^*z$, where $x, y, z \in R$. Then, there exist $\{z_0, z_1, \ldots, z_{k+1}\} \subseteq R$ and $\{t_0, t_1, \ldots, t_{r+1}\} \subseteq R$, where $z_0 = x, z_{k+1} = y = t_0, t_{r+1} = z$ such that

$$(x - z_1) \cap I \neq \emptyset, \ (z_1 - z_2) \cap I \neq \emptyset, \ \ldots, \ (z_k - y) \cap I \neq \emptyset,$$
$$(y - t_1) \cap I \neq \emptyset, \ (t_1 - t_2) \cap I \neq \emptyset, \ \ldots, \ (t_r - z) \cap I \neq \emptyset.$$

We take $\{z_0, z_1, \ldots, z_{k+1}, t_1, t_2, \ldots, t_{r+1}\} \subseteq R$ which satisfies the condition $x\sigma^*z$, and so σ^* is transitive.

Therefore σ^* is an equivalence relation. ∎

We denote $\sigma^*(x)$ the equivalence class with representative x.

Theorem 4.47. *Let R be an H_v-near-ring. If I is a bilaterally H_v-ideal of R, then on the set $R/I = \{\sigma^*(x) \mid x \in R\}$ we define the hyperoperation $\oplus$ and the multiplication $\odot$ as follows:*

$$\sigma^*(x) \oplus \sigma^*(y) = \{\sigma^*(z) \mid z \in \sigma^*(x) + \sigma^*(y)\},$$
$$\sigma^*(x) \odot \sigma^*(y) = \sigma^*(x \cdot y),$$

what gives the factor H_v-near ring $(R/I, \oplus, \odot)$.

Proof. The proof is similar to the proof of Proposition 4.17. ∎

Definition 4.28. If φ is a strong H_v-homomorphism from R_1 into R_2, then the *kernel* of φ is defined by $ker\varphi = \{x \in R_1 \mid \varphi(x) = 0\}$.

It is easy to see that $ker\varphi$ is a weak subhypergroup of R_1 but in general is not normal.

Theorem 4.48. *Let φ be a strong H_v-homomorphism from R_1 into R_2 such that $\varphi(x - x) = 0$, for all $x \in R_1$. Then,*

$$R_1/ker\varphi \cong R_2.$$

Proof. We define $f : R_1/ker\varphi \to R_2$ as follows:

$$f(\sigma^*(x)) = \varphi(x), \text{ for all } x \in R_1.$$

This mapping is well defined, because if $\sigma^*(x) = \sigma^*(y)$, then there exists $\{z_0, z_1, \ldots, z_{k+1}\} \subseteq R$ where $z_0 = x, \ z_{k+1} = y$ such that

$$(x - z_1) \cap ker\varphi \neq \emptyset, \ (z_1 - z_2) \cap ker\varphi \neq \emptyset, \ \ldots, \ (z_k - y) \cap ker\varphi \neq \emptyset.$$

Thus, $0 \in \varphi(x - z_1)$, $0 \in \varphi(z_1 - z_2)$, $\ldots$, $0 \in \varphi(z_k - y)$ or $0 \in \varphi(x) - \varphi(z_1)$, $0 \in \varphi(z_1) - \varphi(z_2)$, $\ldots$, $0 \in \varphi(z_k) - \varphi(y)$ and so $\varphi(x) = \varphi(y)$.

Now, for every $\sigma^*(x), \sigma^*(y) \in R_1/ker\varphi$, we have

$$
\begin{aligned}
f(\sigma^*(x) \oplus \sigma^*(y)) &= f(\{\sigma^*(z) \mid z \in \sigma^*(x) + \sigma^*(y)\}) \\
&= \{\varphi(z) \mid z \in \sigma^*(x) + \sigma^*(y)\} \\
&= \varphi(\sigma^*(x) + \sigma^*(y)) \\
&= \varphi(\sigma^*(x)) + \varphi(\sigma^*(y)) \\
&= \varphi(x) + \varphi(y) \\
&= f(\sigma^*(x)) + f(\sigma^*(y)).
\end{aligned}
$$

Also, we have

$$ f(\sigma^*(x) \odot \sigma^*(y)) = f(\sigma^*(x \cdot y)) = \varphi(x \cdot y) = \varphi(x) \cdot \varphi(y) = f(\sigma^*(x)) \cdot f(\sigma^*(y)), $$

and $f(ker\varphi) = 0$. Therefore, f is a strong homomorphism.

Furtheremore if $f(\sigma^*(x)) = f(\sigma^*(y))$, then $\varphi(x) = \varphi(y)$ which implies $x - y \subseteq ker\varphi$, consequently $(x - y) \cap ker\varphi \neq \emptyset$ and so $\sigma^*(x) = \sigma^*(y)$. Thus, f is a one to one mapping.   ∎

Let $(R, +, \cdot)$ be an H_v-near ring. We define the relation γ^* as the smallest equivalence relation on R such that the quotient R/γ^*, the set of all equivalence classes, is a near-ring. In this case, γ^* called the *fundamental equivalence relation* on R and R/γ^* is called the *fundamental near-ring*. Let $\mathcal{U}$ be the set of all finite sums of products of elements of R. We define the relation γ on R as follows:

$$ a\gamma b \iff \{a, b\} \subseteq u \text{ for some } u \in \mathcal{U}. $$

Let us denote $\widehat{\gamma}$ the transitive closure of γ. Then we can rewrite the definition of $\widehat{\gamma}$ on R as follows:
$a\widehat{\gamma}b$ if and only if there exist $z_1, z_2, \ldots, z_{n+1} \in R$ with $z_1 = a$, $z_{n+1} = b$ and $u_1, \ldots, u_n \in \mathcal{U}$ such that $\{z_i, z_{i+1}\} \subseteq u_i$ $(i = 1, \ldots, n)$.

Theorem 4.49. *The fundamental relation γ^* is the transitive closure of the relation γ.*

Proof. The proof is similar to the proof of Theorem 4.3.   ∎

The kernel of the canonical map $\varphi : R \to R/\gamma^*$ is called the *core* of R and is denoted by ω_R. Here we also denote by ω_R the zero element of R/γ^*. It is easy to prove that the following statements:

(1) $\omega_R = \gamma^*(0)$,

(2) $\gamma^*(-x) = -\gamma^*(x)$ for all $x \in R$.

Lemma 4.10. *Let R_1, R_2 be two H_v-near rings and let γ_1^*, γ_2^* and γ^* be the fundamental equivalence relations on R_1, R_2 and $R_1 \times R_2$ respectively. Then*

$$(x_1, y_1)\gamma^*(x_2, y_2) \;\Leftrightarrow\; x_1\gamma^*x_2, \; y_1\gamma^*y_2,$$

for all $(x_1, y_1), (x_2, y_2) \in R_1 \times R_2$.

Theorem 4.50. *Let R_1, R_2 be two H_v-near rings and let γ_1^*, γ_2^* and γ^* be the fundamental equivalence relations on R_1, R_2 and $R_1 \times R_2$ respectively. Then,*

$$(R_1 \times R_2)/\gamma^* \cong R_1/\gamma_1^* \times R_2/\gamma_2^*.$$

Proof. We consider the map

$$f : R_1/\gamma_1^* \times R_2/\gamma_2^* \to (R_1 \times R_2)/\gamma^*$$

with $f(\gamma_1^*(x), \gamma_2^*(y)) = \gamma^*(x, y)$. It is easy to see that f is an isomorphism. ∎

4.12   Fuzzy H_v-ideals

The concept of a fuzzy ideal of a ring is introduced by Liu in [80]. Then, in [22], Davvaz introduced the concept of fuzzy H_v-ideal of an H_v-ring.

Definition 4.29. If R is a ring and $\mu : R \to [0, 1]$ is a fuzzy subset of R, then μ is called a *left* (respectively, *right*) *fuzzy ideal* if it satisfies the following conditions:

(1) $\min\{\mu(x), \mu(y)\} \leq \mu(x - y)$, for all $x, y \in R$.

(2) $\mu(y) \leq \mu(x \cdot y)$ (respectively, $\mu(x) \leq \mu(x \cdot y)$), for all $x, y \in R$.

The fuzzy subset μ of R is called a *fuzzy ideal* if it is a left and right fuzzy ideal.

In this section, we define the concept of fuzzy H_v-ideal of an H_v-ring and present some of their properties.

Definition 4.30. Let R be an H_v-ring and μ be a fuzzy subset of R. Then, μ is said to be a *left* (respectively, *right*) *fuzzy H_v-ideal* of R if the following axioms hold.

A Walk Through Weak Hyperstructures: H_v-Structures

(1) $\min\{\mu(x), \mu(y)\} \leq \inf\limits_{\alpha \in x+y} \{\mu(\alpha)\}$, for all $x, y \in R$;

(2) for all $x, a \in R$ there exists $y \in R$ such that $x \in a + y$ and

$$\min\{\mu(a), \mu(x)\} \leq \mu(y);$$

(3) for all $x, a \in R$ there exists $z \in R$ such that $x \in z + a$ and

$$\min\{\mu(a), \mu(x)\} \leq \mu(z);$$

(4) $\mu(y) \leq \inf\limits_{\alpha \in x \cdot y} \{\mu(\alpha)\}$ (respectively, $\mu(x) \leq \inf\limits_{\alpha \in x \cdot y} \{\mu(\alpha)\}$), for all $x, y \in R$.

Here, we present all the proofs for the left H_v-ideals. For the right H_v-ideals similar results hold as well.

Example 4.25. Let $(R, +, \cdot)$ be an H_v-ring and μ be a fuzzy H_v-ideal of R. The set $I = \{x \in R \mid \mu(x) = 1\}$ is either the empty-set or an H_v-ideal of R.

Example 4.26. Let $(R, +, \cdot)$ be a ring and μ be a fuzzy subset of R. We define hyperoperations $\oplus, \odot, *$ on R as follows:

$$x \oplus y = \{t \mid \mu(t) = \mu(x + y)\},$$
$$x \odot y = \{t \mid \mu(t) = \mu(x \cdot y)\},$$
$$x * y = y * x = \{t \mid \mu(x) \leq \mu(t) \leq \mu(y)\} \text{ (if } \mu(x) \leq \mu(y)).$$

Then, $(R, *, *), (R, *, \odot), (R, *, \oplus), (R, \oplus, *), (R, \oplus, \odot), (R, \odot, *)$ are H_v-rings. Because for all x, y, z in R we have $[(x+y)+z \in (x \oplus y) \oplus z,\ x+(y+z) \in x \oplus (y \oplus z)]; [(x \cdot y) \cdot z \in (x \odot y) \odot z,\ x \cdot (y \cdot z) \in x \odot (y \odot z)]$ and $[x \in (x*y)*z,\ x \in x*(y*z)]$. Thus, $\oplus, \odot, *$ are weak associative. Moreover, it is clear that reproduction axioms are valid, i.e., $a \oplus R = R \oplus a = R$, $a \odot R = R \odot a = R$ and $a * R = R * a = R$. We also have $[x \cdot (y+z) \in x \odot (y \oplus z),\ (x \cdot y) + (x \cdot z) \in (x \odot y) \oplus (x \odot z)]; [y + z \in x * (y \oplus z),\ y + z \in (x * y) \oplus (x * z)]; [y \cdot z \in x * (y \odot z),\ y \cdot z \in (x * y) \odot (x * z)]; [x + y \in x \oplus (y * z),\ x + y \in (x \oplus y) * (x \oplus z)]; [x \cdot y \in x \odot (y * z),\ x \cdot y \in (x \odot y) * (x \odot z)]$ and $[x \in x*(y*z),\ x \in (x*y)*(x*z)]$. Therefore, $(R, *, *), (R, *, \odot), (R, *, \oplus), (R, \oplus, *), (R, \oplus, \odot)$ and $(R, \odot, *)$ are H_v-rings.

Example 4.27. In Example 4.26, if μ is a fuzzy ideal of R, then μ is a fuzzy H_v-ideal of $(R, \oplus, \odot)$.

Example 4.28. In Example 4.26, if μ is a fuzzy ideal of R, then μ is a fuzzy H_v-ideal of $(R, *, \odot)$.

Now, let μ_t be the level set of μ.

Theorem 4.51. *Let R be an H_v-ring and μ be a fuzzy subset of R. Then, μ is a fuzzy H_v-ideal of R if and only if for every $0 \leq t \leq 1$, $\mu_t \neq \emptyset$ is an H_v-ideal of R.*

Proof. Let μ be a fuzzy H_v-ideal of R. For every x, y in μ_t we have $\mu(x) \geq t$, $\mu(y) \geq t$. Hence, $\min\{\mu(x), \mu(y)\} \geq t$ and so $\inf_{\alpha \in x+y}\{\mu(\alpha)\} \geq t$. Therefore, for every $\alpha \in x + y$ we get $\mu(\alpha) \geq t$ which implies that $\alpha \in \mu_t$, so $x + y \subseteq \mu_t$. Thus, for every $a \in \mu_t$ we have $a + \mu_t \subseteq \mu_t$ and to prove this part of the theorem it is enough to prove that $\mu_t \subseteq a + \mu_t$.

Since μ is a fuzzy H_v-ideal of R, it follows that for every $x \in \mu_t$ there exists $y \in R$ such that $x \in a + y$ and $min\{\mu(a), \mu(x)\} \leq \mu(y)$. From $x \in \mu_t$ and $a \in \mu_t$ we obtain $\min\{\mu(x), \mu(a)\} \geq t$ and so $y \in \mu_t$. Therefore, we have proved that, for every $x \in \mu_t$ there exists $y \in \mu_t$ such that $x \in a + y$ implying that $x \in a + \mu_t$ and this proves $\mu_t \subseteq a + \mu_t$. Therefore, $\mu_t = a + \mu_t$, i.e., the left reproduction axiom is valid for $(\mu_t, +)$. The proof of the right reproduction axiom is similar.

Now, we prove that $R \cdot \mu_t \subseteq \mu_t$. For every $x \in \mu_t$ and $r \in R$ we show that $r \cdot x \subseteq \mu_t$. Since μ is a left fuzzy H_v-ideal we have $t \leq \mu(x) \leq \inf_{\alpha \in r \cdot x}\{\mu(\alpha)\}$. Therefore, for every $\alpha \in r \cdot x$ we obtain $\mu(\alpha) \geq t$ which implies that $\alpha \in \mu_t$. So, $r \cdot x \subseteq \mu_t$.

Conversely, assume that for every t, $0 \leq t \leq 1$, $\mu_t \neq \emptyset$ is an H_v-ideal of R. For every x, y in R we can write $\mu(x) \geq t_0$, $\mu(y) \geq t_0$ where $t_0 = \min\{\mu(x), \mu(y)\}$. Then, $x \in \mu_{t_0}$, $y \in \mu_{t_0}$. Since μ_{t_0} is an H_v-ideal, $x + y \subseteq \mu_{t_0}$. Therefore, for every $\alpha \in x + y$ we have $\mu(\alpha) \geq t_0$ implying that $\inf_{\alpha \in x+y}\{\mu(\alpha)\} \geq t_0$ and so

$$\min\{\mu(x), \mu(y)\} \leq \inf_{\alpha \in x+y}\{\mu(\alpha)\},$$

and in this way the first condition of Definition 4.30 is verified. To verify the second condition, if for every $a, x \in R$ we put $t_1 = min\{\mu(a), \mu(x)\}$ then $x \in \mu_{t_1}$ and $a \in \mu_{t_1}$. Since we have $a + \mu_{t_1} = \mu_{t_1}$, so there exists $y \in \mu_{t_1}$ such that $x \in a + y$. On the other hand, since $y \in \mu_{t_1}$, then $\mu(y) \geq t_1$. Hence, $\min\{\mu(a), \mu(x)\} \leq \mu(y)$ and the second condition of Definition 4.30 is satisfied. In the similar way, the third condition of Definition 4.30 is valid.

Now, we prove the fourth condition of Definition 4.30. For every $x, y \in R$, we put $t_2 = \mu(y)$. Then, $y \in \mu_{t_2}$. Since μ_{t_2} is an H_v-ideal of R, $x \cdot y \subseteq \mu_{t_2}$.

Therefore, for every $\alpha \in x \cdot y$ we have $\alpha \in \mu_{t_2}$ which implies that $\mu(\alpha) \geq t_2$. Hence, $\inf\limits_{\alpha \in x \cdot y} \{\mu(\alpha)\} \geq t_2$ implying that $\mu(y) \leq \inf\limits_{\alpha \in x \cdot y} \{\mu(\alpha)\}$. ∎

The following two corollaries are exactly obtained from Theorem 4.51.

Corollary 4.9. *Let $(R, +, \cdot)$ be an H_v-ring and μ be a fuzzy H_v-ideal of R. Let $0 \leq t_1 < t_2 \leq 1$. Then, $\mu_{t_1} = \mu_{t_2}$ if and only if there is no x in R such that $t_1 \leq \mu(x) < t_2$.*

Corollary 4.10. *Let $(R, +, \cdot)$ be an H_v-ring and μ be a fuzzy H_v-ideal of R. If the range of μ is the finite set $\{t_1, t_2, \ldots, t_n\}$, then the set $\{\mu_{t_i} \mid 1 \leq i \leq n\}$ contains all the level H_v-ideals μ. Moreover if $t_1 < t_2 < \ldots < t_n$, then all the level H_v-ideals μ_{t_i} form the following chain $\mu_{t_1} \subseteq \mu_{t_2} \subseteq \ldots \subseteq \mu_{t_n}$.*

Let μ be a fuzzy subset of the set R. Consider the relation $\sim$ in R defined by $x \sim y$ if and only if $\mu(x) = \mu(y)$. Then, $\sim$ is an equivalence relation in R.

Definition 4.31. Suppose that $\overline{x}$ is the equivalence class containing x. The fuzzy subset $\mu_{\overline{x}}^*$ of R defined by

$$\mu_{\overline{x}}^*(r) = \max\{\mu(x), \mu(r)\}, \text{ for all } r \in R$$

is called the *fuzzy max determined by $\overline{x}$ and μ*, and we denote $\mathcal{M}$, the set of all fuzzy max of μ in R.

Now let $(R, +, \cdot)$ be an H_v-ring and μ be a fuzzy H_v-ideal of R. We define two hyperoperations on $\mathcal{M}$ as follows:

$$\mu_{\overline{x}}^* \uplus \mu_{\overline{y}}^* = \mu_{\overline{x+y}}^* \ , \quad \mu_{\overline{x}}^* \otimes \mu_{\overline{y}}^* = \mu_{\overline{x \cdot y}}^*, \text{ for all } x, y \in R,$$

where $\mu_A^* = \{\mu_{\overline{\alpha}}^* \mid \alpha \in A\}$, for all $A \subseteq R$ and $\overline{x+y} = \overline{x} + \overline{y}$, $\overline{x \cdot y} = \overline{x} \cdot \overline{y}$.

Lemma 4.11. *If $\mu_{\overline{x}}^* = \mu_{\overline{y}}^*$, then $\overline{x} = \overline{y}$.*

Proof. We have $\mu_{\overline{x}}^*(x) = \mu_{\overline{y}}^*(x)$. Thus, $\mu(x) = \max\{\mu(x), \mu(y)\}$ which implies that $\mu(y) \leq \mu(x)$. Similarly, we get $\mu(x) \leq \mu(y)$. So, $\mu(x) = \mu(y)$. Therefore, $\overline{x} = \overline{y}$. ∎

Corollary 4.11. *The hyperoperations $\uplus$ and $\otimes$ are well defined.*

Theorem 4.52. *$(\mathcal{M}, \uplus, \otimes)$ is an H_v-ring.*

Proof. It is straightforward. ∎

Theorem 4.53. *If μ is any fuzzy H_v-ideal of an H_v-ring R, then the map $\psi : R \to \mathcal{M}$ defined by $\psi(x) = \mu_{\overline{x}}^*$, for all $x \in R$ is an inclusion homomorphism.*

Proof. It is straightforward.   ∎

Now, let R be an H_v-ring and γ^* be the fundamental relation on R. Suppose $\gamma^*(a)$ is the equivalence class containing $a \in R$. Then both the sum $\oplus$ and the product $\odot$ on R/γ^*, the set of all equivalence classes, are defined as follows:

$$\gamma^*(a) \oplus \gamma^*(b) = \gamma^*(c), \text{ for all } c \in \gamma^*(a) + \gamma^*(b),$$
$$\gamma^*(a) \odot \gamma^*(b) = \gamma^*(d), \text{ for all } d \in \gamma^*(a) \cdot \gamma^*(b).$$

If $\varphi : R \to R/\gamma^*$ is the canonical map and μ is a fuzzy subset of R, then $\varphi(\mu)$ is fuzzy subset of R/γ^*. In this case, we use μ_{γ^*} instead of $\varphi(\mu)$.

Theorem 4.54. *Let R be an H_v-ring and μ be a fuzzy H_v-ideal of R. Then, μ_{γ^*} is a fuzzy ideal of the ring R/γ^*.*

Proof. Since μ is a fuzzy H_v-ideal of R, it follows that

(1)

$$\min\{\mu_{\gamma^*}(\gamma^*(x)), \mu_{\gamma^*}(\gamma^*(y))\}$$
$$\leq \inf_{\gamma^*(\alpha) \in \gamma^*(x) \oplus \gamma^*(y)} \{\mu_{\gamma^*}(\gamma^*(\alpha))\}, \text{ for all } \gamma^*(x), \gamma^*(y) \in R/\gamma^*.$$

(2) For all $\gamma^*(x), \gamma^*(a) \in R/\gamma^*$, there exists $\gamma^*(y) \in R/\gamma^*$ such that
$$\gamma^*(x) = \gamma^*(a) \oplus \gamma^*(y) \text{ and } \min\{\mu_{\gamma^*}(\gamma^*(x)), \mu_{\gamma^*}(\gamma^*(a))\} \leq \mu_{\gamma^*}(\gamma^*(y)).$$

From (1), we obtain $\min\{\mu_{\gamma^*}(\gamma^*(x)), \mu_{\gamma^*}(\gamma^*(y))\} \leq \mu_{\gamma^*}(\gamma^*(x) \oplus \gamma^*(y))$. Now, for all $\gamma^*(x)$ in R/γ^* we prove that $\mu_{\gamma^*}(\gamma^*(x)) \leq \mu_{\gamma^*}(-\gamma^*(x))$. Since $\gamma^*(x) \in R/\gamma^*$, by considering $\gamma^*(a) = \gamma^*(x)$ which is obtained from the second condition above there exists $\gamma^*(y_1)$ in R/γ^* such that $\gamma^*(x) = \gamma^*(x) \odot \gamma^*(y_1)$ and $\min\{\mu_{\gamma^*}(\gamma^*(x)), \mu_{\gamma^*}(\gamma^*(x))\} \leq \mu_{\gamma^*}(\gamma^*(y_1))$. From $\gamma^*(x) = \gamma^*(x) \odot \gamma^*(y_1)$ we obtain $\omega_H = \gamma^*(y_1)$, where ω_H denotes the unit of the group $(R/\gamma^*, \oplus)$. Therefore,

$$\mu_{\gamma^*}(\gamma^*(x)) \leq \mu_{\gamma^*}(\omega_H). \tag{4.5}$$

Now, by considering $\gamma^*(x), \omega_H$ in R/γ^*, by (2) above there exists $\gamma^*(y_2)$ in R/γ^* such that $\omega_H = \gamma^*(x) \oplus \gamma^*(y_2)$ and $\min\{\mu_{\gamma^*}(\omega_H), \mu_{\gamma^*}(\gamma^*(x))\} \leq \mu_{\gamma^*}(\gamma^*(y_2))$. From $\omega_H = \gamma^*(x) \oplus \gamma^*(y_2)$ we obtain $\gamma^*(y_2) = -\gamma^*(x)$, so

$$\min\{\mu_{\gamma^*}(\omega_H), \mu_{\gamma^*}(\gamma^*(x))\} \leq \mu_{\gamma^*}(-\gamma^*(x)). \tag{4.6}$$

By (4.5) and (4.6) the inequality $\mu_{\gamma^*}(\gamma^*(x)) \leq \mu_{\gamma^*}(-\gamma^*(x))$ is obtained. Now, from $\min\{\mu_{\gamma^*}(\gamma^*(x)), \mu_{\gamma^*}(\gamma^*(y))\} \leq \mu_{\gamma^*}(\gamma^*(x) \oplus \gamma^*(y))$ and $\mu_{\gamma^*}(\gamma^*(x)) \leq \mu_{\gamma^*}(-\gamma^*(x))$ we get

$$\{\mu_{\gamma^*}(\gamma^*(x)), \mu_{\gamma^*}(\gamma^*(y))\} \leq \mu_{\gamma^*}(\gamma^*(x) - \gamma^*(y)).$$

Now, we prove the second condition of the definition of fuzzy ideal. For all $\gamma^*(x), \gamma^*(y) \in R/\gamma^*$, we have $\mu_{\gamma^*}(\gamma^*(x) \odot \gamma^*(y)) = \mu_{\gamma^*}(\gamma^*(x) \odot \gamma^*(a))$, for all $a \in \gamma^*(y)$ and so

$$
\begin{aligned}
\mu_{\gamma^*}(\gamma^*(x) \odot \gamma^*(y)) &= \mu_{\gamma^*}(\gamma^*(x \cdot a)) \\
&= \sup_{\alpha \in \gamma^*(x \cdot a)} \{\mu(\alpha)\} \\
&\geq \sup_{\alpha \in x \cdot a} \{\mu(\alpha)\} \\
&\geq \inf_{\alpha \in x \cdot a} \{\mu(\alpha)\} \\
&\geq \mu(a).
\end{aligned}
$$

Therefore, $\mu_{\gamma^*}(\gamma^*(x) \odot \gamma^*(y)) \geq \sup_{a \in \gamma^*(y)} \{\mu(a)\}$ which implies that $\mu_{\gamma^*}(\gamma^*(x) \odot \gamma^*(y)) \geq \gamma^*(y)$, and the theorem is proved.   ■

Let γ_μ^* be the fundamental equivalence relation on $\mathcal{M}$, the H_v-ring defined in Theorem 4.52, and $\mathcal{U}_\mu$ denotes the set of finite polynomials of elements of $\mathcal{M}$ over $\mathbb{N}$. We denote sum and product on $\mathcal{M}/\gamma_\mu^*$ by $\oplus, \odot$.

Proposition 4.18. *Let $(R, +, \cdot)$ be an H_v-ring. Then, there exists a strong homomorphism $f : R/\gamma^* \to \mathcal{M}/\gamma_\mu^*$.*

Proof.   We define f as follows: $f(\gamma^*(r)) = \gamma_\mu^*(\mu_{\overline{r}}^*)$, for all $r \in R$. We prove that f is well defined. Suppose that $\gamma^*(r) = \gamma^*(r_1)$. Then, $r\gamma^*r_1$. We have $r\gamma^*r_1$ if and only if there exist $x_1, \ldots, x_{m+1}$, where $x_1 = r$, $x_{m+1} = r_1$, $u_1, \ldots, u_m \in \mathcal{U}$ such that $\{x_i, x_{i+1}\} \subseteq u_i$, $i = 1, \ldots, m$. Therefore, $\{\mu_{\overline{x_i}}^*, \mu_{\overline{x_{i+1}}}^*\} \subseteq \mu_{\overline{u_i}}^*$, where $\mu_{\overline{u_i}}^* \in \mathcal{U}_\mu$, $i = 1, \ldots, m$, which implies that $\mu_{\overline{r}}^* \gamma_\mu^* \mu_{\overline{r_1}}^*$ implying $\gamma_\mu^*(\mu_{\overline{r}}^*) = \gamma_\mu^*(\mu_{\overline{r_1}}^*)$.

Now, we show that f is strong homomorphism. For every $\gamma^*(a), \gamma^*(b)$ in R/γ^*, we can write:

$$f(\gamma^*(a) \oplus \gamma^*(b)) = f(\gamma^*(c)) = \gamma_\mu^*(\mu_{\overline{c}}^*), \text{ for all } c \in \gamma^*(a) + \gamma^*(b),$$

$$f(\gamma^*(a) \oplus \gamma^*(b)) = \gamma_\mu^*(\mu_{\overline{a}}^*) \oplus \gamma_\mu^*(\mu_{\overline{b}}^*) = \gamma_\mu^*(\mu_{\overline{d}}^*), \text{ for all } \mu_{\overline{d}}^* \in \gamma_\mu^*(\mu_{\overline{a}}^*) \uplus \gamma_\mu^*(\mu_{\overline{b}}^*).$$

If we put $c \in a + b$ and $\mu_{\overline{d}}^* \in \mu_{\overline{a}}^* \uplus \mu_{\overline{b}}^*$, then $f(\gamma^*(a) \oplus \gamma^*(b)) = f(\gamma^*(a)) \oplus f(\gamma^*(b))$. In the similar way, we have $f(\gamma^*(a) \odot \gamma^*(b)) = f(\gamma^*(a)) \odot f(\gamma^*(b))$.   ■

Corollary 4.12. *The following diagram is commutative:*

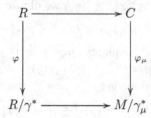

Let $\{R_\alpha \mid \alpha \in \Gamma\}$ be a collection of H_v-rings and
$$\prod_{\alpha \in \Gamma} R_\alpha = \{< x_\alpha > \mid x_\alpha \in R_\alpha\},$$
be the Cartesian product of R_α $(\alpha \in \Gamma)$. We can define two hyperoperations as follows:
$$< x_\alpha > \oplus < y_\alpha >= \{< z_\alpha > \mid z_\alpha \in x_\alpha + y_\alpha, \ \alpha \in \Gamma\},$$
$$< x_\alpha > \odot < y_\alpha >= \{< z_\alpha > \mid z_\alpha \in x_\alpha \cdot y_\alpha, \ \alpha \in \Gamma\}.$$
It follows that $\prod_{\alpha \in \Gamma} R_\alpha$ is an H_v-ring. We call $\prod_{\alpha \in \Gamma} R_\alpha$ the *external direct product* of R_α $(\alpha \in \Gamma)$. Let μ and λ be fuzzy subsets of a non-empty set X. The *Cartesian product* $\mu \times \lambda$ is usually defined by:
$$(\mu \times \lambda)(x,y) = \min\{\mu(x), \lambda(y)\}, \ \text{for all } x, y \in X.$$
Let $\{X_\alpha \mid \alpha \in \Gamma\}$ be a collection of non-empty sets and let μ_α be a fuzzy subset of X_α for all $\alpha \in \Gamma$. Define the Cartesian product of the μ_α by $(\prod_{\alpha \in \Gamma} \mu_\alpha)(x) = \inf_{\alpha \in \Gamma}\{\mu_\alpha(x_\alpha)\}$ where $x =< x_\alpha >$ and $< x_\alpha >$ denotes an element of the Cartesian product $\prod_{\alpha \in \Gamma} X_\alpha$.

Proposition 4.19. *Let* $\{R_\alpha \mid \alpha \in \Gamma\}$ *be a collection of H_v-rings and let* μ_α *be a fuzzy H_v-ideal of R_α. Then* $\prod_{\alpha \in \Gamma} \mu_\alpha$ *is a fuzzy H_v-ideal of* $\prod_{\alpha \in \Gamma} R_\alpha$.

Proof. Let $x =< x_\alpha >$, $y =< y_\alpha > \ \in \prod_{\alpha \in \Gamma} R_\alpha$. Then, for every $z = < z_\alpha >\in x + y =< x_\alpha > \oplus < y_\alpha >$ we have
$$\begin{aligned}
(\prod_{\alpha \in \Gamma} \mu_\alpha)(z) &= \inf_{\alpha \in \Gamma}\{\mu_\alpha(z_\alpha)\} \\
&\geq \inf_{\alpha \in \Gamma}\{\min\{\mu_\alpha(x_\alpha), \mu_\alpha(y_\alpha)\}\} \\
&= \min\left\{\inf_{\alpha \in \Gamma}\{\mu_\alpha(x_\alpha)\}, \inf_{\alpha \in \Gamma}\{\mu_\alpha(y_\alpha)\}\right\} \\
&= \min\left\{(\prod_{\alpha \in \Gamma} \mu_\alpha)(x), (\prod_{\alpha \in \Gamma} \mu_\alpha)(y)\right\}.
\end{aligned}$$
Therefore, the first condition of the definition of an H_v-ideal is satisfied. Now, we prove the second condition as follows. For every $x =< x_\alpha >$ and $a =< a_\alpha >$ in $\prod_{\alpha \in \Gamma} R_\alpha$ there exists $y =< y_\alpha >$ in $\prod_{\alpha \in \Gamma} R_\alpha$ such that $\min\{\mu_\alpha(x_\alpha), \mu_\alpha(a_\alpha)\} \leq \mu_\alpha(y_\alpha)$. Therefore, we have $< x_\alpha >\in< a_\alpha > \oplus < y_\alpha >$ and
$$\begin{aligned}
(\prod_{\alpha \in \Gamma} \mu_\alpha)(y) &= \inf_{\alpha \in \Gamma}\{\mu_\alpha(y_\alpha)\} \geq \inf_{\alpha \in \Gamma}\{\min\{\mu_\alpha(x_\alpha), \mu_\alpha(a_\alpha)\}\} \\
&= \min\left\{\inf_{\alpha \in \Gamma}\{\mu_\alpha(x_\alpha)\}, \inf_{\alpha \in \Gamma}\{\mu_\alpha(a_\alpha)\}\right\} \\
&= \min\left\{(\prod_{\alpha \in \Gamma} \mu_\alpha)(x), (\prod_{\alpha \in \Gamma} \mu_\alpha)(a)\right\}.
\end{aligned}$$

The proof of third condition is similar to that of second condition. To verify the fourth condition, for every $z = < z_\alpha > \in x \odot y = < x_\alpha > \odot < y_\alpha >$ we have

$$(\prod_{\alpha \in \Gamma} \mu_\alpha)(z) = \inf_{\alpha \in \Gamma} \{\mu_\alpha(z_\alpha)\} \geq \inf_{\alpha \in \Gamma} \{\mu_\alpha(y_\alpha)\} = (\prod_{\alpha \in \Gamma} \mu_\alpha)(y).$$

Hence, $(\prod_{\alpha \in \Gamma} \mu_\alpha)(y) \leq \inf_{z \in x \cdot y} \left\{ (\prod_{\alpha \in \Gamma} \mu_\alpha)(z) \right\}.$ ■

The following corollary is exactly obtained from the above proposition.

Corollary 4.13. *Let μ be a fuzzy subset of an H_v-ring R. Then, $\mu \times \mu$ is a fuzzy left (right) H_v-ideal of $R \times R$ if and only if μ is a fuzzy left (right) H_v-ideal of R.*

Definition 4.32. Let μ be a fuzzy H_v-ideal of $R_1 \times R_2$ and let (x_1, x_2), $(a_1, a_2) \in R_1 \times R_2$. Then, there exists $(y_1, y_2) \in R_1 \times R_2$ such that $(x_1, x_2) \in (a_1, a_2) \oplus (y_1, y_2)$ and

$$\min\{\mu(x_1, x_2), \mu(a_1, a_2)\} \leq \mu(y_1, y_2).$$

Now, if for every $r_1, s_1 \in R_1$ there exists $t_1 \in R_1$ such that $(r_1, x_2) \in (s_1, a_2) \oplus (t_1, y_2)$ and

$$\min\{\mu(r_1, x_2), \mu(s_1, a_2)\} \leq \mu(t_1, y_2),$$

and for every $r_2, s_2 \in R_2$ there exists $t_2 \in R_2$ such that $(x_1, r_2) \in (a_1, s_2) \oplus (y_1, t_2)$ and

$$\min\{\mu(x_1, r_2), \mu(a_1, s_2)\} \leq \mu(y_1, t_2),$$

then we say that μ satisfies in the *regular left fuzzy reproduction axiom*. Similarly, we can define the *regular right fuzzy reproduction axiom*. μ is called the *regular fuzzy H_v-ideal* of R if μ satisfies the regular left and right fuzzy reproduction axioms.

Theorem 4.55. *Let R_1, R_2 be H_v-rings with scalar units and μ be a regular fuzzy H_v-ideal of $R_1 \times R_2$. Then, μ_i, $i = 1, 2$ is a fuzzy H_v-ideal of R_i, $i = 1, 2$, respectively, where $\mu_1(x) = \sup_{a \in R_2} \{\mu(x, a)\}$ and $\mu_2(y) = \sup_{b \in R_1} \{\mu(b, y)\}$.*

Proof. We show that μ_1 is a fuzzy H_v-ideal of R_1. Suppose that $x, y \in R_1$. Then, for every $\alpha \in x + y$ we have $\mu_1(\alpha) = \sup_{a \in R_2} \{\mu(\alpha, a)\}$. For every

$a \in R_2$ there exist $r_a, s_a \in R_2$ such that $a \in a + r_a$ and $a \in s_a + a$. Now, we have

$$\mu_1(\alpha) \geq \sup_{a\in R_2} \{\min\{\mu(x,a), \mu(y,r_a)\}\}$$
$$\geq \sup_{a\in R_2} \{\min\{\mu(x,a), \inf_{b\in R_2}\{\mu(y,b)\}\}\}$$
$$= \min\left\{ \sup_{a\in R_2}\{\mu(x,a)\}, \inf_{b\in R_2}\{\mu(y,b)\}\right\}$$
$$= \min\left\{ \mu_1(x), \inf_{b\in R_2}\{\mu(y,b)\}\right\}$$

and also we have

$$\mu_1(\alpha) \geq \sup_{a\in R_2} \{\min\{\mu(x,s_a), \mu(y,a)\}\}$$
$$\geq \sup_{a\in R_2}\left\{ \min\{\inf_{c\in R_2}\{\mu(x,c)\}, \mu(y,a)\}\right\}$$
$$= \min\left\{ \inf_{c\in R_2}\{\mu(x,c)\}, \sup_{a\in R_2}\{\mu(y,a)\}\right\}$$
$$= \min\left\{ \inf_{c\in R_2}\{\mu(x,c)\}, \mu_1(y)\right\}.$$

Therefore,

$$\mu_1(\alpha) \geq \max\left\{ \min\left\{\mu_1(x), \inf_{b\in R_2}\{\mu(y,b)\}\right\}, \min\left\{\inf_{c\in R_2}\{\mu(x,c)\}, \mu_1(y)\right\}\right\}$$
$$= \min\left\{ \max\left\{\mu_1(x), \inf_{c\in R_2}\{\mu(x,c)\}\right\}, \max\left\{\inf_{b\in R_2}\{\mu(y,b)\}, \mu_1(y)\right\}\right\}$$
$$\geq \min\{\mu_1(x), \mu_1(y)\},$$

Now, if $x, a \in R_1$, then for every $r, s \in R_2$ there exists $(y, y_{r,s}) \in R_1 \times R_2$ such that $(x,r) \in (a,s) \oplus (y, y_{r,s})$ and $\min\{\mu(x,r), \mu(a,s)\} \leq \mu(y, y_{r,s})$. Thus,

$$\min\{\mu_1(x), \mu_1(a)\} = \min\left\{ \sup_{r\in R_2}\{\mu(x,r)\}, \sup_{s\in R_2}\{\mu(a,s)\}\right\}$$
$$= \sup_{\substack{r\in R_2\\s\in R_2}} \{\min\{\mu(x,r), \mu(a,s))\}\}$$
$$\leq \sup_{\substack{r\in R_2\\s\in R_2}} \{\mu(y, y_{r,s})\}$$
$$\leq \sup_{z\in R_2} \{\mu(y,z)\} = \mu_1(y).$$

The proof of the third condition is similar to the second condition. Now, we verify the fourth condition of the definition. Suppose that 1 be the unit scalar of R_2. For every $\alpha \in x \cdot y$, we have $(\alpha, a) \in (x,1) \odot (y,a)$. Since μ is a fuzzy H_v-ideal of $R_1 \times R_2$, we obtain $\mu(\alpha, a) \geq \mu(y,a)$ which implies that $\sup_{a\in R_2}\{\mu(\alpha,a)\} \geq \sup_{a\in R_2}\{\mu(y,a)\}$. Therefore $\mu_1(\alpha) \geq \mu_1(y)$ for every $\alpha \in x \cdot y$, and so $\inf_{\alpha\in x\cdot y}\{\mu_1(\alpha)\} \geq \mu_1(y)$. Hence, μ_1 is a fuzzy H_v-ideal of R_1. Similarly, we can prove that μ_2 is a fuzzy H_v-ideal of R_2. ∎

Corollary 4.14. *Let R_1, R_2 be two H_v-rings with scalar units and μ, λ be fuzzy subsets of R_1, R_2, respectively. If $\sup\limits_{x \in R_1}\{\mu(x)\} = \sup\limits_{y \in R_2}\{\lambda(y)\} = 1$ and $\mu \times \lambda$ is a strong fuzzy H_v-ideal of $R_1 \times R_2$, then μ, λ are fuzzy H_v-ideals of R_1, R_2, respectively.*

Proof. Suppose that

$$\mu_1(x) = \sup_{a \in R_2}\{(\mu \times \lambda)(x, a)\} \text{ and } \mu_2(y) = \sup_{b \in R_1}\{(\mu \times \lambda)(b, y)\},$$

then it is enough to show that $\mu(x) = \mu_1(x)$ and $\lambda(y) = \lambda_2(y)$. Let $x \in R_1$ then

$$\begin{aligned}
\mu(x) &= \min\{\mu(x), 1\} \\
&= \min\left\{\mu(x), \sup_{y \in R_2}\{\lambda(y)\}\right\} \\
&= \sup_{y \in R_2}\{\min\{\mu(x), \lambda(y)\}\} \\
&= \sup_{y \in R_2}\{(\mu \times \lambda)(x, y)\} \\
&= \mu_1(x).
\end{aligned}$$

∎

Chapter 5

H_v-modules

5.1 H_v-modules and fundamental relations

Definition 5.1. Let M be a non-empty set. Then, M is called an H_v-module over an H_v-ring R if $(M, +)$ is a weak commutative H_v-group and there exists a map $\cdot : R \times M \to \mathcal{P}^*(M)$ denoted by $(r, m) \mapsto rm$ such that for every $r_1, r_2 \in R$ and every $m_1, m_2 \in M$, we have

(1) $r_1(m_1 + m_2) \cap (r_1 m_1 + r_1 m_2) \neq \emptyset$,
(2) $(r_1 + r_2)m_1 \cap (r_1 m_1 + r_2 m_1) \neq \emptyset$,
(3) $(r_1 r_2)m_1 \cap r_1(r_2 m_1) \neq \emptyset$.

Definition 5.2. Let M_1 and M_2 be two H_v-modules over an H_v-ring R. A mapping $f : M_1 \to M_2$ is called a *strong H_v-homomorphism* if for every $x, y \in M_1$ and every $r \in R$, we have

$$f(x + y) = f(x) + f(y) \text{ and } f(rx) = rf(x).$$

The H_v-modules M_1 and M_2 are called *isomorphic* if the H_v-homomorphism f is one to one and onto. It is denoted by $M_1 \cong M_2$.

Similar to H_v-groups and H_v-rings, the fundamental equivalence relation ε^* on H_v-modules is introduced by Vougiouklis [120]. The *fundamental equivalence relation ε^** on an H_v-module M can be defined as follows.

Definition 5.3. Consider the H_v-module M over an H_v-ring R. If ϑ denotes the set of all expressions consisting of finite hyperoperations of either on R and M or of the external hyperoperations applying on finite sets of elements of R and M. A relation ε can be defined on M whose transitive closure is the fundamental relation ε^*. The relation ε is defined as follows:

for every $x, y \in M$, $x \ \varepsilon \ y$ if and only if $\{x, y\} \subseteq u$ for some $u \in \vartheta$; *i.e*:

$$x \varepsilon y \Leftrightarrow x, y \in \sum_{i=1}^{n} m_i', \quad m_i' = m_i \text{ or } m_i' = \sum_{j=1}^{n_i} (\prod_{k=1}^{k_{ij}} r_{ijk}) m_i,$$

where $m_i \in M$, $r_{ijk} \in R$.

Suppose that $\gamma^*(r)$ is the equivalence class containing $r \in R$ and $\varepsilon^*(x)$ is the equivalence class containing $x \in M$. On M/ε^* the $\oplus$ and the external product $\odot$ using the γ^* classes in R, are defined as follows:

For every $x, y \in M$, and for every $r \in R$,

$$\varepsilon^*(x) \oplus \varepsilon^*(y) = \varepsilon^*(c), \text{ for all } c \in \varepsilon^*(x) + \varepsilon^*(y),$$

$$\gamma^*(r) \odot \varepsilon^*(x) = \varepsilon^*(d), \text{ for all } d \in \gamma^*(r) \cdot \varepsilon^*(x).$$

The kernel of canonical map $\phi : M \to M/\varepsilon_M^*$ is called *heart* of M and it is denoted by ω_M, i.e., $\omega_M = \{x \in M \mid \phi(x) = 0\}$, where 0 is the unit element of the group $(M/\varepsilon^*, \oplus)$. One can prove that the unit element of the group $(M/\varepsilon^*, \oplus)$ is equal to ω_M. By the definition of ω_M, we have

$$\omega_{\omega_M} = Ker(\phi : \omega_M \to \omega_M/\varepsilon_{\omega_M}^* = 0) = \omega_M.$$

The *kernel* of a strong H_v-homomorphism $f : A \to B$ is defined as follows

$$Ker(f) = \{a \in A \mid f(a) \in \omega_B\}.$$

Lemma 5.1. *Let M_1 and M_2 be two H_v-modules over an H_v-ring R and let $\varepsilon_{M_1}^*$, $\varepsilon_{M_2}^*$ and $\varepsilon_{M_1 \times M_2}^*$ be the fundamental relations on M_1, M_2 and $M_1 \times M_2$ respectively. Then,*

$$(x_1, x_2)\varepsilon_{M_1 \times M_2}^*(y_1, y_2) \Leftrightarrow x_1 \varepsilon_{M_1}^* y_1 \text{ and } x_2 \varepsilon_{M_2}^* y_2,$$

for all $(x_1, x_2), (y_1, y_2) \in M_1 \times M_2$.

Proof. It is straightforward.   ∎

Theorem 5.1. *Let M_1 and M_2 be two H_v-modules over an H_v-ring R and let $\varepsilon_{M_1}^*$, $\varepsilon_{M_2}^*$ and $\varepsilon_{M_1 \times M_2}^*$ be the fundamental relations on M_1, M_2 and $M_1 \times M_2$ respectively. Then,*

$$(M_1 \times M_2)/\varepsilon_{M_1 \times M_2}^* \cong M_1/\varepsilon_{M_1}^* \times M_2/\varepsilon_{M_2}^*.$$

Proof. It is straightforward.   ∎

Definition 5.4. Let M be an H_v-module and X, Y are non-empty subsets of M. We say X is *weak equal* to Y and write $X \overset{w}{=} Y$ if and only if for every $x \in X$ there exists $y \in Y$ such that $\varepsilon_M^*(x) = \varepsilon_M^*(y)$ and for every $y \in Y$ there exists $x \in X$ such that $\varepsilon_M^*(x) = \varepsilon_M^*(y)$.

Definition 5.5. Let $M_0 \overset{f_1}{\to} M_1 \overset{f_2}{\to} M_2 \to \cdots \to M_{n-1} \overset{f_n}{\to} M_n$ be a sequence of H_v-modules and strong H_v-homomorphisms. We say this sequence is *exact* if for every $2 \le i \le n$, $Im(f_{i-1}) \overset{w}{=} Ker(f_i)$.

Definition 5.6. A function $f : M_1 \to M_2$ is called *weak-monic* if for every $m_1, m_1' \in M_1$, $f(m_1) = f(m_1')$ implies $\varepsilon_{M_1}^*(m_1) = \varepsilon_{M_1}^*(m_1')$ and f is called *weak-epic* if for every $m_2 \in M_2$ there exists $m_1 \in M_1$ such that $\varepsilon_{M_2}^*(m_2) = \varepsilon_{M_1}^*(f(m_1))$. Finally f is called *weak-isomorphism* if f is weak-monic and weak-epic.

We present the following example for the above definitions.

Example 5.1. Let R be an H_v-ring. Consider the following H_v-modules on R.

(1) $M = \{a, b\}$ together with the following hyperoperations:

$*_M$	a	b
a	a	b
b	b	a

and $\quad \cdot_M : R \times M \to \mathcal{P}^*(M)$
$\qquad\qquad (r,m) \mapsto \{a\}$

(2) $M_1 = \{0, 1, 2\}$ together with the following hyperoperations:

$*_{M_1}$	0	1	2
0	0	1	2
1	1	0,2	1
2	2	1	0

and $\quad \cdot_{M_1} : R \times M_1 \to \mathcal{P}^*(M_1)$
$\qquad\qquad (r,m_1) \mapsto \{0\}$

(3) $M_2 = \{\bar{0}, \bar{1}, \bar{2}\}$ together with the following hyperoperations:

$*_{M_2}$	$\bar{0}$	$\bar{1}$	$\bar{2}$
$\bar{0}$	$\bar{0}$	$\bar{1}$	$\bar{2}$
$\bar{1}$	$\bar{1}$	$\bar{2}$	$\bar{0}$
$\bar{2}$	$\bar{2}$	$\bar{0}$	$\bar{1}$

and $\quad \cdot_{M_2} : R \times M_2 \to \mathcal{P}^*(M_2)$
$\qquad\qquad (r,m_2) \mapsto M_2$

Since $\{0, 2\} \subseteq 1 *_{M_1} 1$, $r \cdot m_1 = 0$ for every $r \in R$ and every $m_1 \in M_1$ and $0 *_{M_1} 0 = 0$, we obtain $M_1/\varepsilon_{M_1}^* = \{\varepsilon_{M_1}^*(0) = \varepsilon_{M_1}^*(2) = \{0, 2\}, \varepsilon_{M_1}^*(1) = \{1\}\}$. Also, since $\varepsilon_{M_1}^*(0) + \varepsilon_{M_1}^*(1) = \varepsilon_{M_1}^*(1)$, it follows that $\omega_{M_1} = \varepsilon_{M_1}^*(0) = \{0, 2\}$. Since $r \cdot_{M_2} m_2 = M_2$ for every $r \in R$ and every $m_2 \in M_2$, we obtain $M_2/\varepsilon_{M_2}^* = \{\{\bar{0}, \bar{1}, \bar{2}\}\}$ and $\omega_{M_2} = \varepsilon_{M_2}^*(\bar{0}) = \varepsilon_{M_2}^*(\bar{1}) = \varepsilon_{M_2}^*(\bar{2}) = M_2$.

Since $(M_1 \times M_2)/\varepsilon^*_{M_1 \times M_2} \cong M_1/\varepsilon^*_{M_1} \times M_2/\varepsilon^*_{M_2}$, it follows that

$$M_1 \times M_2/$$
$$\varepsilon^*_{M_1 \times M_2} = \{\{(0,\bar{0}), (0,\bar{1}), (0,\bar{2}), (2,\bar{0}), (2,\bar{1}), (2,\bar{2})\}, \{(1,\bar{0}), (1,\bar{1}), (1,\bar{2})\}\}.$$

Note that $\omega_{M_1 \times M_2} = \omega_{M_1} \times \omega_{M_2}$. The subsets $X = \{(2,\bar{1}), (2,\bar{2}), (1,\bar{1}), (1,\bar{2})\}$ and $Y = \{(0,\bar{2}), ((1,\bar{0})\}$ of $M_1 \times M_2$ are weakly equal. Now, consider $f \in M[M_1 \times M_2]$, where $f(a) = (2,\bar{2})$, $f(b) = (1,\bar{0})$ and $g \in M_1[M_1 \times M_2]$, where $g(0) = (1,\bar{1})$, $g(1) = (2,\bar{2})$, $g(2) = (1,\bar{1})$. Then, f is weak-epic and g is weak-monic.

5.2 H_v-module of fractions

In Section 4.9, we introduced the concept of H_v-ring of fractions $S^{-1}R$ of a commutative hyperring. The construction of $S^{-1}R$ can be carried through with a hypermodule M over a hyperring R in place of the hyperring R. In this section, we introduce the set of fractions $S^{-1}M$ and define addition and multiplication by elements of $S^{-1}R$. Then we show that $S^{-1}M$ is an H_v-module over the H_v-ring $S^{-1}R$ and is called H_v-module of fractions. The main reference for this part is [21].

Let X be the set of all ordered pairs (m, s) where $m \in M$, $s \in S$. For $A \subseteq M$ and $B \subseteq S$, we denote the set $\{(a, b) \mid a \in A, b \in B\}$ by (A, B). The relation $\sim$ is defined as follows:

$(A, B) \sim (C, D) \Leftrightarrow$ there exists a subset T of S such that $T \cdot (B \cdot C) = T \cdot (D \cdot A)$.

Lemma 5.2. $\sim$ *is an equivalence relation.*

Proof.   The proof is similar to the proof of Lemma 4.4.   ∎

The equivalence class containing (A, B) is denoted by $\| A, B \|$.

If we consider the relation $\sim$ on X, we obtain the following two corollaries.

Corollary 5.1. *For* (m_1, s_1), $(m_2, s_2) \in X$, *we have* $(m_1, s_1) \sim (m_2, s_2)$ *if and only if there exists* $T \subseteq S$ *such that* $T \cdot (s_1 \cdot m_2) = T \cdot (s_2 \cdot m_1)$.

Corollary 5.2. $\sim$ *is an equivalence relation on* X.

In X, the equivalence class containing (m, s) is denoted by $[m, s]$ and we let $S^{-1}M$ to be the set of all the equivalence classes.
We define
$$\ll A, B \gg = \bigcup_{(C,D) \in \| A,B \|} \{[c, d] \mid c \in C, d \in D\}.$$

Corollary 5.3. *For all $m \in M$, $s \in S$, we have $\ll m, s \gg = \ll sm, ss \gg$.*

Now we define addition and multiplication by elements of $S^{-1}R$, as follows:

$$[m_1, s_1] \oplus [m_2, s_2] = \bigcup_{(A,B) \in \|s_1 \cdot m_2 + s_2 \cdot m_1, s_1 \cdot s_2\|} \{[a, b] \mid a \in A, b \in B\}$$
$$= \ll s_1 m_2 + s_2 m_1, s_1 s_2 \gg,$$

$$[r, s] \odot [m_1, s_1] = \bigcup_{(A,B) \in \|rm_1, ss_1\|} \{[a, b] \mid a \in A, b \in B\} = \ll rm_1, ss_1 \gg,$$

for all $[m_1, s_1], [m_2, s_2] \in S^{-1}M$ and $[r, s] \in S^{-1}R$.

Theorem 5.2. *The above definitions are independent of the choices of representatives $[m_1, s_1]$, $[m_2, s_2]$ and $[r, s]$ and that $S^{-1}M$ satisfies the axioms of an H_v-module over $S^{-1}R$.*

If we define

$$r \odot [m_1, s_1] = \ll rm_1, s_1 \gg,$$

then $S^{-1}M$ is an H_v-module over R.

Theorem 5.3. *Let M_1 and M_2 be two hypermodules over a hyperring R and let $f : M_1 \to M_2$ be a strong R-H_v-homomorphism. Then the map $S^{-1}(f) : S^{-1}M_1 \to S^{-1}M_2$ defined by $S^{-1}(f)[m, s] = [f(m), s]$, is a strong $S^{-1}R$-H_v-homomorphism.*

Proof. Suppose that $[m_1, s_1], [m_2, s_2] \in S^{-1}M$ and $[r, s] \in S^{-1}R$. First, we show that $S^{-1}(f)$ is well defined. If $[m_1, s_1] = [m_2, s_2]$, then there exists $T \subseteq S$ such that $T \cdot (s_1 \cdot m_2) = T \cdot (s_2, m_1)$ which implies $f(T \cdot (s_1 \cdot m_2)) = f(T \cdot (s_2 \cdot m_1))$ and so $T \cdot (s_1 \cdot f(m_2)) = T \cdot (s_2 \cdot f(m_1))$ or $[f(m_1), s_1] = [f(m_2), s_2]$. Therefore, $S^{-1}(f)$ is well defined.

Moreover, $S^{-1}(f)$ is an $S^{-1}R$-H_v-homomorphism because, we have

$$S^{-1}(f)([m_1, s_1] \oplus [m_2, s_2])$$
$$= S^{-1}(f)(\bigcup_{(A,B) \in \|s_1 m_2 + s_2 m_1, s_1 s_2\|} \{[a, b] \mid a \in A, b \in B\})$$
$$= \bigcup_{(A,B) \in \|s_1 m_2 + s_2 m_1, s_1 s_2\|} S^{-1}(f)(\{[a, b] \mid a \in A, b \in B\})$$
$$= \bigcup_{(A,B) \in \|s_1 m_2 + s_2 m_1, s_1 s_2\|} \{[f(a), b] \mid a \in A, b \in B\}$$

and

$$S^{-1}(f)([m_1, s_1]) \oplus S^{-1}(f)([m_2, s_2])$$
$$= [f(m_1), s_1] \oplus [f(m_2), s_2]$$
$$= \bigcup_{(A,B) \in \|s_1 f(m_2) + s_2 f(m_1), s_1 s_2\|} \{[a, b] \mid a \in A, b \in B\}$$
$$= \bigcup_{(A,B) \in \|f(s_1 m_2 + s_2 m_1), s_1 s_2\|} \{[a, b] \mid a \in A, b \in B\}.$$

Therefore, we have

$$\{[f(a), b] \mid a \in s_1m_2 + s_2m_1, b \in s_1s_2\} \subseteq S^{-1}(f)([m_1, s_1] \oplus [m_2, s_2]),$$

$$\{[a, b] \mid a \in f(s_1m_2+s_2m_1), b \in s_1s_2\} \subseteq S^{-1}(f)([m_1, s_1]) \oplus S^{-1}(f)([m_2, s_2]).$$

Thus,

$$S^{-1}(f)([m_1, s_1] \oplus [m_2, s_2]) \cap S^{-1}(f)([m_1, s_1]) \oplus S^{-1}(f)([m_2, s_2]) \neq \emptyset.$$

Similarly, we obtain

$$\{[f(a), b] \mid a \in rm_1, b \in ss_1\} \subseteq S^{-1}(f)([r, s] \odot [m_1, s_1]),$$
$$\{[a, b] \mid a \in rf(m_1), b \in ss_1\} \subseteq [r, s] \odot S^{-1}(f)([m_1, s_1]).$$

So,

$$S^{-1}(f)([r, s] \odot [m_1, s_1]) \cap [r, s] \odot S^{-1}(f)([m_1, s_1]) \neq \emptyset,$$

which proves that $S^{-1}(f)$ is an $S^{-1}R$-H_v-homomorphism. ∎

Lemma 5.3. *The natural mapping $\Psi : M \to S^{-1}M$, where $\Psi(m) = [m, 1]$, is an inclusion R-H_v-homomorphism.*

Proof. For every $m_1, m_2 \in M$, we have

$$\begin{aligned}
\Psi(m_1 + m_2) &= \{[\alpha, 1] \mid \alpha \in m_1 + m_2\} \\
&\subseteq \bigcup_{(A,B) \in \|m_1+m_2, 1\|} \{[a, b] \mid a \in A, b \in B\} \\
&= [m_1, 1] \oplus [m_2, 1] \\
&= \Psi(m_1) \oplus \Psi(m_2),
\end{aligned}$$

and for every $r \in R$ and $m \in M$ we have

$$\Psi(rm) = \{[\alpha, 1] \mid \alpha \in rm\} \subseteq \ll rm, 1 \gg = r \odot \Psi(m).$$

Therefore, Ψ is an inclusion R-H_v-homomorphism. ∎

Let ε_s^* be the fundamental equivalence relation on $S^{-1}M$ and $\mathcal{U}_s$ is the set of all expressions consisting of finite hyperoperations of either on $S^{-1}R$ and $S^{-1}M$ or of external hyperoperation. In this case $S^{-1}M/\varepsilon_s^*$ is an $S^{-1}R/\gamma_s^*$-module.

Theorem 5.4. $S^{-1}M/\varepsilon_s^*$ *is an R/γ^*-module.*

Proof. We can define

$$\gamma^*(r) * \varepsilon_s^*([m, s]) = \gamma_s^*([r, 1]) \diamond \varepsilon^*([m, s]).$$

Then, it is clear that $S^{-1}M/\varepsilon_s^*$ is a module over the ring R/γ^*. ∎

Theorem 5.5. *There is an R/γ^*-homomorphism $h : M/\varepsilon^* \to S^{-1}M/\varepsilon_s^*$.*

Proof. We define $h(\varepsilon^*(m)) = \varepsilon_s^*([m,1])$. First we prove that h is well defined. Suppose that $\varepsilon^*(m_1) = \varepsilon^*(m_2)$. So, $m_1\varepsilon^*m_2$. $m_1\varepsilon^*m_2$ if and only if there exist $x_1, \ldots, x_{m+1}; u_1, \ldots, u_m \in \mathcal{U}$ with $x_1 = m_1$, $x_{m+1} = m_2$ such that $\{x_i, x_{i+1}\} \subseteq u_i, i = 1, \ldots, m$ which implies $\{[x_i,1],[x_{i+1},1]\} \subseteq \ll u_i, 1 \gg \in \mathcal{U}_s$. Therefore, $[m_1,1]\varepsilon_s^*[m_2,1]$ and so $\varepsilon_s^*([m_1,1]) = \varepsilon_s^*([m_2,1])$. Thus, h is well defined. Moreover, h is a homomorphism because

$$h(\varepsilon^*(a) \circ \varepsilon^*(b)) = h(\varepsilon^*(c)) = \varepsilon_s^*([c,1]), \text{ for all } c \in \varepsilon^*(a) + \varepsilon^*(b),$$

and

$$h(\varepsilon^*(a)) \circ h(\varepsilon^*(b)) = \varepsilon_s^*([a,1]) \circ \varepsilon_s^*([b,1]) = \varepsilon_s^*([d,s]),$$

for all $[d,s] \in \varepsilon_s^*([a,1]) \oplus \varepsilon_s^*([b,1])$. Thus, setting $d = c \in a+b, s = 1$. Then, it is proved that

$$h(\varepsilon^*(a) \circ \varepsilon^*(b)) = h(\varepsilon^*(a) \circ h(\varepsilon^*(b)).$$

Also, we have

$$h(\gamma^*(r) \diamond \varepsilon^*(m)) = h(\varepsilon^*(r \cdot m)) = \varepsilon^*([a,1]), \text{ for all } a \in r \cdot m,$$

$$\gamma^*(r) * h(\varepsilon^*(m)) = \gamma^*(r) * \varepsilon_s^*([m,1])$$
$$= \gamma_s^*([r,1]) \diamond \varepsilon_s^*([m,s])$$
$$= \varepsilon_s^*([b,s]), \text{ for all } [b,s] \in [r,1] \odot [m,1].$$

Hence, we obtain

$$h(\gamma^*(r) \diamond \varepsilon^*(m)) = \gamma^*(r) * h(\varepsilon^*(m)).$$

Therefore, h is a homomorphism of modules. ∎

5.3 Direct system and direct limit of H_v-modules

The construction of the direct system and direct limit is similar to the usual module theory (see [58, 79, 91]). In [58], Ghadiri and Davvaz considered the category of H_v-modules and prove that the direct limit always exists in this category. Direct limits are defined by a universal property, and so are unique. Also, already Leoreanu [78, 79] and Romeo [88] studied the notions of direct limits of hyperstructures.

A partially ordered set I is said to be a direct set, if for each $i, j \in I$ there exists $k \in I$ such that $i \leq k$ and $j \leq k$. Let I be a direct set and ϑ the category of H_v-modules. Let $(M_i)_{i \in I}$ be a family of H_v-modules indexed by I. For each pair $i, j \in I$ such that $i \leq j$, let $\phi_j^i : M_i \to M_j$ be a homomorphism and suppose that the following axioms are satisfied:

(1) ϕ_i^i is the identity for all $i \in I$,

(2) $\phi_k^i = \phi_k^j \phi_j^i$ whenever $i \leq j \leq k$.

Then, the H_v-modules M_i and strong H_v-homomorphisms ϕ_j^i are said to be a *direct system* $M = (M_i, \phi_j^i)$ over the direct set I. Let $M = (M_i, \phi_j^i)$ be a direct system in ϑ. The *direct limit* of this system, denoted by $\varinjlim M_i$, is an H_v-module and a family of strong H_v-homomorphisms $\alpha_i : M_i \to \varinjlim M_i$, with $\alpha_i = \alpha_j \phi_j^i$ whenever $i \leq j$ satisfying the following universal mapping property:

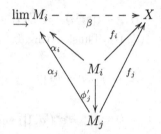

for every H_v-module X and every family of strong H_v-homomorphism $f_i : M_i \to X$ with $f_i = f_j \phi_j^i$, whenever $i \leq j$, there is a unique strong homomorphism $\beta : \varinjlim M_i \longrightarrow X$ making the above diagram commute.

Let X be the disjoint union $\bigcup M_i$. We define an equivalence relation on X by

$a_i \rho a_j$, $a_i \in M_i$, $a_j \in M_j \Leftrightarrow$ there exists an index $k \geq i, j$ with $\phi_k^i a_i = \phi_k j a_j$.

The equivalence class of a_i is denoted by $\rho(a_i)$. Suppose that X/ρ is the set of all equivalent classes. It is clear that $a_1 \rho \phi_j^i a_j$ for $j \geq i$. Now, for $r \in R$ and $\rho(a_i), \rho(a_j) \in X/\rho$, we define

$\rho(a_i) \oplus \rho(a_j) = \{\rho(x) \mid a_k + a_k', \text{ where } a_k = \phi_k^i a_i, \ a_k' = \phi_k^j a_j, \text{ for some } k \geq i, j\}$,

$r \circ \rho(a_i) = \{\rho(x) \mid x \in r a_i\}$.

Proposition 5.1. $(X/\rho, \oplus, \circ)$ *is an H_v-module over R.*

Proof. It is straightforward. ∎

Theorem 5.6. *Let (M_i, ϕ_j^i) be a direct system of H_v-modules indexed by I. Then, the H_v-module X/ρ is $\varinjlim M_i$.*

Proof. We define $\alpha_i : M_i \to X/\rho$ by $a_i \mapsto \rho(a_i)$ and consider the following diagram.

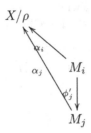

Then, $\alpha_j(\phi^i_j a_i) = \rho(\phi^i_j a_i) = \alpha_i(a_i)$. Thus, $\alpha_j \phi^i_j = \alpha_i$, and so the diagram is commutative. Now, suppose that M is an H_v-module and $\{f_i \mid f_i : M_i \to M\}$ be a family of strong H_v-homomorphisms with $f_i = f_j \phi^i_j$. Define $\zeta : X/\rho \to M$ by $\rho(a_i) \mapsto f_i a_i$. We show that ζ is a strong H_v-homomorphism and so the universal mapping property holds. First, we show that *zeta* is well defined. Suppose that $\rho(a_i) = \rho(b_j)$. Then, there exists $k \geq i, j$ such that $\phi^i_k a_i = \phi^j_k b_j$. Hence, $f_k \phi^i_k a_i = f_k \phi^j_k b_j$ which implies that $f_i a_i = f_j b_j$. Therefore, *zeta* is well defined. Now, let $\rho(a_i), \rho(b_j) \in X/\rho$ and $r \in R$. Then, we obtain

$\zeta(\rho(a_i) \oplus \rho(b_j))$
$= \{\zeta(\rho(x)) \mid x \in a_k + a'_k,$ where $a_k = \phi^i_k a_i,\ a'_k = \phi^j_k b_j\ for\ some\ k \geq i, j\}$
$= \{f_k x \mid x \in a_k + a'_k,$ where $a_k = \phi^i_k a_i,\ a'_k = \phi^j_k b_j\ for\ some\ k \geq i, j\}$
$= f_k(a_k + a'_k),$ where $a_k = \phi^i_k a_i,\ a'_k = \phi^j_k b_j$
$= f_k a_k + f_k a'_k$
$= f_k \phi^i_k a_i + f_k \phi^j_k b_j$
$= f_i a_i + f_j b_j$
$= \zeta(\rho(a_i)) + \zeta(\rho(b_i))$

and

$$\zeta(r \circ \rho(a_i)) = \zeta(\rho(ra_i)) = f_i(ra_i) = r\zeta(\rho(a_i)).$$

Therefore, ζ is a strong H_v-homomorphism and $\zeta \alpha_i = f_i$. ∎

Lemma 5.4. *Let M_1, M_2 be two H_v-modules over an H_v-ring R and $f : M_1 \to M_2$ be a strong H_v-homomorphism. Let $\varepsilon^*_1, \varepsilon^*_2$ and γ^* be the fundamental relations on M_1, M_2 and R. Then, the map $F : M_1/\varepsilon^*_1 \to M_2/\varepsilon^*_2$ defined by $F(\varepsilon^*_1) = \varepsilon^*_2(f(a))$ is an R/γ^*-homomorphism of modules.*

Proof. By hypotheses M_1/ε^*_1 and M_2/ε^*_2 are modules. First, we show that F is well defined. If $\varepsilon^*_1(a) = \varepsilon^*_1(b)$, then there exist $x_1, ..., x_{m+1} \in M_1$

and $u_1, ..., u_m \in \mathcal{U}_{M_1}$ with $x_1 = a$, $x_{m+1} = b$ such that $\{x_i, x_{i+1}\} \subseteq u_i$, $i = 1, ..., m$. Since f is a strong H_v-homomorphism, it follows that $f(u_i) \in \mathcal{U}_{M_2}$. Therefore, $f(a)\varepsilon_2^* f(b)$ which implies that $\varepsilon_2^*(f(a)) = \varepsilon_2^*(f(b))$, and so $F(\varepsilon_1^*(a)) = F(\varepsilon_1^*(b))$. Thus, F is well defined. Now, we have

$$F(\varepsilon_1^*(a) \oplus \varepsilon_1^*(b)) = F(\varepsilon_1^*(a + b)) = \varepsilon_2^*(f(a + b)) = \varepsilon_2^*(f(a) + f(b))$$
$$= \varepsilon_2^*(f(a)) \oplus \varepsilon_2^*(f(b) = F(\varepsilon_1^*(a)) \oplus F(\varepsilon_1^*(b))$$

and

$$F(\gamma^* \odot \varepsilon_1^*(a)) = F(\varepsilon_1^*(ra)) = \varepsilon_2^*(f(ra)) = \varepsilon_2^*(rf(a)) = \gamma^* \odot F(\varepsilon_1^*(a)). \blacksquare$$

Proposition 5.2. *Let (M_i, ϕ_j^i) be a direct system of H_v-modules over an H_v-ring R indexed by a directed set I. Then, $(M_i/\varepsilon_{M_i}^*, \phi_j^{i*})$ is a direct system of modules over the ring R/γ^*, where*

$$\phi_j^{i*} : M_i/\varepsilon_{M_i}^* \to M_j/\varepsilon_{M_j}^*$$
$$\varepsilon_{M_i}^*(a_i) \mapsto \varepsilon_{M_j}^*(\phi_j^i a_i).$$

Proof.   By Lemma 5.4, $(M_i/\varepsilon_{M_i}^*, \phi_j^{*i})$ is a family of R/γ^*-modules and R/γ-homomorphisms. It is clear that ϕ_i^{i*} is the identity for all $i \in I$. Now, for $i \leq j \leq k$, we have

$$(\phi_k^j \phi_j^i)^*(\varepsilon_{M_i}^*(a_i)) = \phi_k^{i*}(\varepsilon_{M_i}^*(a_i)) = \varepsilon_{M_k}^*(\phi_k^i a_i) = \varepsilon_{M_k}^*(\phi_k^j \phi_j^i)a_i)$$
$$= \varepsilon_{M_k}^*(\phi_k^j(\phi_j^i a_i)) = \phi_k^{j*}(\varepsilon_{M_j}^*(\phi_j^i a_i)) = \phi_k^{j*}\phi_j^{i*}(\varepsilon_{M_i}^*(a_i)).$$

Therefore, $(\phi_k^j \phi_j^i)^* = \phi_k^{i*} = \phi_k^{j*}\phi_j^{i*}.$ ∎

Proposition 5.3. *Let $\rho(a_i)$, $\rho(b_j) \in X/\rho$. We define*

$$\rho(a_i) \; \theta \; \rho(b_j) \iff \text{there exists } k \geq i, j \text{ such that } \phi_k^i a_i \varepsilon_{M_k} \phi_k^j b_j.$$

Then, $\theta = \varepsilon_{X/\rho}$.

Proof.   Suppose that $\rho(a_i) \; \varepsilon_{X/\rho} \; \rho(b_j)$. Then, there exist $r_1, ..., r_n \in R$ and $t_1, ..., t_m \in \bigcup M_i$ such that $\{\rho(a_i), \rho(b_j)\} \subseteq r_1\rho(t_1) + ... + r_m\rho(t_m)$. Suppose that $t_1 \in M_{i_1}, ..., t_m \in M_{i_m}$ and $k \geq i_1, ..., i_m, i, j$. Hence,

$$r_1\rho(t_1) + ... + r_m\rho(t_m) = r_1\rho(\phi_k^{i_1} t_1) + ... + r_m\rho(\phi_k^{i_m} t_m)$$
$$= \{\rho(x) \mid x \in \phi_k^{i_1} r_1 t_1 + ... + \phi_k^{i_m} r_m t_m\},$$

and so $\{\rho(\phi_k^i a_i), \rho(\phi_k^j b_j)\} \subseteq \{\rho(x) \mid x \in \phi_k^{i_1} r_1 t_1 + ... + \phi_k^{i_m} r_m t_m\}$. Now, for $n \geq k$, we obtain $\{\phi_n^i a_i, \phi_n^j b_j\} \subseteq \phi_n^{i_1} r_1 t_1 + ... + \phi_n^{i_m} r_m t_m$ which implies that $\phi_n^i a_i \varepsilon_{M_n} \phi_n^j b_j$. Hence, $\rho(a_i) \; \theta \; \rho(b_j)$.

Conversely, if $\rho(a_i) \; \theta \; \rho(b_j)$, then there exists $k \geq i, j$ such that

$\phi_k^i a_i \varepsilon_{M_k} \phi_k^j b_j$. So, there exist $t_1, ..., t_m \in M_k \subseteq \bigcup M_i$ and $r_1, ..., r_m \in R$ such that

$$\{\phi_k^i a_i, \phi_k^j b_j\} \subseteq r_1 t_1 + ... + r_m t_m.$$

Thus, $\{\rho(\phi_k^i a_i), \rho(\phi_k^j b_j)\} \subseteq r_1 \rho(t_1) + ... + r_m \rho(t_m)$ which implies that $\rho(a_i) \ \varepsilon_{X/\rho} \ \rho(b_j)$.

Therefore, $\theta = \varepsilon_{X/\rho}$. ∎

Theorem 5.7. *Let* (M_i, ϕ_j^i) *be a direct system of* H_v*-modules over an* H_v*-ring* R *indexed by a directed set* I, *and let* ε^* *be the fundamental relation of* $\varinjlim M_i$. *Then,*

$$\varinjlim(M_i/\varepsilon_{M_i}^*) \cong (\varinjlim M_i)/\varepsilon^*.$$

Proof. It is straightforward. ∎

5.4   M[-] and -[M] Functors

Vaziri, Ghadiri and Davvaz introduced the H_v-module $M[A]$ and determined it's heart. They studied the connection between equivalence relations $\varepsilon_{M[A]}^*$ and ε_A^*. Moreover, they introduced $M[-]$ and $-[M]$ functors. The main reference for this section is [109, 110].

Let $f : A \to B$ be a strong H_v-homomorphism of H_v-modules over an H_v-ring R. Then, $F : A/\varepsilon_A^* \to B/\varepsilon_B^*$ where $F(\varepsilon_A^*(a)) = \varepsilon_B^*(f(a))$ is an R/γ^*-homomorphism of R/γ^*-modules. Let R be a weak-commutative H_v-ring and **H** be the set of all H_v-modules and all strong R-homomorphisms. One can show that **H** is a category. Also, set **H*** the category of R/γ^*-modules and R/γ^*-homomorphisms. Then we have the following theorem.

Theorem 5.8. *Let* $T :$ **H** $\to$ **H***, *defined by* $T(M_1) = M_1/\varepsilon_1^*$ *and when* $f : M_1 \to M_2$ *is a strong* H_v*-homomorphism*

$$T(f) : M_1/\varepsilon_1^* \to M_2/\varepsilon_2^*,$$
$$\varepsilon_1^*(a) \mapsto \varepsilon_2^*(f(a))$$

where ε_1^* *and* ε_2^* *are fundamental relations on* M_1 *and* M_2, *respectively. Then,* T *is a functor, and is called natural functor.*

Proof. It is easy to see that T is well defined. We have the following:

(1) If $M_1 \in obj$ **H**, then $M_1/\varepsilon_1^* \in obj$ **H***;

(2) If $f : M_1 \to M_2$ is a strong H_v-homomorphism, then by Lemma 5.4, $F : M_1/\varepsilon_1^* \to M_2/\varepsilon_2^*$ is an R/γ^*-homomorphism of modules.

(3) Suppose that ε_3^* is the fundamental relation on M_3. If $M_1 \xrightarrow{f} M_2 \xrightarrow{g} M_3$ is an exact sequence of H_v-modules and strong H_v-homomorphisms in **H**, then

$$T(gf)(\varepsilon_1^*(a)) = \varepsilon_3^*(gf(a)) = T(g)\varepsilon_2^*(f(a))$$
$$= T(g)T(f)(\varepsilon_1^*(a)),$$

so $T(gf) = T(g)T(f)$;

(4) For every $M_1 \in obj$ **H**, we have

$$T(1_{M_1})(\varepsilon_1^*(a)) = \varepsilon_1^*(1_{M_1}(a)) = \varepsilon_1^*(a),$$

so $T(1_{M_1}) = 1_{T(M_1)}$.

Therefore, T is a functor.   ∎

Now, we want to introduce $M[-]$ and $-[M]$ functors and investigate some related concepts.

Suppose that M and N are two H_v-modules and $M[N]$ is the set of all functions on M with values in N. First we equip $M[N]$ to appropriate hyperoperations to be an H_v-module. Then, we introduce the functors $M[-]$ and $-[M]$ and investigate some related concepts. Throughout this paper, the hyperoperations in M, N and $M[N]$ will be shown with same symbols.

Theorem 5.9. *The $M[N]$ with the following hyperoperations is an H_v-module.*

$$f + g = \{h \in M[N] \mid h(x) \in f(x) + g(x), \ \forall x \in M\},$$
$$r \cdot f = \{k \in M[N] \mid k(x) \in r \cdot f(x), \ \forall x \in M\}.$$

Proof. The hyperoperations $+$ and $\cdot$ in $M[N]$ are well-define, for $+$ and $\cdot$ in N are well-define. Let $f, g, h \in M[N]$. We have

$$(f+g)+h = \{l \in M[N] \mid l(x) \in f(x)+g(x), \ \forall x \in M\}+h = \bigcup_{l \in f+g} l+h$$
$$= \{L \in M[N] \mid L(x) \in l(x)+h(x), \ \forall x \in M, \ l(x) \in f(x)+g(x)\}$$

and

$$f+(g+h) = f+\{k \in M[N] \mid k(x) \in g(x)+h(x), \ \forall x \in M\} = \bigcup_{k \in g+h} f+k$$
$$= \{K \in M[N] \mid K(x) \in f(x)+k(x), \ \forall x \in M, \ k(x) \in g(x)+h(x)\}.$$

Since N is an H_v-group, for all $x \in M$ there exists $n_x \in [(f(x) + g(x)) + h(x)] \cap [f(x) + (g(x) + h(x))]$. We define $u \in M[N]$ by $u(x) = n_x$, according to the choice axiom. Then, $u \in [(f+g)+h] \cap [f+(g+h)]$ and associativity is satisfied.

For the reproduction axiom let $f, g \in M[A]$. Then, for all $x \in M$, $f(x), g(x) \in N$ and so, there exists $y_x \in N$ such that $f(x) \in g(x) + y_x$. We define $h \in M[N]$ by $h(x) = y_x$, then $f \in g + h$. Similarly, there exists $h' \in M[N]$ such that $f \in h' + g$. Since N is an H_v-module, the conditions of H_v-modules satisfy in $M[N]$. We check only one of the H_v-module conditions. Let $r_1, r_2 \in R$ and $f \in M[N]$. Since N is an H_v-module, it follows that for every $x \in M$ there exists $n_x \in [(r_1 + r_2)f(x)] \cap [r_1 f(x) + r_2 f(x)]$. We define $h \in M[N]$ by $h(x) = n_x$. Obviously, $h \in [(r_1 + r_2)f] \cap [(r_1 f + r_2 f)] \neq \emptyset$. ∎

Lemma 5.5. *Let $f : A \to B$ be a strong H_v-homomorphism and M be an H_v-module. Then,*

(1) *The map $\overline{f} \colon M[A] \to M[B]$ defined by $\overline{f}(\phi) = f \circ \phi$ is a strong H_v-homomorphism.*

(2) *The map $\overline{f} \colon B[M] \to A[M]$ defined by $\overline{f}(\phi) = \phi \circ f$ is a strong H_v-homomorphism.*

Proof. (1) Let $\phi_1, \phi_2 \in M[A]$. Then,

$$\overline{f}(\phi_1 + \phi_2) = \{ f \circ h \mid h \in M[A],\ h(m) \in \phi_1(m) + \phi_2(m), \forall m \in M \},$$

$$\overline{f}(\phi_1) + \overline{f}(\phi_2) = f \circ \phi_1 + f \circ \phi_2 = \{ h' \in M[B] \mid h'(m) \in f \circ \phi_1(m) + f \circ \phi_2(m) \}.$$

Suppose that $f \circ h \in \overline{f}(\phi_1 + \phi_2)$, where $h \in M[A]$ and $h(m) \in \phi_1(m) + \phi_2(m)$ for every $m \in M$. Then, $f(h(m)) \in f(\phi_1(m) + \phi_2(m)) = f(\phi_1(m)) + f(\phi_2(m))$. Therefore, $\overline{f}(\phi_1 + \phi_2) \subseteq \overline{f}(\phi_1) + \overline{f}(\phi_2)$.

Conversely, suppose that $h' \in \overline{f}(\phi_1) + \overline{f}(\phi_2)$. We need to find an $h \in M[A]$ such that $h' = f \circ h$ and $h(m) \in \phi_1(m) + \phi_2(m)$. By hypothesis for $m \in M$, we have

$$h'(m) = b_m \in f \circ \phi_1(m) + f \circ \phi_2(m) = f(\phi_1(m) + \phi_2(m)) \subseteq Im(f).$$

So, $b_m \in f(\phi_1(m) + \phi_2(m))$. Now, according to the choice axiom we can select $a \in f^{-1}(b_m)$ such that $a \in \phi_1(m) + \phi_2(m)$ and define $h(m) = a$.

Similarly, one can show that $\overline{f}(r\phi) = r\,\overline{f}(\phi)$.

(2) Let $\phi_1, \phi_2 \in B[M]$. Then,

$$\overline{f}(\phi_1 + \phi_2) = \{ h \circ f \mid h \in B[M], h(b) \in \phi_1(b) + \phi_2(b) \},$$

$$\overline{f}(\phi_1) + \overline{f}(\phi_2) = \phi_1 \circ f + \phi_2 \circ f = \{ h' \in A[M] \mid h'(a) \in \phi_1 \circ f(a) + \phi_2 \circ f(a) \}.$$

Suppose that $h \circ f \in \overline{f}$ $(\phi_1 + \phi_2)$, where $h \in B[M]$ and $h(b) \in \phi_1(b) + \phi_2(b)$ for every $b \in B$. Since $Im(f) \subseteq B$, we have $h(f(a)) \in \phi_1(f(a)) + \phi_2(f(a))$ for every $a \in A$. Therefore, $\overline{f}$ $(\phi_1 + \phi_2) \subseteq \overline{f}$ $(\phi_1) + \overline{f}$ (ϕ_2).

Conversely, suppose that $h' \in \overline{f}$ $(\phi_1) + \overline{f}$ (ϕ_2). We need to find an $h \in B[M]$ such that $h' = h \circ f$ and $h(b) \in \phi_1(b) + \phi_2(b)$. For every $b \in Im(f) \subseteq B$ we define $h(b) = h'(a)$, where $f(a) = b$ and for every $b \in B \backslash Im(f)$ according to the choice axiom we select an m_b in $\phi_1(b) + \phi_2(b) \subseteq M$ and define $h(b) = m_b$. Then, h satisfies the requirement conditions.

Similarly, one can show that $\overline{f}$ $(r\phi) = r\, \overline{f}$ (ϕ).   ∎

Lemma 5.6. *Let M be an H_v-module and $f : A \to B$ be a morphism in the category $\mathbf{H}$. Then,*

(1) $M[-]$ $:\mathbf{H} \to \mathbf{H}$ *define by* $M[-](A)$ $=$ $M[A]$ *and* $M[-](f)$ $=$ $\overline{f} \colon M[A] \to M[B]$, *where* $\overline{f}$ $(\phi) = f \circ \phi$, *is a covariant functor.*

(2) $-[M]$ $:\mathbf{H} \to \mathbf{H}$ *define by* $-[M](A)$ $=$ $A[M]$ *and* $-[M](f)$ $=$ $\overline{f} \colon B[M] \to A[M]$, *where* $\overline{f}$ $(\phi) = \phi \circ f$, *is a contravariant functor.*

Proof.   (1) By Theorem 5.9 if A is an H_v-module, then $M[-](A) = M[A]$ is an H_v-module. By Lemma 5.5 if $f : A \to B$ is a strong H_v-homomorphism, then $M[-](f) = \overline{f}$ is a strong H_v-homomorphism. Now, let $A \xrightarrow{f} B \xrightarrow{g} C$ be a strong H_v-homomorphism in $\mathbf{H}$. Then,

$$M[-](g \circ f)(\phi) = g \circ f \circ \phi = g(f \circ \phi) = M[-](g)(f \circ \phi) = M[-](g) \circ M[-](f)(\phi)$$

and for every $A \in obj\mathbf{H}$ we have $M[-](1_A)(\phi) = 1_A \circ \phi = \phi$. Then, $M[-](1_A) = 1_{M[-](A)}$ and so, $M[-]$ is a covariant functor.

(2) By Theorem 5.9 if A is an H_v-module, then $-[M](A) = A[M]$ is an H_v-module. By Lemma 5.5 if $f : A \to B$ is a strong H_v-homomorphism, then $-[M](f) = \overline{f}$ is a strong H_v-homomorphism. Now, let $A \xrightarrow{f} B \xrightarrow{g} C$ be a strong H_v-homomorphism in $\mathbf{H}$. Then,

$$-[M](g \circ f)(\phi) = \phi \circ g \circ f = (\phi \circ g)f = -[M](f)(\phi \circ g) = -[M](f) \circ -[M](g)(\phi),$$

and for every $A \in obj\mathbf{H}$ we have $-[M](1_A)(\phi) = \phi \circ 1_A = \phi$. Then, $-[M](1_A) = 1_{-[M](A)}$ and so, $-[M]$ is a contravariant functor.   ∎

Lemma 5.7. *Let*

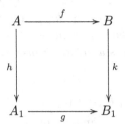

be a commutative diagram of H_v-modules and strong H_v-homomorphisms. Then, the following diagrams are commutative.

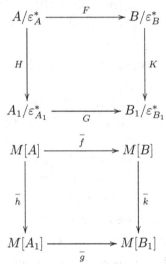

Proof. We have $T(A) = A/\varepsilon_A^*$ and $T(f : A \to B) = F : A/\varepsilon^* \to B/\varepsilon_B^*$, where $F(\varepsilon_A^*(a)) = \varepsilon_B^*(f(a))$. Therefore,

$$K \circ F = T(k) \circ T(f) = T(k \circ f) = T(g \circ h) = T(g) \circ T(h) = G \circ H.$$

We have $M[-](A) = M[A]$, $M[-](f : A \to B) = \bar{f}: M[A] \to M[B]$ where $\bar{f}(\phi) = f \circ \phi$. Therefore,

$$\bar{k} \circ \bar{f} = M[-](k) \circ M[-](f) = M[-](k \circ f) = M[-](g \circ h)$$
$$= M[-](g) \circ M[-](h) = \bar{g} \circ \bar{h}. \qquad \blacksquare$$

We know that the combination of two covariant functors is a covariant functor. Therefore, the map $S = T \circ M[-]:\mathbf{H} \to \mathbf{H^*}$ is a covariant functor, where

$$S(A) = M[A]/\varepsilon_{M[A]}^* \text{ and } S(f : A \to B) = \bar{F}: M[A]/\varepsilon_{M[A]}^* \to M[B]/\varepsilon_{M[B]}^*,$$

where $\bar{F}(\varepsilon_{M[A]}^*(\phi)) = \varepsilon_{M[B]}^*(f \circ \phi)$.

Lemma 5.8. *For every $A \in objH$, $\tau_A : T(A) \to S(A)$ defined by $\tau_A(\varepsilon_A^*(a)) = \varepsilon_{M[A]}^*(\phi_a)$ is a R/γ^*-homomorphism, where $\phi_a : M \to A$ defined by $\phi_a(m) = a$ for every $m \in M$. Then, the family $\tau = (\tau_A : T(A) \to S(A))_{A \in objH}$ is a natural transformation from T to S.*

Proof. We have

$$\tau_A(\varepsilon_A^*(a) \oplus \varepsilon_A^*(b)) = \tau_A(\varepsilon_A^*(a+b)) = \varepsilon_{M[A]}^*(\phi_t),$$

where $t \in a + b$. On the other hand, we obtain

$$\begin{aligned}
\tau_A(\varepsilon_A^*(a)) \oplus \tau_A(\varepsilon_A^*(b)) &= \varepsilon_{M[A]}^*(\phi_a) \oplus \varepsilon_{M[A]}^*(\phi_b) = \varepsilon_{M[A]}^*(\phi_a + \phi_b) \\
&= \varepsilon_{M[A]}^*(\{\phi \in M[A] \mid \phi(m) \in \phi_a(m) + \phi_b(m), \forall m \in M\}) \\
&= \varepsilon_{M[A]}^*(\{\phi \in M[A] \mid \phi(m) \in a+b, \forall m \in M\}) \\
&= \varepsilon_{M[A]}^*(\phi_t),
\end{aligned}$$

where $t \in a + b$. Therefore, $\tau_A(\varepsilon_A^*(a) \oplus \varepsilon_A^*(b)) = \tau_A(\varepsilon_A^*(a)) \oplus \tau_A(\varepsilon_A^*(b))$. Similarly, we have

$$\begin{aligned}
\tau_A(\gamma^*(r) \odot \varepsilon_A^*(a)) &= \tau_A(\varepsilon_A^*(d)), \text{ for some } d \in \gamma^*(r) \cdot \varepsilon_A^*(a) \\
&= \varepsilon_{M[A]}^*(\phi_d), \text{ for some } d \in r \cdot a
\end{aligned}$$

and

$$\begin{aligned}
\gamma^*(r) \odot \tau_A(\varepsilon_A^*(a)) &= \gamma^*(r) \odot \varepsilon_{M[A]}^*(\phi_a) \\
&= \varepsilon_{M[A]}^*(h) \text{ for some } h \in r \cdot \phi_a \\
&= \varepsilon_{M[A]}^*(h),
\end{aligned}$$

where for every $m \in M$, $h(m) \in r \cdot \phi_a(m) = r \cdot a$. Therefore,

$$\tau_A(\gamma^*(r) \odot \varepsilon_A^*(a)) = \gamma^*(r) \odot \tau_A(\varepsilon_A^*(a)).$$

Now, let $f : A \to B$ be a morphism in H and consider the following diagram.

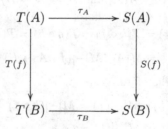

We have

$$\begin{aligned}
S(f) \circ \tau_A(\varepsilon_A^*(a)) &= S(f)(\varepsilon_{M[A]}^*(\phi_a)) = \varepsilon_{M[B]}^*(f \circ \phi_a), \\
\tau_B \circ T(f)(\varepsilon_A^*(a)) &= \tau_B(\varepsilon_B^*(f(a))) = \varepsilon_{M[B]}^*(\phi_{f(a)}).
\end{aligned}$$

Obviously, $f \circ \phi_a = \phi_{f(a)}$ and so, $S(f) \circ \tau_A = \tau_B \circ T(f)$ and $\tau : T \to S$ is a natural transformation. ∎

Lemma 5.9. *Let H_1 and H_2 be two H_v-modules. Then, $H_1 \times H_2$ is a product object in $\mathbf{H}$ category.*

Proof. The proof is straightforward. ∎

Notice that Lemma 5.9 can be generalized to the cartesian product of n arbitrary H_v-modules.

Theorem 5.10. *Let M be an H_v-module. Then,*

$$M[H_1 \times H_2] \cong M[H_1] \times M[H_2].$$

Proof. It is easy to see that the map $\phi : M[H_1] \times M[H_2] \to M[H_1 \times H_2]$ defined by $\phi(f_1, f_2) = f : M \to H_1 \times H_2$, where $f(m) = (f_1(m), f_2(m))$ is well-defined. Now, we have

$$\phi((f_1, g_1) + (f_2, g_2)) = \phi(\{(f, g) \mid f \in f_1 + f_2, g \in g_1 + g_2\})$$
$$= \{h \mid h(m) = (f(m), g(m)), f(m) \in f_1(m) + f_2(m), g(m) \in g_1(m) + g_2(m)\}.$$

On the other hand, we have

$$\phi((f_1, g_1)) = h \in M[H_1 \times H_2] \text{ such that } h(m) = (f_1(m), g_1(m)),$$
$$\phi((f_2, g_2)) = k \in M[H_1 \times H_2] \text{ such that } k(m) = (f_2(m), g_2(m)).$$

And

$$h + k = \{l \mid l(m) \in h(m) + k(m) = (f_1(m), g_1(m)) + (f_2(m), g_2(m))\}$$
$$= \{l \mid l(m) = (f(m), g(m)), f(m) \in f_1(m) + f_2(m), g(m) \in g_1(m) + g_2(m)\}.$$

Therefore, $\phi((f_1, g_1) + (f_2, g_2)) = \phi((f_1, g_1)) + \phi((f_2, g_2))$.
Similarly, one can show that $\phi(r(f, g)) = r\phi((f, g))$.
Now, let $f \in M[H_1 \times H_2]$ where $f(m) = (h_{1m}, h_{2m})$. We define $f_1 \in M[H_1]$ by $f(m) = h_{1m}$ and $f_2 \in M[H_2]$ by $f(m) = h_{2m}$. Obviously, $\phi((f_1, f_2)) = f$.
Finally, suppose that $\phi((f_1, f_2)) = \phi((g_1, g_2))$. Then, for every $m \in M$, we obtain $(f_1(m), f_2(m)) = (g_1(m), g_2(m))$ and so, $(f_1, f_2) = (g_1, g_2)$. ∎

Notice that in finite mode in Theorem 5.10 we have

$$|M[H_1] \times M[H_2]| = |M[H_1]| \times |M[H_2]| = |H_1|^{|M|} \times |H_2|^{|M|}$$
$$= |H_1 \times H_2|^{|M|} = |M[H_1 \times H_2]|.$$

So, it is enough to show that ϕ is one to one or onto.

Corollary 5.4. *Let $M, H_1, H_2, \ldots, H_n$ be H_v-modules. Then,*

$$M[H_1 \times H_2 \times H_3 \times \cdots \times H_n] \cong M[H_1] \times M[H_2] \times M[H_3] \times \cdots \times M[H_n].$$

5.5   Five short lemma and snake lemma in H_v-modules

In this section, we investigate the five short lemma and snake lemma in H_v-modules. The main reference is [109, 110].

Lemma 5.10. *If* $\varepsilon^*_{M[A]}(f) = \varepsilon^*_{M[A]}(g)$, *then* $\varepsilon^*_A(f(m)) = \varepsilon^*_A(g(m))$, *for every* $m \in M$; *i.e. if for some* $m \in M$, $\varepsilon^*_A(f(m)) \neq \varepsilon^*_A(g(m))$ *then* $\varepsilon^*_{M[A]}(f) \neq \varepsilon^*_{M[A]}(g)$.

Proof.   Suppose that $f \; \varepsilon^*_{M[A]} \; g$. Then, there exist $f_0 = f, f_1, \cdots, f_n = g$ in $M[A]$ such that $f_i \; \varepsilon_{M[A]} \; f_{i+1}$ for $i = 0, 1, \ldots, n-1$. So, $\{f_i, f_{i+1}\} \subseteq$

$$\sum_{j=1}^{n_i} g'_{ij}, \text{ for } i = 0, 1, \ldots, n-1, \text{ where } g'_{ij} = g_{ij} \text{ or } g'_{ij} = \sum_{k=1}^{n_{ij}} (\prod_{l=1}^{l_{ijk}} r_{ijkl}) g_{ij} \text{ for}$$

$g_{ij} \in M[A]$ and $r_{ijkl} \in R$. Now, since

$$\sum_{j=1}^{n_i} g'_{ij} = \{h \in M[N] \mid h(m) \in g'_{i1}(m) + g'_{i2}(m) + \cdots + g'_{in_i}(m), \; \forall m \in M\},$$

we have $\{f_i(m), f_{i+1}(m)\} \subseteq \sum_{j=1}^{n_i} g'_{ij}(m)$ for every $m \in M$ and so, there exist $a_0 = f_0(m) = f(m), a_1 = f_1(m), \ldots, a_n = f_n(m) = g(m) \in A$ such that a_i $\varepsilon_A \; a_{i+1}$, for $i = 0, 1, \ldots, n-1$. Therefore, for every $m \in M$, we have $f(m)$ $\varepsilon^*_A \; g(m)$.   ∎

In the following example we show that the converse of Lemma 5.10 is not true in general.

Example 5.2. Consider $f, g \in M[M_1 \times M_2]$ as in Example 5.1 and define $f(a) = (2,\bar{2})$, $f(b) = (1,\bar{0})$ and $g(a) = (0,\bar{1})$, $g(b) = (1,\bar{2})$. By Example 5.1 we have $\varepsilon^*_{M_1 \times M_2}(f(a)) = \varepsilon^*_{M_1 \times M_2}(g(a))$ and $\varepsilon^*_{M_1 \times M_2}(f(b)) = \varepsilon^*_{M_1 \times M_2}(g(b))$. Since for every $r \in R$ and every $m_1 \in M_1$, $rm_1 = \{0\}$ and on the other hand, for every two elements m_2 and m'_2 of M_2, $m_2 *_{M_2} m'_2$ is singleton, it follows $\varepsilon^*_{M[M_1 \times M_2]}(f) \neq \varepsilon^*_{M[M_1 \times M_2]}(g)$.

In the following lemma, we determine the heart of $M[A]$.

Lemma 5.11. *Let* M *and* A *be two* H_v-*modules. Then,* $\omega_{M[A]} = M[\omega_A]$.

Proof.   Suppose that $f \in \omega_{M[A]}$. Then, for every $g \in M[A]$ we have

$$\varepsilon^*_{M[A]}(g) = \varepsilon^*_{M[A]}(f) \oplus \varepsilon^*_{M[A]}(g) \left(= \varepsilon^*_{M[A]}(f+g) \right).$$

Now, by Lemma 5.10 for every $m \in M$ we obtain

$$\varepsilon^*_A((f+g)(m)) = \varepsilon^*_A(g(m)).$$

But for every $m \in M$, we have $(f+g)(m) = \{l(m) \mid l \in f+g\} = f(m) + g(m)$. So,

$$\varepsilon_A^*((f+g)(m)) = \varepsilon_A^*(f(m) + g(m)) = \varepsilon_A^*(f(m)) \oplus \varepsilon_A^*(g(m)) = \varepsilon_A^*(g(m)).$$

Therefore, for every $m \in M$, we obtain $\varepsilon_A^*(f(m)) \in \omega_A$ and so, $f \in M[\omega_A]$.

Conversely, suppose that $f \in M[\omega_A]$. Then, for every $g \in M[A]$ and all $m \in M$, we have

$$\varepsilon_A^*(f(m) + g(m)) = \varepsilon_A^*(f(m)) \oplus \varepsilon_A^*(g(m)) = \varepsilon_A^*(g(m)).$$

So, for every $g \in M[A]$ and all $m \in M$, we have $f(m) + g(m) \in \varepsilon_A^*(g(m))$ and we obtain

$$\varepsilon_{M[A]}^*(f) \oplus \varepsilon_{M[A]}^*(g) = \varepsilon_{M[A]}^*(f+g) = \varepsilon_{M[A]}^*(\{l \mid l(m) \in f(m)+g(m)\}) = \varepsilon_{M[A]}^*(g)$$

and consequently $f \in \omega_{M[A]}$. ∎

In the following, we want to investigate the exactness of $-[M]$ and $M[-]$ functors.

Let $A \xrightarrow{f} B \xrightarrow{g} C$ be an exact sequence. Then, for every $a \in A$, we have $f(a) \in Im(f) \overset{w}{=} Ker(g)$ and so, $\varepsilon_B^*(f(a)) = \varepsilon_B^*(b)$ for some $b \in Ker(g)$. Now, we obtain

$$\varepsilon_C^*(g(f(a))) = G(\varepsilon_B^*(f(a))) = G(\varepsilon_B^*(b)) = \varepsilon_C^*(g(b)) = \omega_C.$$

Therefore, for every $a \in A$ we have $g(f(a)) \in \omega_C$.

Now, by considering $-[M]$ functor on the exact sequence $A \xrightarrow{f} B \xrightarrow{g} C$, we obtain

$$C[M] \xrightarrow{\bar{g}} B[M] \xrightarrow{\bar{f}} A[M].$$

We want to check the exactness of this sequence. We have

$$Im(\bar{g}) = \{\bar{g}\ (\phi) \mid \phi \in C[M]\} = \{\phi \circ g \mid \phi \in C[M]\},$$
$$Ker(\bar{f}) = \{\psi \in B[M] \mid \bar{f}\ (\psi) = \psi \circ f \in \omega_{A[M]} = A[\omega_M]\}.$$

Let ϕ be a function in $C[M]$ such that $\phi(\omega_C) \cap \omega_M = \emptyset$ (note that it is necessary to $\omega_M \neq M$). Then, for every $a \in A$ since $g \circ f(a) \in \omega_C$ and $\phi(\omega_C) \cap \omega_M = \emptyset$, we obtain $\varepsilon_M^*(\phi(g(f(a)))) \neq \omega_M$. On the other hand, for every $\psi \in Ker(\bar{f})$ and every $a \in A$, $\varepsilon_M^*(\psi(f(a))) = \omega_M$. So, by Lemma 5.10 for $\phi \circ g \in Im(\bar{g})$ there is no member of $Ker(\bar{g})$ such that its class be equal to class of $\phi \circ g$. Therefore, in general the $-[M]$ functor is not exact. The same discussion is establish for the $M[-]$ functor.

Example 5.3. Consider the H_v-modules M, M_1 and M_2 as Example 1 and the sequence $M \xrightarrow{f} M_1 \xrightarrow{i} M_1$ where $f(a) = 0$, $f(b) = 2$ and i be identity. It is easy to see that the sequence $M \xrightarrow{f} M_1 \xrightarrow{i} M_1$ is exact. But the sequence

$$M_1[M_1 \times M_2] \xrightarrow{\bar{i}} M_1[M_1 \times M_2] \xrightarrow{\bar{f}} M[M_1 \times M_2]$$

is not exact, because for $\phi \in M_1[M_1 \times M_2]$ defined by $\phi(0) = (1,\bar{1})$, $\phi(1) = (2,\bar{1})$ and $\phi(2) = (1,\bar{2})$ there is no member of $Ker(\bar{f})$ such that its class be equal to class of ϕ.

In the following theorem we show that if the converse of Lemma 5.10 is establish, then the functors $M[-]$ and $-[M]$ are exact.

Theorem 5.11. *Let $A \xrightarrow{f} B \xrightarrow{g} C$ be an exact sequence of H_v-modules and strong H_v-homomorphisms. If the converse of Lemma 5.10 is establish, then the sequences*

$$C[M] \xrightarrow{\bar{g}} B[M] \xrightarrow{\bar{f}} A[M] \tag{5.1}$$

$$M[A] \xrightarrow{\bar{f}} M[B] \xrightarrow{\bar{g}} M[C] \tag{5.2}$$

are exact sequences.

Proof. We prove (2). The proof of (1) is similar. Suppose that $h \in Im(\bar{f})$. Then, there exists $\phi \in M[A]$ such that $h = \bar{f}(\phi) = f \circ \phi \in M[B]$. For every $m \in M$, $f \circ \phi(m) \in Im(f)$ and so, there exists $b_m \in Ker(g)$ such that $\varepsilon_B^*(f \circ \phi(m)) = \varepsilon_B^*(b_m)$. Now, we define $k \in M[B]$ by $k(m) = b_m$. Since $\bar{g}(k) = g \circ k \in M[\omega_C] = \omega_{M[C]}$, we obtain $k \in Ker\ \bar{g}$. Finally, by the converse of Lemma 5.10 we have $\varepsilon_{M[B]}^*(h) = \varepsilon_{M[B]}^*(k)$.

Conversely, let $k \in Ker(\bar{g})$, then $\bar{g}(k) = g \circ k \in \omega_{M[C]} = M[\omega_C]$. So, for all $m \in M$, $g \circ k(m) \in \omega_C$ and $k(m) \in Ker(g)$. Then, there exists $b_m = f(a) \in Im(f)$ for some $a \in A$ such that $\varepsilon_B^*(b_m) = \varepsilon_B^*(k(m))$. We define $\psi \in M[A]$ by $\psi(m) = a$ and set $\phi = f \circ \psi = \bar{f}(\psi) \in Im(\bar{f})$. Now, by the converse of Lemma 5.10 we obtain $\varepsilon_{M[B]}^*(k) = \varepsilon_{M[B]}^*(\phi)$. ∎

Lemma 5.12. *Let A, B and C are H_v-modules. Then,*

(1) $\omega_A \xrightarrow{i} A \xrightarrow{f} B$ *is exact if and only if f is weak-monic.*

(2) $B \xrightarrow{g} C \xrightarrow{j} \omega_C$ is exact if and only if g is weak-epic.

(3) $\omega_A \xrightarrow{i} A \xrightarrow{f} B \xrightarrow{g} C \xrightarrow{j} \omega_C$ is exact if and only if f is weak-monic, g is weak-epic and $Im(f) \overset{w}{=} Ker(g)$.

Proof. (1) Suppose that the given sequence is exact. It is enough to show that $Ker(f) = \omega_A$. Always, we have $\omega_A \subseteq Ker(f)$. On the other hand, if $a \in Ker(f)$, then there exists $a_1 \in Im(i) = \omega_A$ such that $\varepsilon_A^*(a) = \varepsilon_A^*(a_1) = \omega_A$ and so, $a \in \omega_A$. Therefore, $Ker(f) = \omega_A$ and f is weak-monic.

Conversely, suppose that f is weak-monic. Then, $Ker(f) = \omega_A = Im(i)$ and consequently $Ker(f) \overset{w}{=} Im(i)$.

(2) Suppose that the given sequence is exact. Then, $Im(g) \overset{w}{=} Ker(j)$ and so, for every $c \in Ker(j) (= C$ since $\omega_{\omega_C} = \omega_C)$ there exists $b \in B$ such that $\varepsilon_C^*(g(b)) = \varepsilon_C^*(c)$. Therefore, g is weak-epic.

Conversely, suppose that g is weak-epic. Then, for every $c \in C (= Ker(j))$ there exists $b \in B$ such that $\varepsilon_C^*(g(b)) = \varepsilon_C^*(c)$. On the other hand, for all $g(b) \in Im(g) \subseteq C$ there exist some $t \in B$ such that $\varepsilon_C^*(g(b)) = \varepsilon_C^*(g(t))$, where $g(t) \in C = Ker(j)$ and consequently $Im(g) \overset{w}{=} Ker(j)$.

(3) It follows from (1), (2) and the definition of exactness. ∎

Lemma 5.13. Let $f : A \to B$ be a strong H_v-homomorphism of H_v-modules. Then, f is weak-epic if and only if F is onto. Moreover f is weak-monic if and only if F is one to one. Finally, f is weak-isomorphism if and only if F is isomorphism.

Proof. Suppose that f is weak-epic and $\varepsilon_B^*(b) \in B/\varepsilon_B^*$. Since f is weak-epic, there exists $a \in A$ such that $\varepsilon_B^*(f(a)) = \varepsilon_B^*(b)$. But $\varepsilon_B^*(f(a)) = F(\varepsilon_A^*(a))$. So, $F(\varepsilon_A^*(a)) = \varepsilon_B^*(b)$ and consequently F is onto.

Conversely, let F is onto. Then, for every $b \in B$ there exists $\varepsilon_A^*(a) \in A/\varepsilon_A^*$ such that $F(\varepsilon_A^*(a)) = \varepsilon_B^*(b)$. But $F(\varepsilon_A^*(a)) = \varepsilon_B^*(f(a))$. So, there exists $a \in A$ such that $\varepsilon_B^*(f(a)) = \varepsilon_B^*(b)$ and consequently f is weak-epic. The second part is proved in [41]. The third part is a obvious result. ∎

Theorem 5.12. Let $\omega_A \xrightarrow{i} A \xrightarrow{f} B \xrightarrow{g} C \xrightarrow{j} \omega_C$ be an exact sequence of H_v-modules and strong H_v-homomorphisms over an H_v-ring R. Then,

$$0 = \omega_A/\varepsilon_{\omega_A}^* \xrightarrow{I} A/\varepsilon_A^* \xrightarrow{F} B/\varepsilon_B^* \xrightarrow{G} C/\varepsilon_C^* \xrightarrow{J} \omega_c/\varepsilon_{\omega_c}^* = 0$$

is an exact sequence of R/γ^*-homomorphisms and R/γ^*-modules.

Proof.   It follows from Lemma 5.12, Lemma 5.13 and Theorem 4.8 of [41] that say if $A \xrightarrow{f} B \xrightarrow{g} C$ is an exact sequence, then $A/\varepsilon_A^* \xrightarrow{F} B/\varepsilon_B^* \xrightarrow{G} C/\varepsilon_C^*$ is an exact sequence. ∎

Theorem 5.13. *(Five short lemma in H_v-modules) Let*

$$
\begin{array}{ccccccc}
\omega_A & \longrightarrow & A & \xrightarrow{\;f\;} & B & \xrightarrow{\;g\;} & C & \longrightarrow & \omega_C \\
 & & \downarrow{\scriptstyle h} & & \downarrow{\scriptstyle k} & & \downarrow{\scriptstyle l} & & \\
\omega_{A_1} & \longrightarrow & A_1 & \xrightarrow[f_1]{} & B_1 & \xrightarrow[g_1]{} & C_1 & \longrightarrow & \omega_{C_1}
\end{array}
$$

be a commutative diagram of H_v-modules and H_v-homomorphisms over an H_v-ring R with both rows exact. Then,

(1) *If h and l are weak-monic, then k is weak-monic.*
(2) *If h and l are weak-epic, then k is weak-epic.*
(3) *If h and l are weak-isomorphism, then k is weak-isomorphism.*

Proof.   (1) By Lemma 5.7 and Theorem 5.12 the following diagram of R/γ^*-modules and R/γ^*-homomorphisms is commutative with both rows exact

$$
\begin{array}{ccccccc}
0 = \omega_A/\varepsilon_{\omega_A}^* & \longrightarrow & A/\varepsilon_A^* & \xrightarrow{\;F\;} & B/\varepsilon_B^* & \xrightarrow{\;G\;} & C/\varepsilon_C^* & \longrightarrow & C/\varepsilon_{\omega_C}^* = 0 \\
 & & \downarrow{\scriptstyle H} & & \downarrow{\scriptstyle K} & & \downarrow{\scriptstyle L} & & \\
0 = \omega_{A_1}/\varepsilon_{\omega_{A_1}}^* & \longrightarrow & A_1/\varepsilon_{A_1}^* & \xrightarrow[F_1]{} & B_1/\varepsilon_{B_1}^* & \xrightarrow[G_1]{} & C_1/\varepsilon_{C_1}^* & \longrightarrow & \omega_{C_1}/\varepsilon_{\omega_{C_1}}^* = 0.
\end{array}
$$

By Lemma 5.13, H and L are one to one R/γ^*-homomorphisms. Then, by five short lemma in modules K is one to one R/γ^*-homomorphism. So, by Lemma 5.13, k is weak-monic R-homomorphism.

Alternative Proof. It is enough to show that $Ker(k) = \omega_B$. Always, we have $\omega_B \subseteq Ker(k)$. On the other hand, suppose that $b \in Ker(k)$. Then, $k(b) \in \omega_{B_1}$ and so, $g_1(k(b)) \in g_1(\omega_{B_1})$. Since $g_1(\omega_{B_1}) \subseteq \omega_{C_1}$, we have $g_1(k(b)) \in \omega_{C_1}$. Since $g_1 \circ k = l \circ g$ and l is weak monic, we obtain $g(b) \in Ker(l) = \omega_C$. Then, $b \in Ker(g) \overset{w}{=} Im(f)$ and consequently

$$\varepsilon_B^*(b) = \varepsilon_B^*(f(a)) \text{ for some } a \in A. \tag{5.3}$$

Since k is strong H_v-homomorphism, we have $\varepsilon_{B_1}^*(k(b)) = \varepsilon_{B_1}^*(k(f(a)))$. Since $k \circ f = f_1 \circ h$ and $b \in Ker(k)$, we obtain $\varepsilon_{B_1}^*(k(b)) = \varepsilon_{B_1}^*(f_1(h(a))) = \omega_{B_1}$. So, $f_1(h(a)) \in \omega_{B_1}$. Since f_1 is weak-monic we obtain $h(a) \in \omega_{A_1}$ and since h is weak-monic it follow that $a \in \omega_A$. So, $f(a) \in f(\omega_A) \subseteq \omega_B$ and by Equation (5.3) we obtain $\varepsilon_B^*(b) = \varepsilon_B^*(f(a)) = \omega_B$. Therefore, $b \in \omega_B$ and prove is complete.

(2) It is similar to (1).

(3) It follows from (1) and (2).   ∎

Now, we prove the snake lemma and close this section.

Theorem 5.14. *(Snake lemma in H_v-modules) Let*

$$
\begin{array}{ccccccc}
A & \xrightarrow{\ f\ } & B & \xrightarrow{\ g\ } & C & \longrightarrow & \omega_C \\
{\scriptstyle h}\downarrow & & {\scriptstyle k}\downarrow & & {\scriptstyle l}\downarrow & & \\
\omega_{A_1} & \longrightarrow & A_1 & \xrightarrow[f_1]{} & B_1 & \xrightarrow[g_1]{} & C_1
\end{array}
$$

be a commutative diagram of H_v-modules and strong homomorphisms over an H_v-ring R with both exact rows. If l is weak-monic, then there exists an exact sequence as follows:

$$Ker(h) \xrightarrow{\ \alpha\ } Ker(k) \xrightarrow{\ \beta\ } Ker(l).$$

Proof.   First we want to define α and β. We have

$$
\begin{aligned}
Ker(h) &= \{a \in A \mid h(a) \in \omega_{A_1}\},\\
Ker(k) &= \{b \in B \mid k(b) \in \omega_{B_1}\},\\
Ker(l) &= \{c \in C \mid l(c) \in \omega_{C_1}\}.
\end{aligned}
$$

Now, for $a \in Ker(h)$, $f_1 \circ h(a) \in f_1(\omega_{A_1}) \subseteq \omega_{B_1}$. Since $f_1 \circ h(a) = k \circ f(a)$, we obtain $f(a) \in Ker(k)$. Also, for $b \in Ker(k)$, $g_1 \circ k(b) \in g_1(\omega_{B_1}) \subseteq \omega_{C_1}$. Since $g_1 \circ k(b) = l \circ g(b)$, we obtain $g(b) \in Ker(l)$.

We define α by $\alpha(a) = f(a)$, for every $a \in Ker(h)$ and β by $\beta(b) = g(b)$, for every $b \in Ker(k)$. Since the $Ker(h)$, $Ker(k)$ and $Ker(l)$ are H_v-submodules of A, B and C respectively and f, g are strong homomorphisms, it follows that α and β are strong homomorphisms.

We show that $Im(\alpha) \overset{w}{=} Ker(\beta)$. Let $x \in Im(\alpha)$ then, $x = f(a)$ for some $a \in Ker(h)(\subseteq A)$. The first row is exact, so there exists $b \in Ker(g)$ such that $\varepsilon_B^*(f(a)) = \varepsilon_B^*(b)$, where $g(b) \in \omega_C$. Since l is weak-monic we have $ker(l) = \omega_C$; but $\omega_{ker(l)} = \omega_{\omega_C} = \omega_C$ and so $\beta(b) = g(b) \in \omega_{Ker(l)}$. It is enough to show $b \in Ker(k)$. Since $\varepsilon_B^*(f(a)) = \varepsilon_B^*(b)$ and $f(a) \in Im(\alpha)(\subseteq Ker(k))$, we obtain $b \in Ker(k)$.

Conversely, let $b \in Ker(\beta)$, then $\beta(b) = g(b) \in \omega_{Ker(l)} = \omega_C$ and $b \in Ker(g)$. Since first row is exact, there exists $f(a) \in Im(f)$ for some $a \in A$ such that $\varepsilon_B^*(b) = \varepsilon_B^*(f(a))$. It is enough to show $a \in Ker(h)$. Since k is strong and diagram is commutative, we obtain $\varepsilon^*_{B_1}(k(b)) = \varepsilon^*_{B_1}(k(f(a))) = \varepsilon^*_{B_1}(f_1(h(a)))$. Since $b \in Ker(\beta)(\subseteq Ker(k))$, it follows $f_1(h(a)) \in \omega_{B_1}$ and $h(a) \in Ker(f_1)$. Since f_1 is weak-monic (by exactness) we have $Ker(f_1) = \omega_{A_1}$. Therefore, $a \in Ker(h)$. ∎

5.6   Shanuel's lemma in H_v-modules

Vaziri and Ghadiri studied the concepts of star homomorphism, (star) iso-morph sequences and star projective H_v-modules in order to find a gener-alization of Shanuel's lemma. In this section, we present these concepts. The main reference is [109, 110].

Definition 5.7. A mapping $f : M_1 \to M_2$ of H_v-modules M_1 and M_2 over an H_v-ring R is called a star homomorphism if for every $x, y \in M_1$ and every $r \in R$: $\varepsilon^*_{M_2}(f(x+y)) = \varepsilon^*_{M_2}(f(x) + f(y))$ and $\varepsilon^*_{M_2}(f(rx)) = \varepsilon^*_{M_2}(rf(x))$; i.e. $f(x+y) \overset{w}{=} f(x) + f(y)$ and $f(rx) \overset{w}{=} rf(x)$.

Every strong homomorphism is a star homomorphism but the converse is not true necessarily by the following example.

Example 5.4. Let R be an H_v-ring. Consider the following H_v-modules on R:

(1) $M_1 = \{a, b\}$ together with the following hyperoperations:

$*_{M_1}$	a	b
a	a	b
b	b	a

and $\cdot_{M_1} : R \times M_1 \to \mathcal{P}^*(M_1),$
$(r, m_1) \mapsto \{a\}$

(2) $M_2 = \{0, 1, 2\}$ together with the following hyperoperations:

$*_{M_2}$	0	1	2
0	0	1	2
1	1	$\{0,2\}$	1
2	2	1	0

and $\cdot_{M_2} : R \times M_2 \to \mathcal{P}^*(M_2).$
$(r, m_2) \mapsto \{0\}$

We obtain $M_2/\varepsilon^*_{M_2} = \{\varepsilon^*_{M_2}(0) = \{0,2\}, \varepsilon^*_{M_2}(1) = \{1\}\}$. If $f : M_1 \to M_2$ defined by $f(a) = 0$ and $f(b) = 1$ then f is a star homomorphism but not strong homomorphism because $f(b *_{M_1} b) \neq f(b) *_{M_2} f(b)$.

Definition 5.8. Two mappings $f, g : M \to N$ on H_v modules are called weak equal if for every $m \in M$; $\varepsilon^*_N(f(m)) = \varepsilon^*_N(g(m))$ and denote by

$f \stackrel{w}{=} g$. The following diagram of H_v-modules and strong homomorphisms is called star commutative if $g \circ f \stackrel{w}{=} h$.

$$
\begin{array}{ccc}
 & & A \\
 & {\scriptstyle f}\nearrow & \downarrow {\scriptstyle h} \\
B & \xrightarrow[g]{} & C
\end{array}
$$

Also, it is said commutative if for every $a \in A$, $g \circ f(a) = h(a)$.

Definition 5.9. The sequences

$$\omega_A \to A \xrightarrow{f} B \xrightarrow{g} C \to \omega_C$$

and

$$\omega_{A'} \to A' \xrightarrow{f'} B' \xrightarrow{g'} C' \to \omega_{C'}$$

are called isomorph (star isomorph) if there exist weak-isomorphisms (star homomorphisms) $\alpha : A \to A'$, $\beta : B \to B'$ and $\gamma : C \to C'$ such that the following diagram is commutative (star commutative):

$$
\begin{array}{ccccccccc}
\omega_A & \longrightarrow & A & \xrightarrow{f} & B & \xrightarrow{g} & C & \longrightarrow & \omega_C \\
 & & \downarrow{\scriptstyle \alpha} & & \downarrow{\scriptstyle \beta} & & \downarrow{\scriptstyle \gamma} & & \\
\omega_{A'} & \longrightarrow & A' & \xrightarrow[f']{} & B' & \xrightarrow[g']{} & C' & \longrightarrow & \omega_{C'}.
\end{array}
$$

Definition 5.10. An H_v-module P is called star projective if for every diagram of strong homomorphisms and H_v-modules as follows

$$
\begin{array}{c}
P \\
\downarrow{\scriptstyle f} \\
M \xrightarrow{g} N \longrightarrow \omega_N
\end{array}
$$

that it's row is exact, there exists a strong homomorphism $\phi : P \to M$ such that $g \circ \phi \stackrel{w}{=} f$.

According to [41] for every strong homomorphism $f : M \to N$ there is the R/γ^*-homomorphism $F : M/\varepsilon_M^* \to N/\varepsilon_N^*$ of R/γ^*-modules defined by $F(\varepsilon_M^*(m)) = \varepsilon_N^*(f(m))$.

Lemma 5.14. *Let $f : A \to B$ be a strong homomorphism of H_v-modules. Then, f is weak-epic (weak-monic) if and only if F is onto (one to one). So, f is weak-isomorphism if and only if F is isomorphism.*

Proof.   Suppose that f is weak-epic and $\varepsilon_B^*(b) \in B/\varepsilon_B^*$. Since f is weak-epic, there exists $a \in A$ such that $\varepsilon_B^*(f(a)) = \varepsilon_B^*(b)$. But $\varepsilon_B^*(f(a)) = F(\varepsilon_A^*(a))$. So, $F(\varepsilon_A^*(a)) = \varepsilon_B^*(b)$ and consequently F is onto.

Conversely, let F is onto. Then, for every $b \in B$ there exists $\varepsilon_A^*(a) \in A/\varepsilon_A^*$ such that $F(\varepsilon_A^*(a)) = \varepsilon_B^*(b)$. But $F(\varepsilon_A^*(a)) = \varepsilon_B^*(f(a))$. So, there exists $a \in A$ such that $\varepsilon_B^*(f(a)) = \varepsilon_B^*(b)$ and consequently f is weak-epic. The proof of the remaining part is straightforward.   ∎

Theorem 5.15. *(Shanuel's lemma in H_v-modules) Let P_1 and P_2 are two star projective H_v-modules. Then the following exact sequences are star isomorph.*

$$\omega_K \longrightarrow K \xrightarrow{\ f\ } P_1 \xrightarrow{\ g\ } M \longrightarrow \omega_M \ , \qquad (5.4)$$

$$\omega_L \longrightarrow L \xrightarrow{\ f_1\ } P_2 \xrightarrow{\ g_1\ } M \longrightarrow \omega_M \ . \qquad (5.5)$$

Proof.   Let $\gamma : M \to M$ be identity on M. Since P_1 is a star projective H_v-module, there exists a strong homomorphism $\beta : P_1 \to P_2$ such that for every $p \in P_1$; $\varepsilon_M^*(g_1 \circ \beta(p)) = \varepsilon_M^*(g(p))$. Now, for every $k \in K$; $f(k) \in P_1$ and then by exactness of sequence (5.4) we have $\beta \circ f(k) \in Ker(g_1)$ and so by exactness of sequence (5.5) there exists $l_k \in L$ such that $\varepsilon_{P_2}^*(\beta(f(k))) = \varepsilon_{P_2}^*(f_1(l_k))$. We define $\alpha : K \to L$ by $\alpha(k) = l_k$. Suppose $k_1, k_2 \in K$ and $r \in R$, we have:

$$\begin{aligned}
\varepsilon_{P_2}^*(\beta \circ f(k_1 + k_2)) &= \varepsilon_{P_2}^*(\beta(f(k_1)) + \beta(f(k_2))) \\
&= \varepsilon_{P_2}^*(\beta f(k_1)) \oplus \varepsilon_{P_2}^*(\beta(f(k_2))) \\
&= \varepsilon_{P_2}^*(f_1(l_{k_1})) \oplus \varepsilon_{P_2}^*(f_1(l_{k_2})) \\
&= \varepsilon_{P_2}^*(f_1(l_{k_1}) + f_1(l_{k_2})) \\
&= \varepsilon_{P_2}^*(f_1(l_{k_1} + l_{k_2})) \\
&= \varepsilon_{P_2}^*(f_1(\alpha(k_1) + \alpha(k_2))) \\
&= F_1(\varepsilon_L^*(\alpha(k_1) + \alpha(k_2))),
\end{aligned}$$

on the other hand

$$\begin{aligned}
\varepsilon_{P_2}^*(\beta \circ f(k_1 + k_2)) &= \{\varepsilon_{P_2}^*(\beta(f(t)))|\ t \in k_1 + k_2\} \\
&= \{\varepsilon_{P_2}^*(f_1(l_t))|\ t \in k_1 + k_2;\ \varepsilon_{P_2}^*(\beta(f(t))) = \varepsilon_{P_2}^*(f_1(l_t))\} \\
&= \{\varepsilon_{P_2}^*(f_1(\alpha(t)))|\ t \in k_1 + k_2\} \\
&= \varepsilon_{P_2}^*(f_1(\alpha(k_1 + k_2))) \\
&= F_1(\varepsilon_L^*(\alpha(k_1 + k_2))).
\end{aligned}$$

Thus $F_1(\varepsilon_L^*(\alpha(k_1 + k_2))) = F_1(\varepsilon_L^*(\alpha(k_1) + \alpha(k_2)))$. Now by Lemma 5.14, F_1 is one to one and $\varepsilon_L^*(\alpha(k_1 + k_2)) = \varepsilon_L^*(\alpha(k_1) + \alpha(k_2))$.

Also,

$$\varepsilon_{P_2}^*(\beta \circ f(rk_1)) = \varepsilon_{P_2}^*(r\beta(f(k_1)))$$
$$= \gamma^*(r) \odot \varepsilon_{P_2}^*(\beta(f(k_1)))$$
$$= \gamma^*(r) \odot \varepsilon_{P_2}^*(f_1(l_{k_1}))$$
$$= \varepsilon_{P_2}^*(rf_1(l_{k_1}))$$
$$= \varepsilon_{P_2}^*(rf_1(\alpha(k_1)))$$
$$= \varepsilon_{P_2}^*(f_1(r\alpha(k_1)))$$
$$= F_1(\varepsilon_L^*(r\alpha(k_1)),$$

on the other hand

$$\varepsilon_{P_2}^*(\beta(f(rk_1))) = \{\varepsilon_{P_2}^*(\beta(f(t)))|\ t \in rk_1\}$$
$$= \{\varepsilon_{P_2}^*(f_1(l_t))|\ t \in rk_1;\ \varepsilon_{P_2}^*(\beta(f(t))) = \varepsilon_{P_2}^*(f_1(l_t))\}$$
$$= \{\varepsilon_{P_2}^*(f_1(\alpha(t)))|\ t \in rk_1\}$$
$$= \varepsilon_{P_2}^*(f_1(\alpha(rk_1)))$$
$$= F_1(\varepsilon_L^*(\alpha(rk_1))).$$

Thus $F_1(\varepsilon_L^*(\alpha(rk_1))) = F_1(\varepsilon_L^*(r\alpha(k_1)))$. Now by Lemma 5.14, F_1 is one to one and $\varepsilon_L^*(\alpha(rk_1)) = \varepsilon_L^*(r\alpha(k_1)$ and α is a star homomorphism.
One can check the star commutativity on these star homomorphisms. ■

Theorem 5.16. *(1) Let*

$$\omega_A \longrightarrow A \xrightarrow{f} B \xrightarrow{g} C$$
$$\downarrow{\beta} \qquad \downarrow{\gamma}$$
$$\omega_{A_1} \longrightarrow A_1 \xrightarrow{f_1} B_1 \xrightarrow{g_1} C_1$$

be a star commutative diagram of H_v-modules and strong H_v-homomorphisms with both exact rows. Then there exists a star homomorphism $\alpha : A \to A_1$ such that star-commute the diagram.
(2) Let

$$A \xrightarrow{f} B \xrightarrow{g} C \longrightarrow \omega_C$$
$$\downarrow{\alpha} \qquad \downarrow{\beta}$$
$$A_1 \xrightarrow{f_1} B_1 \xrightarrow{g_1} C_1 \longrightarrow \omega_{C_1}$$

be a star commutative diagram of H_v-modules and strong homomorphisms with both exact rows. Then, there exists a star homomorphism $\gamma : C \to C_1$ such that star commute the diagram.

Proof.   (1) For every $a \in A$ we have $\varepsilon^*_{C_1}(g_1 \circ \beta \circ f(a)) = \varepsilon^*_{C_1}(\gamma \circ g \circ f(a))$. First row is exact and γ is strong homomorphism. Then $g \circ f(a) \in \omega_C$ and $\gamma \circ g \circ f(a) \in \omega_{C_1}$. So, $\beta \circ f(a) \in Ker(g_1)$ and there exists $a_1 \in A_1$ such that $\varepsilon^*_{B_1}(\beta \circ f(a)) = \varepsilon^*_{B_1}(f_1(a_1))$. Now we define $\alpha : A \to A_1$ by $\alpha(a) = a_1$.

Similar to the proof of Theorem 5.15 one can show that α is a star homomorphism. Also, for every $a \in A$ we have

$$\varepsilon^*_{B_1}(f_1 \circ \alpha(a)) = \varepsilon^*_{B_1}(f_1(a_1)) = \varepsilon^*_{B_1}(\beta \circ f(a)).$$

(2) Since g is weak-epic for every $c \in C$ there exists $b_c \in B$ such that $\varepsilon^*_C(c) = \varepsilon^*_C(g(b_c))$. We define $\gamma : C \to C_1$ by $\gamma(c) = g_1 \circ \beta(b_c)$. The remaining of proof is straight forward and similar to the proof of (1).   ∎

5.7   Product and direct sum in H_v-modules

In this section, we present the concepts of product and direct sum of H_v-modules. The main reference is [110].

Definition 5.11. Let M be an H_v-module and H, K are H_v-submodules of M. M is said direct sum of H and K if $H \cap K \subseteq \omega_M$ and $\varepsilon^*(H + K) = \varepsilon^*(M)$. We denote it by $H \oplus K = M$.

Example 5.5. For every H_v-module M we have $M = \omega_M \oplus M$.

Example 5.6. Consider the following weak commutative H_v-group:

$*_M$	0	1	2	3	4	5	6
0	{0,1}	{0,1}	2	3	4	5	6
1	{0,1}	{0,1}	2	3	4	5	6
2	2	2	{0,1}	{5,6}	{5,6}	{2,3,4}	{2,3,4}
3	3	3	{5,6}	{0,1}	{5,6}	{2,3,4}	{2,3,4}
4	4	4	{5,6}	{5,6}	{0,1}	{2,3,4}	{2,3,4}
5	5	5	{2,3,4}	{2,3,4}	{2,3,4}	6	{0,1}
6	6	6	{2,3,4}	{2,3,4}	{2,3,4}	{0,1}	5

One can check that $R = (M, *_M, .)$ is an H_v-ring where $r_1.r_2 = \{0,1\}$ for every $r_1, r_2 \in R$ and M is an H_v-module over the H_v-ring R. Also,

$$M/\varepsilon^*_M = \{\varepsilon^*_M(0), \varepsilon^*_M(2)\}$$

where

$$\varepsilon^*_M(0) = \omega_M = \{0,1,5,6\}, \quad \varepsilon^*_M(2) = \{2,3,4\}.$$

Now, $H = \{0,1,2\}$ and $K = \{0,1,5,6\}$ are H_v-submodules of M and $H \oplus K = M$.

Proposition 5.4. *Let* $f : M \to M$ *be a strong homomorphism of* H_v-*modules such that* $f^2 = f$. *Then,* M *is direct sum of* $Im(f)$ *and* $Ker(f)$. *Moreover* f *is identity on* $Im(f) \cap Ker(f)$.

Proof. Let $m \in Im(f) \cap Ker(f)$ then

$$m = f(m_1) \text{ for some } m_1 \text{ in } M \qquad (5.6)$$

and

$$f(m) \in \omega_M. \qquad (5.7)$$

By apply f on Eq. (5.6) we obtain $f(m) = f^2(m_1) = f(m_1) = m$ as a member of ω_M by Eq. (5.7). So, $Im(f) \cap Ker(f) \subseteq \omega_M$ and f is identity on $Im(f) \cap Ker(f)$. Now, for every $m \in M$ we have:

$$F(F(\varepsilon^*(m))) = F(\varepsilon^*(f(m))) = \varepsilon^*(f^2(m)) = \varepsilon^*(f(m)) = F(\varepsilon^*(m)).$$

So $Im(F) + Ker(F) = M/\varepsilon_M^*$, since F is a R/γ^*-module such that $F^2 = F$. Therefore $\varepsilon^*(Im(f) + Ker(f)) = \varepsilon^*(M)$. ∎

Let $\{M_i\}_{i \in I}$ be a non-empty collection of H_v-modules. The product of this collection

$$\sqcap_{i \in I} \{M_i\} = \{(x_i) | \ x_i \in M; \ \forall \ i \in I\}$$

with the following hyper operations is an H_v-module:

$$(x_i) + (y_i) = \{(z_i) | \ z_i \in x_i + y_i\},$$

$$r(x_i) = \{(w_i) | \ w_i \in rx_i\}.$$

Lemma 5.15. *Let* $\sqcap_{i \in I} M_i$ *be the product of non-empty collection of* H_v-*modules. Then* (1) $P_k : \sqcap M_i \to M_k$ *defined by* $P_k((x_i)) = x_k$ *is a strong homomorphism.*
(2) *For every exact sequence* $M_1 \overset{\phi}{\to} M \overset{\psi}{\to} M_2$ *the mapping:*
$\lambda_1 : M_1 \to M_1 \sqcap M_2$ *defined by* $\lambda_1(x) = (x, \psi\phi(x))$ *is a strong homomorphism. Also,* $\lambda_2 : M_2 \to M_1 \sqcap M_2$ *defined by* $\lambda_2(x) = (a, x)$, *where* a *is an arbitrary member of* ω_{M_1}, *is a star homomorphism. In particular if there exists a* $t \in \omega_{M_1}$ *such that* $t + t = t$, *then* λ_2 *is a strong homomorphism.*
(3) $P_k \lambda_k = I_{M_k}$.

Proof. (1)

$$P_k((x_i) + (y_i)) = P_k(\{(z_i) | \ z_i \in x_i + y_i\}) = \{z_k | \ z_k \in x_k + y_k\}.$$

On the other hand,

$$P_k((x_i)) + P_k((y_i)) = x_k + y_k.$$

Similarly, we obtain $P_k(r(x_i)) = rP_k((x_i))$.

(2) We have

$$\lambda_1(x + y) = \bigcup_{a \in x+y, b \in \psi\phi(x+y)} (a, b)$$
$$= (x, \psi\phi(x)) + (y, \psi\phi(y))$$
$$= \lambda_1(x) + \lambda_1(y).$$

Similarly we obtain $\lambda_1(rx) = r\lambda_1(x)$. Also,

$$\varepsilon^*(\lambda_2(x + y)) = \varepsilon^*(\bigcup_{a \in \omega_{M_1}, b \in x+y} (a, b))$$
$$= \varepsilon^*((a_1, x) + (a_1, y)) \text{ where } a_1 \in \omega_{M_1}$$
$$= \varepsilon^*((a_1, x)) \oplus \varepsilon^*((a_1, y))$$
$$= \varepsilon^*(\lambda_2(x)) \oplus \varepsilon^*(\lambda_2(y))$$
$$= \varepsilon^*(\lambda_2(x) + \lambda_2(y)).$$

Similarly $\varepsilon^*(\lambda_2(rx)) = \varepsilon^*(r\lambda_2(x))$.

(3) The proof of this part is straightforward. ∎

Theorem 5.17. *Let $\{M_i\}$ be a non-empty collection of H_v-modules. For every H_v-module X and every collection of strong homomorphisms $\{f_i : X \to M_i\}$ there exists an unique strong homomorphism $\phi : X \to \sqcap M_i$ defined by $\phi(x) = (f_i(x))$ such that for every $i \in I$ the following diagram is commutative.*

Proof. The proof is straight forward. ∎

We want to define the inverse of a weak-isomorphism to determine the conditions for split an exact sequence.

Lemma 5.16. *Let $f : M \to N$ be a weak-isomorphism. Then $f^{-1} : N \to M$ defined by $f^{-1}(n) = m_n$ for selected $m_n \in F^{-1}(\varepsilon_N^*(n))$, is a star homomorphism such that $f^{-1} \circ f \stackrel{w}{=} I_M$ and $f \circ f^{-1} \stackrel{w}{=} I_N$.*

Proof. Since f is a weak-isomorphism by Lemma 5.14, F is a isomorphism and have an inverse. For every $n_1, n_2 \in N$ we have

$$f^{-1}(n_1 + n_2) = \{m_c \mid m_c \in F^{-1}(\varepsilon_N^*(c)),\ c \in n_1 + n_2\}, \qquad (5.8)$$

On the other hand

$$\begin{aligned}
f^{-1}(n_1) + f^{-1}(n_2) &= m_{n_1} + m_{n_2} \\
&\subseteq F^{-1}(\varepsilon_N^*(n_1)) + F^{-1}(\varepsilon_N^*(n_2)) \qquad (5.9)\\
&= F^{-1}(\varepsilon_N^*(n_1 + n_2)).
\end{aligned}$$

From Eq. (5.8) and Eq. (5.9) we obtain $\varepsilon_M^*(f^{-1}(n_1 + n_2)) = \varepsilon_M^*(f^{-1}(n_1) + f^{-1}(n_2))$ (notice that for every $n_1, n_2 \in N$, $n_1 + n_2 \subseteq \varepsilon_N^*(n)$ for some $n \in n_1 + n_2$).

Similarly, we obtain $\varepsilon_M^*(f^{-1}(rn)) = \varepsilon_M^*(rf^{-1}(n))$.

Finally, for every $m \in M$ we have

$$\begin{aligned}
f^{-1} \circ f(m) &\in F^{-1}(\varepsilon_N^*(f(m))) \\
&= F^{-1}(F(\varepsilon_M^*(m))) \\
&= \varepsilon_M^*(m)
\end{aligned}$$

and for every $n \in N$,

$$f \circ f^{-1}(n) = f(m_n), \text{ where } m_n \in F^{-1}(\varepsilon_N^*(n)).$$

But $f(m_n) \in \varepsilon_N^*(n)$. ∎

Definition 5.12. Let f be a weak-isomorphism, the f^{-1} defined in Lemma 5.16 is called the inverse of f. It is clear that this inverse is not unique necessary.

Theorem 5.18. Let M_1, M_2 and M be three H_v-modules and the sequence

$$\omega_{M_1} {\to} M_1 \overset{\phi}{\to} M \overset{\psi}{\to} M_2 {\to} \omega_{M_2} \qquad (5.10)$$

is exact:

(1) If there exists a star homomorphism $\phi' : M \to M_1$ ($\psi' : M_2 \to M$) such that $\phi'\phi \overset{w}{=} I_{M_1}$ ($\psi\psi' \overset{w}{=} I_{M_2}$), then the sequence (5.10) is star isomorph with the sequence

$$\omega_{M_1} {\to} M_1 \overset{\lambda_1}{\to} M_1 \sqcap M_2 \overset{P_2}{\to} M_2 {\to} \omega_{M_2}. \qquad (5.11)$$

(2) If the sequences (5.10) and (5.11) are isomorph, then there exist a star homomorphisms $\phi' : M \to M_1$ and $\psi' : M_2 \to M$ such that $\phi'\phi \overset{w}{=} I_{M_1}$, $\psi\psi' \overset{w}{=} I_{M_2}$.

Proof. (1) We define $\alpha : M \to M_1 \sqcap M_2$ by $\alpha(x) = (\phi'(x), \psi(x))$. It is easy to see that α is a star homomorphism. Since for every $m_1 \in M_1$ we have $\phi'\phi(m_1) \in \varepsilon^*_{M_1}(m_1)$ and $\psi\phi(m_1) \in \omega_{M_1}$, the following diagram is star commutative with both exact rows.

$$
\begin{array}{ccccccccc}
\omega_{M_1} & \longrightarrow & M_1 & \xrightarrow{\ \phi\ } & M & \xrightarrow{\ \psi\ } & M_2 & \longrightarrow & \omega_{M_2} \,. \\
 & & \downarrow{\scriptstyle 1_{M_1}} & & \downarrow{\scriptstyle \alpha} & & \downarrow{\scriptstyle 1_{M_2}} & & \\
\omega_{M_1} & \longrightarrow & M_1 & \xrightarrow{\ \lambda_1\ } & M_1 \sqcap M_2 & \xrightarrow{\ P_2\ } & M_2 & \longrightarrow & \omega_{M_2}
\end{array}
$$

Now let there exists the star homomorphism $\psi' : M_2 \to M$ such that $\psi\psi' \overset{w}{=} I_{M_2}$. we define the mapping $\beta : M_1 \times M_2 \to M$ by $\beta((m_1, m_2)) = m_{m_1,m_2}$ where m_{m_1,m_2} is a member of $\phi(m_1) + \psi'(m_2)$ (according to choice axiom). we show that β is a star homomorphism. We have:
$\varepsilon^*(\beta((a_1, a_2) + (a_1', a_2'))) = \varepsilon^*(\beta((t_1, t_2)))$, where $t_1 \in a_1 + a_1'$ and $t_2 \in a_2 + a_2'$.
and

$$
\begin{aligned}
\varepsilon^*(\beta((a_1, a_2))) \oplus \varepsilon^*(\beta((a_1', a_2'))) &= \varepsilon^*(\phi(a_1) + \psi'(a_2)) \oplus \varepsilon^*(\phi(a_1') + \psi'(a_2')) \\
&= \varepsilon^*(\phi(a_1) + \psi'(a_1') + \phi(a_2) + \psi'(a_2')) \\
&= \varepsilon^*(\phi(t_1)) \oplus \varepsilon^*(\psi'(t_2)) \\
&= \varepsilon^*(\phi(t_1) + \psi'(t_2)) \\
&= \varepsilon^*(\beta((t_1, t_2)))
\end{aligned}
$$

where $t_1 \in a_1 + a_1'$ and $t_2 \in a_2 + a_2'$. So β is a star homomorphism. One can show that the following diagram is star commutative.

$$
\begin{array}{ccccccccc}
\omega_{M_1} & \longrightarrow & M_1 & \xrightarrow{\ \phi\ } & M & \xrightarrow{\ \psi\ } & M_2 & \longrightarrow & \omega_{M_2} \,. \\
 & & \uparrow{\scriptstyle 1_{M_1}} & & \uparrow{\scriptstyle \beta} & & \uparrow{\scriptstyle 1_{M_2}} & & \\
\omega_{M_1} & \longrightarrow & M_1 & \xrightarrow{\ \lambda_1\ } & M_1 \sqcap M_2 & \xrightarrow{\ P_2\ } & M_2 & \longrightarrow & \omega_{M_2}
\end{array}
$$

(2) By hypothesis there exist weak-isomorphisms $\alpha : M_1 \to M_1$, $\beta : M \to M_1 \sqcap M_2$ and $\gamma : M_2 \to M_2$ such that commute the following diagram

$$
\begin{array}{ccccccccc}
\omega_{M_1} & \longrightarrow & M_1 & \xrightarrow{\ \phi\ } & M & \xrightarrow{\ \psi\ } & M_2 & \longrightarrow & \omega_{M_2} \\
 & & \downarrow{\scriptstyle \alpha} & & \downarrow{\scriptstyle \beta} & & \downarrow{\scriptstyle \gamma} & & \\
\omega_{M_1} & \longrightarrow & M_1 & \xrightarrow{\ \lambda_1\ } & M_1 \sqcap M_2 & \xrightarrow{\ P_2\ } & M_2 & \longrightarrow & \omega_{M_2}
\end{array}
$$

By Lemma 5.16, there exists star homomorphism $\alpha^{-1} : M_1 \to M_1$ such that $\alpha^{-1} \circ \alpha \stackrel{w}{=} I_{M_1}$. Now, we define $\phi' : M \to M_1$ by $\phi' = \alpha^{-1}P_1\beta$. Consequently ϕ' is a star homomorphism and

$$\phi'\phi = \alpha^{-1}P_1\beta\phi = \alpha^{-1}P_1\lambda_1\alpha = \alpha^{-1}1_{M_1}\alpha \stackrel{w}{=} I_{M_1}.$$

Similarly, by hypothesis there exist weak-isomorphisms: $\alpha : M_1 \to M_1$, $\beta : M_1 \sqcap M_2 \to M$ and $\gamma : M_2 \to M_2$ such that the following diagram is commutative.

$$
\begin{array}{ccccccccc}
\omega_{M_1} & \longrightarrow & M_1 & \stackrel{\lambda_1}{\longrightarrow} & M_1 \sqcap M_2 & \stackrel{P_2}{\longrightarrow} & M_2 & \longrightarrow & \omega_{M_2} \\
 & & \alpha \downarrow & & \beta \downarrow & & \gamma \downarrow & & \\
\omega_{M_1} & \longrightarrow & M_1 & \stackrel{}{\underset{\phi}{\longrightarrow}} & M & \stackrel{}{\underset{\psi}{\longrightarrow}} & M_2 & \longrightarrow & \omega_{M_2}
\end{array}
$$

By Lemma 5.16, there exists star homomorphism $\gamma^{-1} : M_2 \to M_2$ such that $\gamma \circ \gamma^{-1} \stackrel{w}{=} I_{M_2}$ Now, we define $\psi' : M_2 \to M$ by $\psi' = \beta\lambda_2\gamma^{-1}$. Obviously ψ' is a star homomorphism and

$$\psi\psi' = \psi\beta\lambda_2\gamma^{-1} = \gamma P_2\lambda_2\gamma^{-1} = \gamma 1_{M_2}\gamma^{-1} \stackrel{w}{=} I_{M_2}. \qquad \blacksquare$$

An exact sequence in Theorem 5.18 is called a *split sequence*.

5.8   Fuzzy and intuitionistic fuzzy H_v-submodules

The concept of fuzzy modules was introduced by Negoita and Ralescu.

Definition 5.13. Let M be a module over a ring R. A fuzzy set μ in M is called a *fuzzy submodule* of M if for every $x, y \in M$ and $r \in R$ the following conditions are satisfied:

(1) $\mu(0) = 1$;
(2) $\min\{\mu(x), \mu(y)\} \leq \mu(x - y)$, for all $x, y \in M$;
(3) $\mu(x) \leq \mu(rx)$, for all $x \in M$ and $r \in R$.

In [25], Davvaz defined the concept of a fuzzy H_v-submodule of an H_v-module which is a generalization of the concept of a fuzzy submodule.

Definition 5.14. Let M be an H_v-module over an H_v-ring R and μ a fuzzy set in M. Then μ is said to be a *fuzzy H_v-submodule* of M if the following axioms hold:

(1) $\min\{\mu(x), \mu(y)\} \leq \inf\{\mu(z) \mid z \in x + y\}$, for all $x, y \in M$;

(2) for all $x, a \in M$ there exists $y \in M$ such that $x \in a + y$ and

$$\min\{\mu(a), \mu(x)\} \leq \mu(y);$$

(3) for all $x, a \in M$ there exists $z \in M$ such that $x \in z + a$ and

$$\min\{\mu(a), \mu(x)\} \leq \mu(z);$$

(4) $\mu(x) \leq \inf\{\mu(z) \mid z \in r \cdot x\}$, for all $x \in M$ and $r \in R$.

Theorem 5.19. *For a fuzzy subset μ of an H_v-module M, the following statements are equivalent.*

(1) μ *is a fuzzy H_v-submodule of M.*
(2) μ_t $(t \in ImA)$ *is an H_v-submodule of M.*

Proof.   The proof is similar to the proof of Theorem 4.51.   ∎

After the introduction of fuzzy sets by Zadeh, there have been a number of generalizations of this fundamental concept. The notion of intuitionistic fuzzy sets introduced by Atanassov is one among them.

Definition 5.15. An *intuitionistic fuzzy set A* in a non-empty set X is an object having the form

$$A = \{(x, \mu_A(x), \lambda_A(x)) \mid x \in X\},$$

where the functions $\mu_A : X \to [0,1]$ and $\lambda_A : X \to [0,1]$ denote the degree of membership (namely $\mu_A(x)$) and the degree of nonmembership (namely $\lambda_A(x)$) of each element $x \in X$ to the set A respectively, and $0 \leq \mu_A(x) + \lambda_A(x) \leq 1$ for all $x \in X$. For the sake of simplicity, we shall use the symbol $A = (\mu_A, \lambda_A)$ for the intuitionistic fuzzy set $A = \{(x, \mu_A(x), \lambda_A(x)) \mid x \in X\}$.

Definition 5.16. Let $A = (\mu_A, \lambda_A)$ and $B = (\mu_B, \lambda_B)$ be intuitionistic fuzzy sets in X. Then

(1) $A \subseteq B$ if and only if $\mu_A(x) \leq \mu_B(x)$ and $\lambda_A(x) \geq \lambda_B(x)$, for all $x \in X$;
(2) $A^c = \{(x, \lambda_A(x), \mu_A(x)) \mid x \in X\}$;
(3) $A \cap B = \{(x, \min\{\mu_A(x), \mu_B(x)\}, \max\{\lambda_A(x), \lambda_B(x)\}) \mid x \in X\}$;
(4) $A \cup B = \{(x, \max\{\mu_A(x), \mu_B(x)\}, \min\{\lambda_A(x), \lambda_B(x)\}) \mid x \in X\}$;
(5) $\Box A = \{(x, \mu_A(x), \mu_A^c(x)) \mid x \in X\}$;
(6) $\Diamond A = \{(x, \lambda_A^c(x), \lambda_A(x)) \mid x \in X\}$.

Biswas applied the concept of intuitionistic fuzzy sets to the theory of groups and studied intuitionistic fuzzy subgroups of a group. Now, we define an intuitionistic fuzzy submodule of a module.

Definition 5.17. Let M be a module over a ring R. An intuitionistic fuzzy set $A = (\mu_A, \lambda_A)$ in M is called an *intuitionistic fuzzy submodule* of M if

(1) $\mu_A(0) = 1$;

(2) $\min\{\mu_A(x), \mu_A(y)\} \leq \mu_A(x - y)$ for all $x, y \in M$,

(3) $\mu_A(x) \leq \mu_A(r \cdot x)$, for all $x \in M$ and $r \in R$;

(4) $\lambda_A(0) = 0$;

(5) $\lambda_A(x - y) \leq \max\{\lambda_A(x), \lambda_A(y)\}$, for all $x, y \in M$;

(6) $\lambda_A(r \cdot x) \leq \lambda_A(x)$, for all $x \in M$ and $r \in R$.

Davvaz et al. applied the concept of intuitionistic fuzzy sets to H_v-modules. They introduced the notion of intuitionistic fuzzy H_v-submodules of an H_v-module and investigate some related properties. The main reference for this section is [40]. In what follows, let M denote an H_v-module over an H_v-ring R unless otherwise specified. We start by defining the notion of intuitionistic fuzzy H_v-submodules.

Definition 5.18. An intuitionistic fuzzy set $A = (\mu_A, \lambda_A)$ in M is called an *intuitionistic fuzzy H_v-submodule* of M if

(1) $\min\{\mu_A(x), \mu_A(y)\} \leq \inf\{\mu_A(z) \mid z \in x + y\}$, for all $x, y \in M$;

(2) for all $x, a \in M$ there exist $y, z \in M$ such that $x \in (a + y) \cap (z + a)$ and

$$\min\{\mu_A(a), \mu_A(x)\} \leq \min\{\mu_A(y), \mu_A(z)\};$$

(3) $\mu_A(x) \leq \inf\{\mu_A(z) \mid z \in r \cdot x\}$, for all $x \in M$ and $r \in R$;

(4) $\sup\{\lambda_A(z) \mid z \in x + y\} \leq \max\{\lambda_A(x), \lambda_A(y)\}$, for all $x, y \in M$;

(5) for all $x, a \in M$ there exist $y, z \in M$ such that $x \in (a + y) \cap (z + a)$ and

$$\max\{\lambda_A(y), \lambda_A(z)\} \leq \max\{\lambda_A(a), \lambda_A(x)\};$$

(6) $\sup\{\lambda_A(z) \mid z \in r \cdot x\} \leq \lambda_A(x)$, for all $x \in M$ and $r \in R$.

Lemma 5.17. *If $A = (\mu_A, \lambda_A)$ is an intuitionistic fuzzy H_v-submodule of M, then so is $\square A = (\mu_A, \mu_A^c)$.*

Proof. It is sufficient to show that μ_A^c satisfies the conditions (4), (5), and (6) of Definition 5.18. For $x, y \in M$ we have

$$\min\{\mu_A(x), \mu_A(y)\} \leq \inf\{\mu_A(z) \mid z \in x + y\}$$

and so $\min\{1 - \mu_A^c(x), 1 - \mu_A^c(y)\} \leq \inf\{1 - \mu_A^c(z) \mid z \in x + y\}$. Hence,

$$\min\{1 - \mu_A^c(x), 1 - \mu_A^c(y)\} \leq 1 - \sup\{\mu_A^c(z) \mid z \in x + y\}$$

which implies $\sup\{\mu_A^c(z) \mid z \in x + y\} \leq 1 - \min\{1 - \mu_A^c(x), 1 - \mu_A^c(y)\}$. Therefore

$$\sup\{\mu_A^c(z) \mid z \in x + y\} \leq \max\{\mu_A^c(x), \mu_A^c(y)\},$$

and thus the condition (4) of Definition 5.18 is valid.

Now, let $a, x \in M$. Then there exist $y, z \in M$ such that $x \in (a + y) \cap (z + a)$ and $\min\{\mu_A(a), \mu_A(x)\} \leq \min\{\mu_A(y), \mu_A(z)\}$. It follows that

$$\min\{1 - \mu_A^c(a), 1 - \mu_A^c(x)\} \leq \min\{1 - \mu_A^c(y), 1 - \mu_A^c(z)\}$$

so that

$$\max\{\mu_A^c(y), \mu_A^c(z)\} \leq \max\{\mu_A^c(a), \mu_A^c(x)\}.$$

Hence the condition (5) of Definition 5.18 is satisfied.

For the condition (6), let $x \in M$ and $r \in R$. Since μ_A is a fuzzy H_v-submodule of M, we have

$$\mu_A(x) \leq \inf\{\mu_A(z) \mid z \in r \cdot x\}$$

and so

$$1 - \mu_A^c(x) \leq \inf\{1 - \mu_A^c(z) \mid z \in r \cdot x\},$$

which implies that

$$\sup\{\mu_A^c(z) \mid z \in r \cdot x\} \leq \mu_A^c(x).$$

Therefore the condition (6) of Definition 5.18 is satisfied. ∎

Lemma 5.18. *If $A = (\mu_A, \lambda_A)$ is an intuitionistic fuzzy H_v-submodule of M, then so is $\Diamond A = (\lambda_A^c, \lambda_A)$.*

Proof. The proof is similar to the proof of Lemma 5.17. ∎

Combining the above two lemmas it is not difficult to verify that the following theorem is valid.

Theorem 5.20. *$A = (\mu_A, \lambda_A)$ is an intuitionistic fuzzy H_v-submodule of M if and only if $\Box A$ and $\Diamond A$ are intuitionistic fuzzy H_v-submodules of M.*

Corollary 5.5. *$A = (\mu_A, \lambda_A)$ is an intuitionistic fuzzy H_v-submodule of M if and only if μ_A and λ_A^c are fuzzy H_v-submodules of M.*

Definition 5.19. For any $t \in [0,1]$ and fuzzy set μ in M, the set

$$U(\mu;t) = \{x \in M \mid \mu(x) \geq t\} \text{ (respectively, } L(\mu;t) = \{x \in M \mid \mu(x) \leq t\})$$

is called an *upper* (respectively, *lower*) *t-level cut* of μ.

Theorem 5.21. *If $A = (\mu_A, \lambda_A)$ is an intuitionistic fuzzy H_v-submodule of M, then the sets $U(\mu_A;t)$ and $L(\lambda_A;t)$ are H_v-submodules of M for every $t \in Im(\mu_A) \cap Im(\lambda_A)$.*

Proof. Let $t \in Im(\mu_A) \cap Im(\lambda_A) \subseteq [0,1]$ and let $x, y \in U(\mu_A;t)$. Then $\mu_A(x) \geq t$ and $\mu_A(y) \geq t$ and so $\min\{\mu_A(x), \mu_A(y)\} \geq t$. It follows from the condition (1) of Definition 5.18 that $\inf\{\mu_A(z) \mid z \in x + y\} \geq t$. Therefore $z \in U(\mu_A;t)$ for all $z \in x + y$, and so $x + y \subseteq U(\mu_A;t)$. Hence $a + U(\mu_A;t) \subseteq U(\mu_A;t)$ and $U(\mu_A;t) + a \subseteq U(\mu_A;t)$ for all $a \in U(\mu_A;t)$. Now, let $x \in U(\mu_A;t)$. Then there exist $y, z \in M$ such that $x \in (a + y) \cap (z + a)$ and $\min\{\mu_A(x), \mu_A(a)\} \leq \min\{\mu_A(y), \mu_A(z)\}$. Since $x, a \in U(\mu_A;t)$, we have $t \leq \min\{\mu_A(x), \mu_A(a)\}$ and so $t \leq \min\{\mu_A(y), \mu_A(z)\}$, which implies $y \in U(\mu_A;t)$ and $z \in U(\mu_A;t)$. This proves that $U(\mu_A;t) \subseteq a + U(\mu_A;t)$ and $U(\mu_A;t) \subseteq U(\mu_A;t) + a$.

Now, for every $r \in R$ and $x \in U(\mu_A;t)$ we show that $r \cdot x \subseteq U(\mu_A;t)$. Since A is an intuitionistic fuzzy H_v-submodule of M, we have

$$t \leq \mu_A(x) \leq \inf\{\mu_A(z) \mid z \in r \cdot x\}.$$

Therefore, for every $z \in r \cdot x$ we get $\mu_A(z) \geq t$ which implies $z \in U(\mu_A;t)$, so $r \cdot x \subseteq U(\mu_A;t)$.

If $x, y \in L(\lambda_A;t)$, then $\max\{\lambda_A(x), \lambda_A(y)\} \leq t$. It follows from the condition (4) of Definition 5.18 that $\sup\{\lambda_A(z) \mid z \in x + y\} \leq t$. Therefore for all $z \in x + y$ we have $z \in L(\lambda_A;t)$, so $x + y \subseteq L(\lambda_A;t)$. Hence for all $a \in L(\lambda_A;t)$ we have $a + L(\lambda_A;t) \subseteq L(\lambda_A;t)$ and $L(\lambda_A;t) + a \subseteq L(\lambda_A;t)$. Now, let $x \in L(\lambda_A;t)$. Then there exist $y, z \in M$ such that $x \in (a+y) \cap (z+a)$ and $\max\{\lambda_A(y), \lambda_A(z)\} \leq \max\{\lambda_A(a), \lambda_A(x)\}$. Since $x, a \in L(\lambda_A;t)$, we have $\max\{\lambda_A(a), \lambda_A(x)\} \leq t$ and so $\max\{\lambda_A(y), \lambda_A(z)\} \leq t$. Thus $y \in L(\lambda_A;t)$ and $z \in L(\lambda_A;t)$. Hence $L(\lambda_A;t) \subseteq a+L(\lambda_A;t)$ and $L(\lambda_A;t) \subseteq L(\lambda_A;t)+a$.

Now, we show that $r \cdot x \subseteq L(\lambda_A;t)$ for every $r \in R$ and $x \in L(\lambda_A;t)$. Since A is an intuitionistic fuzzy H_v-submodule of M, we have

$$\sup\{\lambda_A(z) \mid z \in r \cdot x\} \leq \lambda_A(x) \leq t.$$

Therefore, for every $z \in r \cdot x$ we get $\lambda_A(z) \leq t$, which implies $z \in L(\lambda_A;t)$, so $r \cdot x \subseteq L(\lambda_A;t)$. ∎

Theorem 5.22. *If $A = (\mu_A, \lambda_A)$ is an intuitionistic fuzzy set in M such that all non-empty level sets $U(\mu_A; t)$ and $L(\lambda_A; t)$ are H_v-submodules of M, then $A = (\mu_A, \lambda_A)$ is an intuitionistic fuzzy H_v-submodule of M.*

Proof.   Assume that all non-empty level sets $U(\mu_A; t)$ and $L(\lambda_A; t)$ are H_v-submodules of M. If $t_0 = \min\{\mu_A(x), \mu_A(y)\}$ and $t_1 = \max\{\lambda_A(x), \lambda_A(y)\}$ for $x, y \in M$, then $x, y \in U(\mu_A; t_0)$ and $x, y \in L(\lambda_A; t_1)$. So $x + y \subseteq U(\mu_A; t_0)$ and $x + y \subseteq L(\lambda_A; t_1)$. Therefore for all $z \in x + y$ we have $\mu_A(z) \geq t_0$ and $\lambda_A(z) \leq t_1$, i.e.,

$$\inf\{\mu_A(z) \mid z \in x + y\} \geq \min\{\mu_A(x), \mu_A(y)\}$$

and

$$\sup\{\lambda_A(z) \mid z \in x + y\} \leq \max\{\lambda_A(x), \lambda_A(y)\},$$

which verify the conditions (1) and (4) of Definition 5.18.

Now, if $t_2 = \min\{\mu_A(a), \mu_A(x)\}$ for $x, a \in M$, then $a, x \in U(\mu_A; t_2)$. So there exist $y_1, z_1 \in U(\mu_A; t_2)$ such that $x \in a + y_1$ and $x \in z_1 + a$. Also we have $t_2 \leq \min\{\mu_A(y_1), \mu_A(z_1)\}$. Therefore the condition (2) of Definition 5.18 is verified. If we put $t_3 = \max\{\lambda_A(a), \lambda_A(x)\}$ then $a, x \in L(\lambda_A; t_3)$. So there exist $y_2, z_2 \in L(\lambda_A; t_3)$ such that $x \in a + y_2$ and $x \in z_2 + a$ and we have $\max\{\lambda_A(y_2), \lambda_A(y_2)\} \leq t_3$, and so the condition (5) of Definition 5.18 is verified.

Now, we verify the conditions (3) and (6). Let $t_4 = \mu_A(x)$ and $t_5 = \lambda_A(x)$ for some $x \in M$ and let $r \in R$. Then $x \in U(\mu_A; t_4)$ and $x \in L(\lambda_A, t_5)$. Since $U(\mu_A; t_4)$ and $L(\lambda_A, t_5)$ are H_v-submodules of M, we get $r \cdot x \subseteq U(\mu_A; t_4)$ and $r \cdot x \subseteq L(\lambda_A, t_5)$. Therefore for every $z \in r \cdot x$ we have $z \in U(\mu_A; t_4)$ and $z \in L(\lambda_A, t_5)$ which imply $\mu_A(z) \geq t_4$ and $\lambda_A(z) \leq t_5$. Hence

$$\inf\{\mu_A(z) \mid z \in r \cdot x\} \geq t_4 = \mu_A(x)$$

and

$$\sup\{\lambda_A(z) \mid z \in r \cdot x\} \leq t_5 = \lambda_A(x).$$

This completes the proof.   ∎

Corollary 5.6. *Let S be an H_v-submodule of an H_v-module M. If fuzzy sets μ and λ in M are defined by*

$$\mu(x) = \begin{cases} \alpha_0 & \text{if } x \in S, \\ \alpha_1 & \text{if } x \in M \setminus S, \end{cases} \qquad \lambda(x) = \begin{cases} \beta_0 & \text{if } x \in S, \\ \beta_1 & \text{if } x \in M \setminus S, \end{cases}$$

where $0 \leq \alpha_1 < \alpha_0$, $0 \leq \beta_0 < \beta_1$ and $\alpha_i + \beta_i \leq 1$ for $i = 0, 1$. Then $A = (\mu, \lambda)$ is an intuitionistic fuzzy H_v-submodule of M and $U(\mu; \alpha_0) = S = L(\lambda; \beta_0)$.

Corollary 5.7. *Let χ_S be the characteristic function of an H_v-submodule S of M. Then $A = (\chi_S, \chi_S^c)$ is an intuitionistic fuzzy H_v-submodule of M.*

Theorem 5.23. *If $A = (\mu_A, \lambda_A)$ is an intuitionistic fuzzy H_v-submodule of M, then*

$$\mu_A(x) = \sup\{\alpha \in [0,1] \mid x \in U(\mu_A; \alpha)\}$$

and

$$\lambda_A(x) = \inf\{\alpha \in [0,1] \mid x \in L(\lambda_A; \alpha)\}$$

for all $x \in M$.

Proof. Let $\delta = \sup\{\alpha \in [0,1] \mid x \in U(\mu_A; \alpha)\}$ and let $\varepsilon > 0$ be given. Then $\delta - \varepsilon < \alpha$ for some $\alpha \in [0,1]$ such that $x \in U(\mu_A; \alpha)$. This means that $\delta - \varepsilon < \mu_A(x)$ so that $\delta \le \mu_A(x)$ since ε is arbitrary.

We now show that $\mu_A(x) \le \delta$. If $\mu_A(x) = \beta$, then $x \in U(\mu_A; \beta)$ and so

$$\beta \in \{\alpha \in [0,1] \mid x \in U(\mu_A; \alpha)\}.$$

Hence

$$\mu_A(x) = \beta \le \sup\{\alpha \in [0,1] \mid x \in U(\mu_A; \alpha)\} = \delta.$$

Therefore

$$\mu_A(x) = \delta = \sup\{\alpha \in [0,1] \mid x \in U(\mu_A; \alpha)\}.$$

Now let $\eta = \inf\{\alpha \in [0,1] \mid x \in L(\lambda_A; \alpha)\}$. Then

$$\inf\{\alpha \in [0,1] \mid x \in L(\lambda_A; \alpha)\} < \eta + \varepsilon$$

for any $\varepsilon > 0$, and so $\alpha < \eta + \varepsilon$ for some $\alpha \in [0,1]$ with $x \in L(\lambda_A; \alpha)$. Since $\lambda_A(x) \le \alpha$ and ε is arbitrary, it follows that $\lambda_A(x) \le \eta$.

In order to prove $\lambda_A(x) \ge \eta$, let $\lambda_A(x) = \zeta$. Then $x \in L(\lambda_A; \zeta)$ and thus $\zeta \in \{\alpha \in [0,1] \mid x \in L(\lambda_A; \alpha)\}$. Hence

$$\inf\{\alpha \in [0,1] \mid x \in L(\lambda_A; \alpha)\} \le \zeta,$$

i.e., $\eta \le \zeta = \lambda_A(x)$. Consequently

$$\lambda_A(x) = \eta = \inf\{\alpha \in [0,1] \mid x \in L(\lambda_A; \alpha)\},$$

which completes the proof. ∎

Definition 5.20. A fuzzy set μ in a set X is said to have *sup property* if for every non-empty subset S of X, there exists $x_0 \in S$ such that

$$\mu(x_0) = \sup_{x \in S}\{\mu(x)\}.$$

Proposition 5.5. *Let M_1 and M_2 be two H_v-modules over an H_v-ring R and $f : M_1 \to M_2$ be a surjection. If $A = (\mu_A, \lambda_A)$ is an intuitionistic fuzzy H_v-submodule of M_1 such that μ_A and λ_A have sup property, then*

(1) $f(U(\mu_A; t)) = U(f(\mu_A); t)$,
(2) $f(L(\lambda_A; t)) \subseteq L(f(\lambda_A); t)$.

Proof.   (1) We have

$$
\begin{aligned}
y \in U(f(\mu_A); t) &\Leftrightarrow f(\mu_A)(y) \geq t \\
&\Leftrightarrow \sup_{x \in f^{-1}(y)} \{\mu_A(x)\} \geq t \\
&\Leftrightarrow \exists x_0 \in f^{-1}(y),\ \mu_A(x_0) \geq t \\
&\Leftrightarrow \exists x_0 \in f^{-1}(y),\ x_0 \in U(\mu_A; t) \\
&\Leftrightarrow f(x_0) = y,\ x_0 \in U(\mu_A; t) \\
&\Leftrightarrow y \in f(U(\mu_A; t)).
\end{aligned}
$$

(2) We have

$$
\begin{aligned}
y \in L(f(\lambda_A); t) &\Rightarrow f(\lambda_A)(y) \leq t \\
&\Rightarrow \sup_{x \in f^{-1}(y)} \{\lambda_A(x)\} \leq t \\
&\Rightarrow \lambda_A(x) \leq t,\ \text{for all } x \in f^{-1}(y) \\
&\Rightarrow x \in L(\lambda_A; t),\ \text{for all } x \in f^{-1}(y) \\
&\Rightarrow y \in f(L(\lambda_A; t)). \qquad \blacksquare
\end{aligned}
$$

Proposition 5.6. *Let M_1 and M_2 be two H_v-modules over an H_v-ring R and $f : M_1 \to M_2$ be a map. If $B = (\mu_B, \lambda_B)$ is an intuitionistic fuzzy H_v-submodule of M_2, then*

(1) $f^{-1}(U(\mu_B; t)) = U(f^{-1}(\mu_B); t)$,
(2) $f^{-1}(L(\lambda_B; t)) = L(f^{-1}(\lambda_B); t)$

for every $t \in [0, 1]$.

Proof.   (1) We have

$$
\begin{aligned}
x \in U(f^{-1}(\mu_B); t) &\Leftrightarrow f^{-1}(\mu_B)(x) \geq t \\
&\Leftrightarrow \mu_B(f(x)) \geq t \\
&\Leftrightarrow f(x) \in U(\mu_B; t) \\
&\Leftrightarrow x \in f^{-1}(U(\mu_B; t)).
\end{aligned}
$$

(2) We have

$$
\begin{aligned}
x \in L(f^{-1}(\lambda_B); t) &\Leftrightarrow f^{-1}(\lambda_B)(x) \leq t \\
&\Leftrightarrow \lambda_B(f(x)) \leq t \\
&\Leftrightarrow f(x) \in L(\lambda_B; t) \\
&\Leftrightarrow x \in f^{-1}(L(\lambda_B; t)). \qquad \blacksquare
\end{aligned}
$$

Definition 5.21. Let f be a map from a set X to a set Y. If $B = (\mu_B, \lambda_B)$ is an intuitionistic fuzzy set in Y, then the *inverse image* of B under f is defined by:

$$f^{-1}(B) = (f^{-1}(\mu_B), f^{-1}(\lambda_B)).$$

It is easy to see that $f^{-1}(B)$ is an intuitionistic fuzzy set in X.

Corollary 5.8. *Let M_1 and M_2 be two H_v-modules over an H_v-ring R and $f : M_1 \to M_2$ be a strong epimorphism. If $B = (\mu_B, \lambda_B)$ is an intuitionistic fuzzy H_v-submodule of M_2, then $f^{-1}(B)$ is an intuitionistic fuzzy H_v-submodule of M_1.*

Proof. Assume that $B = (\mu_B, \lambda_B)$ is an intuitionistic fuzzy H_v-submodule of M_2. By Theorem 4.7, we know that the sets $U(\mu_B; t)$ and $L(\lambda_B; t)$ are H_v-submodules of M_2 for every $t \in Im(\mu_B) \cap Im(\lambda_B)$. It follows from Proposition 3.6 that $f^{-1}(U(\mu_B; t))$ and $f^{-1}(L(\lambda_B; t))$ are H_v-submodules of M_1. Using Proposition 4.14, we have

$$f^{-1}(U(\mu_B; t)) = U(f^{-1}(\mu_B); t),$$

$$f^{-1}(L(\lambda_B; t)) = L(f^{-1}(\lambda_B); t).$$

Now, the proof is completed. ∎

Suppose $\gamma^*(r)$ is the equivalence class containing $r \in R$, and $\varepsilon^*(x)$ the equivalence class containing $x \in M$. The kernel of the canonical map $\varphi : M \to M/\varepsilon^*$ is is denoted by ω_M.

Definition 5.22. Let M be an H_v-module over an H_v-ring R and let $A = (\mu_A, \lambda_A)$ be an intuitionistic fuzzy H_v-submodule of M. The intuitionistic fuzzy set $A/\varepsilon^* = (\overline{\mu_A}^{\varepsilon^*}, \underline{\lambda_{A_{\varepsilon^*}}})$ is defined as follows:

$$\overline{\mu_A}^{\varepsilon^*} : M/\varepsilon^* \to [0, 1]$$

$$\overline{\mu_A}^{\varepsilon^*}(\varepsilon^*(x)) = \begin{cases} \sup_{a \in \varepsilon^*(x)} \{\mu_A(a)\} & \text{if } \varepsilon^*(x) \neq \omega_M \\ 1 & \text{if } \varepsilon^*(x) = \omega_M \end{cases}$$

and

$$\underline{\lambda_{A_{\varepsilon^*}}} : M/\varepsilon^* \to [0, 1]$$

$$\underline{\lambda_{A_{\varepsilon^*}}}(\varepsilon^*(x)) = \begin{cases} \inf_{a \in \varepsilon^*(x)} \{\lambda_A(a)\} & \text{if } \varepsilon^*(x) \neq \omega_M \\ 0 & \text{if } \varepsilon^*(x) = \omega_M. \end{cases}$$

In the following we show that

$$0 \le \overline{\mu_A}^{\varepsilon^*}(\varepsilon^*(x)) + \underline{\lambda_{A_{\varepsilon^*}}}(\varepsilon^*(x)) \le 1,$$

for all $\varepsilon^*(x) \in M/\varepsilon^*$.

If $\varepsilon^*(x) = \omega_M$, then the above inequalities are clear. Assume that $x \in H$ and $\varepsilon^*(x) \ne \omega_M$. Since $0 \le \mu_A(a)$ and $0 \le \lambda_A(a)$ for all $a \in \varepsilon^*(x)$, we have

$$0 \le \sup_{a \in \varepsilon^*(x)} \{\mu_A(a)\} + \inf_{a \in \varepsilon^*(x)} \{\lambda_A(a)\}$$

or

$$0 \le \overline{\mu_A}^{\varepsilon^*}(\varepsilon^*(x)) + \underline{\lambda_{A_{\varepsilon^*}}}(\varepsilon^*(x)).$$

On the other hand, we have

$$\mu_A(a) + \lambda_A(a) \le 1 \quad \text{or} \quad \mu_A(a) \le 1 - \lambda_A(a),$$

for all $a \in \varepsilon^*(x)$, and so

$$\begin{aligned}
\overline{\mu_A}^{\varepsilon^*}(\varepsilon^*(x)) &= \sup_{a \in \varepsilon^*(x)} \{\mu_A(a)\} \\
&\le \sup_{a \in \varepsilon^*(x)} \{1 - \lambda_A(a)\} \\
&= 1 - \inf_{a \in \varepsilon^*(x)} \{\lambda_A(a)\} \\
&= 1 - \underline{\lambda_{A_{\varepsilon^*}}}(\varepsilon^*(x)).
\end{aligned}$$

Hence $\overline{\mu_A}^{\varepsilon^*}(\varepsilon^*(x)) + \underline{\lambda_{A_{\varepsilon^*}}}(\varepsilon^*(x)) \le 1$.

Theorem 5.24. *Let M be an H_v-module over an H_v-ring R and let μ be a fuzzy H_v-submodule of M. Then $\overline{\mu_A}^{\varepsilon^*}$ is a fuzzy submodule of the module M/ε^*.*

Proof.   It is similar to the proof of Theorem 4.54.   ∎

Lemma 5.19. *We have*

$$(\overline{\lambda_A^c}^{\varepsilon^*})^c = \underline{\lambda_{A_{\varepsilon^*}}}.$$

Proof.   If $\varepsilon^*(x) = \omega_M$, then

$$(\overline{\lambda_A^c}^{\varepsilon^*})^c(\omega_H) = 1 - (\overline{\lambda_A^c}^{\varepsilon^*})(\omega_M) = 0 = \underline{\lambda_{A_{\varepsilon^*}}}(\omega_M).$$

Now, assume that $\varepsilon^*(x) \ne \omega_M$. Then

$$\begin{aligned}
(\overline{\lambda_A^c}^{\varepsilon^*})^c(\varepsilon^*(x)) &= 1 - (\overline{\lambda_A^c}^{\varepsilon^*})(\varepsilon^*(x)) \\
&= 1 - \sup_{a \in \varepsilon^*(x)} \{\lambda_A^c(a)\} \\
&= 1 - \sup_{a \in \varepsilon^*(x)} \{1 - \lambda_A(a)\} \\
&= \inf_{a \in \varepsilon^*(x)} \{\lambda_A(a)\} \\
&= \underline{\lambda_{A_{\varepsilon^*}}}(\varepsilon^*(x)).
\end{aligned}$$

    ∎

Theorem 5.25. *Let M be an H_v-module over an H_v-ring R and let $A = (\mu_A, \lambda_A)$ be an intuitionistic fuzzy H_v-submodule of M. Then $A/\varepsilon^* = (\overline{\mu_A}^{\varepsilon^*}, \underline{\lambda_{A_{\varepsilon^*}}})$ is an intuitionistic fuzzy submodule of the fundamental module M/ε^*.*

Proof.   It is straightforward.   ∎

Chapter 6

Hyperalgebra and Lie-Santilli theory

6.1 H_v-vector space

In this section, we present the notions of H_v-vector space and H_v-Lie algebra. These concepts are introduced by Vougiouklis [119, 120, 136].

Definition 6.1.
Let $(F, +, \cdot)$ be an H_v-field, $(V, +)$ a weak commutative H_v-group and there exists an external hyperoperation

$$\cdot : F \times V \to \mathcal{P}^*(V)$$
$$(a, x) \mapsto ax$$

such that for all $a, b \in F$ and $x, y \in V$ we have

(1) $a(x + y) \cap (ax + ay) \neq \emptyset$,
(2) $(a + b)x \cap (ax + bx) \neq \emptyset$,
(3) $(ab)x \cap a(bx) \neq \emptyset$.

Then V is called an H_v-*vector space* over F. In the case of an H_v-ring instead of H_v-field, we have an H_v-module.

In the above cases the fundamental relation ε^* is the smallest equivalence such that the quotient V/ε^* is a vector space over the fundamental field F/γ^*. Indeed, similar to H_v-groups, H_v-rings and H_v-modules, we have the following theorem.

Theorem 6.1. *Let $(V, +)$ be an H_v-vector space over the H_v-field F. Denote by ϑ the set of all expressions consisting of finite hyperoperations either on F and V or the external hyperoperation applied on finite sets of elements of F and V. We define the relation ε in V as follows:*

$$x \varepsilon y \iff \{x, y\} \subset u \text{ where } u \in \vartheta.$$

Then the relation ε^* is the transitive closure of the relation ε.

Proof. Let $\widehat{\varepsilon}$ be the transitive closure of ε and denote $\widehat{\varepsilon}(x)$ the class of the element x. First, we prove that the quotient set $V/\widehat{\varepsilon}$ is a vector space over the field F/γ^*. In $V/\widehat{\varepsilon}$ the sum $\oplus$ and the external product $\odot$, using the γ^* classes in F, is defined in the usual manner for all $A \in F$ and $x, y \in V$

$$\widehat{\varepsilon}(x) \oplus \widehat{\varepsilon}(y) = \{\widehat{\varepsilon}(z) \mid z \in \widehat{\varepsilon}(x) + \widehat{\varepsilon}(y)\},$$
$$\gamma^*(a) \odot \widehat{\varepsilon}(y) = \{\widehat{\varepsilon}(z) \mid z \in \gamma^*(a) \cdot \widehat{\varepsilon}(y)\}.$$

Take $x' \in \widehat{\varepsilon}(x)$ and $y' \in \widehat{\varepsilon}(y)$. Then, we have $x'\widehat{\varepsilon}x$ if and only if there exist $x_1, ..., x_{m+1}$ with $x_1 = x'$, $x_{m+1} = x$ and $u_1, ..., u_m \in \vartheta$ such that $\{x_i, x_{i+1} \subseteq u_i$ for $i = 1, ..., m$, and $y'\widehat{\varepsilon}y$ if and only if there exist $y_1, ..., y_{n+1}$ with $y_1 = y'$, $y_{n+1} = y$ and $v_1, ..., v_m \in \vartheta$ such that $\{y_j, y_{j+1} \subseteq v_j$ for $j = 1, ..., n$. From the above we obtain

$$\{x_i, x_{i+1}\} + y_1 \subseteq u_i + v_1, \text{ for } i = 1, ..., m-1, \tag{6.1}$$

$$x_{m+1} + \{y_j, y_{j+1}\} \subseteq u_m + v_j, \text{ for } j = 1, ..., n. \tag{6.2}$$

Thus, the sums

$$u_i + v_1 = t_i, \text{ for } i = 1, ..., m-1 \text{ and } u_m + v_j = t_{m+j-1}, \text{ for } j = 1, ..., n,$$

are also the elements of ϑ. Therefore, $t_k \in \vartheta$, for all $k \in \{1, ..., m+n-1\}$. Now, pick up any elements $z_1, ..., z_{m+n}$ such that

$$z_i \in x_i + y_1, \text{ for } i = 1, ..., m \text{ and } z_{m+j} \in x_{m+1} + y_{j+1}, \text{for } j = 1, ..., n,$$

and by using Equations 6.1 and 6.2, we have

$$\{z_k, z_{k+1}\} \subseteq t_k, \text{ for } k = 1, ..., m+n-1.$$

Therefore, every element $z_1 \in x_1 + y_1 = a' + b'$ is $\widehat{\varepsilon}$ equivalent to every element $z_{m+n} \in x_{m+1} + y_{n+1} = x + y$. Thus, $\widehat{\varepsilon}(x) + \widehat{\varepsilon}(y)$ is singleton. Hence, we can write

$$\widehat{\varepsilon}(x) \oplus \widehat{\varepsilon}(y) = \widehat{\varepsilon}(z), \text{ for all } z \in \widehat{\varepsilon}(x) + \widehat{\varepsilon}(y).$$

In a similar way, using the properties of γ^* in F, one can prove that

$$\gamma^*(a) \odot \widehat{\varepsilon}(x) = \widehat{\varepsilon}(z), \text{ for all } z \in \gamma^*(a) \cdot \widehat{\varepsilon}(x).$$

The weak associativity and the weak distributivity in F and V guarantee that the associativity and distributivity in the quotient set $V/\widehat{\varepsilon}$ over F/γ^*. Therefore, $V/\widehat{\varepsilon}$ is a vector space over the field F/γ^*.

Now, let σ be an equivalence relation in V such that V/σ is a vector space over F/γ^*. Denote $\sigma(x)$ the class of x. Then, $\sigma(x) \oplus \sigma(y)$ and $\gamma^*(a) \odot \sigma(x)$ are singletons for all $a \in F$ and $x, y \in V$, i.e.,

$$\sigma(x) \oplus \sigma(y) = \sigma(z), \text{ for all } z \in \sigma(x) + \sigma(y),$$
$$\gamma^*(a) \odot \sigma(x) = \sigma(z), \text{ for all } z \in \gamma^*(a) \cdot \sigma(x).$$

Thus, we can write, for every $a \in F$, $x, y \in V$ and $A \subseteq \gamma^*(a)$, $X \subseteq \sigma(x)$, $X \subseteq \sigma(y)$,

$$\sigma(x) \oplus \sigma(y) = \sigma(x + y) = \sigma(X + Y),$$
$$\gamma^*(a) \odot \sigma(x) = \sigma(ax) = \sigma(A \cdot X).$$

By the induction, we extend these relations on finite sums and external products. Thus, for every $u \in \vartheta$, we have

$$\sigma(x) = \sigma(u), \text{ for all } x \in u.$$

Consequently, for every $x \in V$, we obtain

$$x' \in \varepsilon(x) \;\Rightarrow\; x' \in \sigma(x).$$

But σ is transitively closed, so we obtain

$$x' \in \widehat{\varepsilon}(x) \;\Rightarrow\; x' \in \sigma(x).$$

That means that the relation $\widehat{\varepsilon}$ is the smallest equivalence relation in V such that $V/\widehat{\varepsilon}$ is a vector space over the field F/γ^*, i.e., $\widehat{\varepsilon} = \varepsilon^*$.   ∎

Remark 6.1. Let $(V, +)$ be an H_v-vector space over F. Denote $\sum a_i v_i$ where $a_i v_i$ is either the the product $a_i v_i$ if $a_i \in F$ or the element v_i if a_i does not appeared. Then, from the fundamental property it is immediate that

$$\varepsilon^*\left(\sum a_i v_i \right) = \sum \gamma^*(a_i) \varepsilon^*(v_i).$$

Denoting by $\beta^*(v)$ the fundamental class of v in $(V, +)$. Then, obviously, $\beta^*(u) \subseteq \varepsilon^*(v)$, for all $v \in V$.

Too elements $v_1, v_2 \in V$, which are not β^*-equivalent, can be ε^*-equivalent if they belong to a product av with $a \in F$ and $v \in V$.

Corollary 6.1. *If for all $a \in F$ and $v \in V$, there exists an element $v \in V$ such that $\gamma^*(a)\beta^*(v) \subseteq \beta^*(u)$, then $\varepsilon^* = \beta^*$.*

Definition 6.2. Let $[U]$ be the set of elements which belong to the sets of the form $\sum a_i v_i$, for all $a_i \in F$ and $v_i \in U$. If $[U] = V$, then U generates V. If there is no element $u \in U$ such that $u \in [U \setminus \{u\}]$, then U is called *linearly independent* subset of V.

Theorem 6.2. *If the vectors $\varepsilon^*(v_1), ..., \varepsilon^*(v_s)$ are independent over F/γ^*, then for all $v_1', ..., v_s' \in V$, where $v_i' \in \varepsilon^*(v_i)$, $i = 1, ..., s$, are independent over F.*

Proof. Suppose that there exists $v_t' \in [U_t]$, where $U_t = \{v_1', ..., v_s'\} \setminus \{v_t'\}$. Therefore there exists $a_i' s$ such that $v_t' \in \sum a_i v_i'$, where $v_i' \in U_i$. So, from Remark 6.1, it is obtained that $\varepsilon^* \left(\sum a_i v_i' \right) = \sum \gamma^*(a_i) \varepsilon^*(v_i')$ which is a contradiction. ∎

Definition 6.3. Let $(L, +)$ be an H_v-vector space over the H_v-field $(F, +, \cdot)$, $\phi : F \to F/\gamma^*$ the canonical map and $\omega_F = \{x \in F \mid \phi(x) = 0\}$, where 0 is the zero of the fundamental field F/γ^*. Similarly, let ω_L be the core of the canonical map $\phi' : L \to L/\varepsilon^*$ and denote by the same symbol 0 the zero of L/ε^*. Consider the *bracket (commutator) hhyperoperation*:

$$[\,,\,] : L \times L \to P(L) : (x, y) \to [x, y].$$

Then L is an H_v-*Lie algebra* over F if the following axioms are satisfied.

(L1) The bracket hyperoperation is bilinear, i.e.,
$$[\lambda_1 x_1 + \lambda_2 x_2, y] \cap (\lambda_1 [x_1, y] + \lambda_2 [x_2, y]) \neq \varnothing$$
$$[x, \lambda_1 y_1 + \lambda_2 y_2] \cap (\lambda_1 [x, y_1] + \lambda_2 [x, y_2]) \neq \varnothing,$$
for all $x, x_1, x_2, y, y_1, y_2 \in L, \lambda_1, \lambda_2 \in F$,

(L2) $[x, x] \cap \omega_L \neq \varnothing$, for all $x \in L$,

(L3) $([x, [y, z]] + [y, [z, x]] + [z, [x, y]]) \cap \omega_L \neq \varnothing$, for all $x, y, z \in L$.

This is a general definition thus one can use special cases in order to face problems in applied sciences.

6.2   e-hyperstructures

e-hyperstructures are a special kind of hyperstructures and, in what follows, we shall see that they can be interpreted as a generalization of two important concepts for physics: Isotopies and Genotopies. On the other hand, biological systems such as cells or organisms at large are open and irreversible because they grow. The representation of more complex systems, such as neural networks, requires more advances methods, such as hyperstructures. In this manner, e-hyperstructures can play a significant role for the representation of complex systems in physics and biology, such as nuclear fusion, the reproduction of cells or neural systems.

These applications were investigated by Santilli and Vougiouklis and we mention here some of their results and examples (see [97], [95]). Firstly, we shall define and analyze several types of e-hyperstructures.

Definition 6.4. A hypergroupoid $(H, \cdot)$ is called an e-*hypergroupoid* if H contains a scalar identity (also called unit) e, which means that for all $x \in H$, $x \cdot e = e \cdot x = x$.

In an e-hypergroupoid, an element x' is called *inverse* of a given element $x \in H$ if $e \in x \cdot x' = x' \cdot x$.

Clearly, if a hypergroupoid contains a scalar unit, then it is unique, while the inverses are not necessarily unique. In what follows, we use some examples which are obtained as follows: Take a set where an operation "$\cdot$" is defined, then we "enlarge" the operation putting more elements in the products of some pairs. Thus a hyperoperation "$\circ$" can be obtained, for which we have $x \cdot y \in x \circ y$, for all $x, y \in H$. Recall that the hyperstructures obtained in this way are H_b-structures.

Example 6.1. Consider the usual multiplication on the subset $\{1, -1, i, -i\}$ of complex numbers. Then we can consider the hyperoperation $\circ$ defined in the following table:

$\circ$	1	-1	i	$-i$
1	1	-1	i	$-i$
-1	-1	1	$-i$	$\{i, -i\}$
i	i	$-i$	-1	1
$-i$	$-i$	i	$\{1, i\}$	$\{-1, i\}$

Notice that we enlarged the products $(-1) \cdot (-i)$, $(-i) \cdot i$ and $(-i) \cdot (-i)$ by setting $(-1) \circ (-i) = \{i, -i\}$, $(-i) \circ i = \{1, i\}$ and $(-i) \circ (-i) = \{-1, i\}$. We obtain an e-hypergroupoid, with the scalar unit 1. The inverses of the elements $-1, i, -i$ are $-1, -i, i$ respectively. Moreover, the above structure is an H_v-abelian group, which means that the hyperoperation $\circ$ is weak associative, weak commutative and the reproductive axiom holds.

Example 6.2. Consider the set $H = \{f_i \mid i \in \{1, 2, 3, 4, 5, 6\}\}$ of real functions, defined from the real open interval $(0, 1)$ to $(0, 1)$, where $f_1(x) = x$, $f_2(x) = (1-x)^{-1}$, $f_3(x) = 1 - x^{-1}$, $f_4(x) = x^{-1}$, $f_5(x) = 1 - x$, $f_6(x) = x(1-x)^{-1}$. Let the multiplication on H be the usual composition of functions. We can obtain a hyperoperation ○, given by the following table:

○	f_1	f_2	f_3	f_4	f_5	f_6
f_1	f_1	f_2	f_3	f_4	f_5	f_6
f_2	f_2	f_3	f_1	$\{f_6,f_5\}$	$\{f_4,f_6\}$	$\{f_5,f_4\}$
f_3	f_3	f_1	f_2	$\{f_5,f_6\}$	$\{f_6,f_4\}$	$\{f_4,f_5\}$
f_4	f_4	f_5	f_6	f_1	f_2	f_3
f_5	f_5	f_6	f_4	$\{f_3,f_2\}$	f_1	f_2
f_6	f_6	f_4	f_5	$\{f_2,f_3\}$	$\{f_3,f_2\}$	f_1

We obtain an e-hypergroupoid, with the scalar unit f_1. The inverses of the elements f_2, f_3, f_4, f_5, f_6 are f_3, f_2, f_4, f_5, f_6 respectively. Moreover, the above structure is an H_v-abelian group.

Example 6.3. Consider now the finite noncommutative quaternion group

$$Q = \{1,\ -1,\ i,\ -i,\ j,\ -j,\ k,\ -k\},$$

for which the multiplication is given by the following table:

○	1	-1	i	$-i$	j	$-j$	k	$-k$
1	1	-1	i	$-i$	j	$-j$	k	$-k$
-1	-1	1	$-i$	i	$-j$	j	$-k$	$-k$
i	i	$-i$	-1	1	k	$-k$	$-j$	j
$-i$	$-i$	i	1	-1	$-k$	k	j	$-j$
j	j	$-j$	$-k$	k	-1	1	i	$-i$
$-j$	$-j$	j	k	$-k$	1	-1	$-i$	i
k	k	$-k$	j	$-j$	$-i$	i	-1	1
$-k$	$-k$	k	$-j$	j	i	$-i$	1	-1

Denote $\bar{i} = \{i, -i\}, \bar{j} = \{j, -j\}$ and $\bar{k} = \{k, -k\}$. We can obtain a hyperoperation ○, given by the following table:

○	1	-1	i	$-i$	j	$-j$	k	$-k$
1	1	-1	i	$-i$	j	$-j$	k	$-k$
-1	-1	1	$-i$	i	$-j$	j	$-k$	$-k$
i	i	$-i$	-1	1	k	$\bar{k}$	$\bar{j}$	j
$-i$	$-i$	i	1	-1	$\bar{k}$	k	j	$\bar{j}$
j	j	$-j$	$\bar{k}$	k	-1	1	i	$\bar{i}$
$-j$	$-j$	j	k	$\bar{k}$	1	-1	$\bar{i}$	i
k	k	$-k$	j	$\bar{j}$	$\bar{i}$	i	-1	1
$-k$	$-k$	k	$\bar{j}$	j	i	$\bar{i}$	1	-1

We obtain an e-hypergroupoid, with the scalar unit 1. The inverses of the elements $-1, i, -i, j, -j, k, -k$ are $-1, -i, i, -j, j, -k, k$ respectively. Moreover, the above structure is an H_v-abelian group, too.

It is immediate the following basic result, that holds for all the above examples.

Theorem 6.3. *The weak associativity is valid for all H_b-structures with associative basic operations.*

We are interested now in another kind of an e-hyperstructure, which is the e-hyperfield.

Definition 6.5. A set F, endowed with an operation $+$, which we call addition, and a hyperoperation $\cdot$, called multiplication, is said to be an e-hyperfield if the following axioms are valid.

(1) $(F, +)$ is an abelian group where 0 is the additive unit;
(2) the multiplication $\cdot$ is weak associative;
(3) the multiplication $\cdot$ is weak distributive with respect to $+$, i.e., for all $x, y, z \in F$,
$$x(y + z) \cap (xy + xz) \neq \emptyset, \quad (x + y)z \cap (xz + yz) \neq \emptyset;$$
(4) 0 is an absorbing element, i.e., for all $x \in F$, $0 \cdot x = x \cdot 0 = 0$;
(5) there exists a multiplicative scalar unit 1, i.e., for all $x \in F$, $1 \cdot x = x \cdot 1 = x$;
(6) for every element $x \in F$ there exists an inverse x^{-1} such that $1 \in x \cdot x^{-1} \cap x^{-1} \cdot x$.

The elements of an e-hyperfield $(F, +, \cdot)$ are called e-hypernumbers.

Example 6.4.

(1) Starting with the ring $\mathbb{Z}_3 = \{\bar{0}, \bar{1}, \bar{2}\}$, we can obtain a hyperring by enlarging the product $\bar{2} \circ \bar{2} = \{\bar{1}\}$ to $\bar{2} \circ \bar{2} = \{\bar{1}, \bar{2}\}$. In other words, we obtain the following table:

$\circ$	$\bar{0}$	$\bar{1}$	$\bar{2}$
$\bar{0}$	$\bar{0}$	$\bar{0}$	$\bar{0}$
$\bar{1}$	$\bar{0}$	$\bar{1}$	$\bar{2}$
$\bar{2}$	$\bar{0}$	$\bar{2}$	$\{\bar{1}, \bar{2}\}$

The above structure is an e-hyperfield.

(2) In the above example, only a hyperproduct is not a singleton. These hyperstructures, for which only a hyperproduct is not a singleton, are called *very thin* and they are useful to the theory of representations of H_v-groups by hypermatrices, see [117]. Hence, a way to obtain

a very thin hyperstructure is the following one: we take a classical structure and we choose two elements a, b, then we can enlarge the product $a \cdot b$. Therefore, in order to obtain a very thin e-hyperfield we can take a field and enlarge only one product of two, nonzero and non-unit elements. This simple change of the operation leads to enormous changes to the algebraic hyperstructure, so it looks like a chain reaction in physics.

(3) Another large class of e-hyperfields can be obtained by using H_b-structures. For instance, we can take the field of real numbers $\mathbb{R}$, or the field of complex numbers $\mathbb{C}$ or the field of quaternions Q and then we can enlarge all products of nonzero and nonunit elements by adding nonzero elements and so we obtain e-hyperfields.

(4) We can use the above method starting from an e-hyperfield or a ring, not necessarily a field. For instance, we can take the ring $\mathbb{Z}_6 = \{\bar{0}, \bar{1}, \bar{2}, \bar{3}, \bar{4}, \bar{5}\}$ of integers modulo 6. We consider a hyperoperation $\circ$ given by the following table:

$\circ$	$\bar{0}$	$\bar{1}$	$\bar{2}$	$\bar{3}$	$\bar{4}$	$\bar{5}$
$\bar{0}$	$\bar{0}$	$\bar{0}$	$\bar{0}$	$\bar{0}$	$\bar{0}$	$\bar{0}$
$\bar{1}$	$\bar{0}$	$\bar{1}$	$\bar{2}$	$\bar{3}$	$\bar{4}$	$\bar{5}$
$\bar{2}$	$\bar{0}$	$\bar{2}$	$\bar{4}$	$\{\bar{0},\bar{1}\}$	$\{\bar{2},\bar{3}\}$	$\{\bar{4},\bar{5}\}$
$\bar{3}$	$\bar{0}$	$\bar{3}$	$\{\bar{0},\bar{1}\}$	$\{\bar{3},\bar{2}\}$	$\{\bar{0},\bar{5}\}$	$\{\bar{3},\bar{4}\}$
$\bar{4}$	$\bar{0}$	$\bar{4}$	$\{\bar{2},\bar{3}\}$	$\{\bar{0},\bar{5}\}$	$\{\bar{4},\bar{1}\}$	$\bar{2}$
$\bar{5}$	$\bar{0}$	$\bar{5}$	$\{\bar{4},\bar{5}\}$	$\{\bar{3},\bar{4}\}$	$\bar{2}$	$\bar{1}$

Then $(\mathbb{Z}_6, +, \circ)$ is an e-hyperfield, for which the multiplication is not closed in $\mathbb{Z}_6 - \{\bar{0}\}$.

For the following example, we recall a *P-hyperstructure* notion. Let $(G, \cdot)$ be a semigroup and $P \subseteq G$, $P \neq \emptyset$. The following hyperoperations are called *P-hyperoperations*:

$$xP^*y = xPy, \quad xP_r^*y = xyP, \quad xP_l^*y = Pxy,$$

for all $x, y \in G$. If in a set they are defined P-hyperoperations, then we obtain P-hyperstructures.

The P-hyperoperation P^* is associative, so (G, P^*) is a semihypergroup. If $P \subseteq Z(G)$, where $Z(G)$ is the center of G, then the above three hyperoperations coincide. P-hyperoperations can be defined in groupoids or hypergroupoids as well, and so we obtain a large class of hyperstructures. We can also define P-hyperoperations in sets with partial operations. Moreover, in structures with more than one operations, we can define more P-hyperoperations. In a P-hypergroup the set of left or right units is P. The set of left inverses of x with respect to the unit p_0^{-1} is $p_0^{-1}x^{-1}P^{-1}$ and similarly the set of right inverses of x with respect to the unit p_0^{-1} is $P^{-1}x^{-1}p_0^{-1}$.

Definition 6.6. An *e-hypermatrix* is a matrix with entries elements of an e-hyperfield.

We can define the product of two e-matrices in an usual manner: the elements of product of two e-matrices (a_{ij}), (b_{ij}) are $c_{ij} = \sum a_{ik} \circ b_{kj}$, where the sum of products is the usual sum of sets.

If we consider the e-hyperfield given in Example 6.4(1), then we have:

$$
\begin{bmatrix} \bar{2} & \bar{1} \\ \bar{2} & \bar{0} \end{bmatrix} \circ \begin{bmatrix} \bar{2} & \bar{1} \\ \bar{1} & \bar{1} \end{bmatrix} = \begin{bmatrix} \bar{2}\circ\bar{2}+\bar{1}\circ\bar{1} & \bar{2}\circ\bar{1}+\bar{1}\circ\bar{1} \\ \bar{2}\circ\bar{2}+\bar{0}\circ\bar{1} & \bar{2}\circ\bar{1}+\bar{0}\circ\bar{1} \end{bmatrix}
$$

$$
= \begin{bmatrix} \{\bar{1},\bar{2}\}+\bar{1} & \bar{2}+\bar{1} \\ \{\bar{1},\bar{2}\}+\bar{0} & \bar{2}+\bar{0} \end{bmatrix}
$$

$$
= \begin{bmatrix} \{\bar{2},\bar{0}\} & \bar{0} \\ \{\bar{1},\bar{2}\} & \bar{2} \end{bmatrix}
$$

$$
= \left\{ \begin{bmatrix} \bar{2} & \bar{0} \\ \bar{1} & \bar{2} \end{bmatrix}, \begin{bmatrix} \bar{2} & \bar{0} \\ \bar{2} & \bar{2} \end{bmatrix}, \begin{bmatrix} \bar{0} & \bar{0} \\ \bar{1} & \bar{2} \end{bmatrix}, \begin{bmatrix} \bar{0} & \bar{0} \\ \bar{2} & \bar{2} \end{bmatrix} \right\}.
$$

Moreover, notice that the product of an e-hypernumber with an e-hypermatrix is also a hyperoperation. For instance, again on the above hyperfield, we have

$$
\bar{2} \circ \begin{bmatrix} \bar{2} & \bar{1} \\ \bar{2} & \bar{2} \end{bmatrix} = \begin{bmatrix} \bar{2}\circ\bar{2} & \bar{2}\circ\bar{1} \\ \bar{2}\circ\bar{2} & \bar{2}\circ\bar{2} \end{bmatrix}
$$

$$
= \left\{ \begin{bmatrix} \bar{1} & \bar{2} \\ \bar{1} & \bar{1} \end{bmatrix}, \begin{bmatrix} \bar{1} & \bar{2} \\ \bar{1} & \bar{2} \end{bmatrix}, \begin{bmatrix} \bar{1} & \bar{2} \\ \bar{2} & \bar{1} \end{bmatrix}, \begin{bmatrix} \bar{1} & \bar{2} \\ \bar{2} & \bar{2} \end{bmatrix}, \begin{bmatrix} \bar{2} & \bar{2} \\ \bar{1} & \bar{1} \end{bmatrix}, \begin{bmatrix} \bar{2} & \bar{2} \\ \bar{1} & \bar{2} \end{bmatrix}, \begin{bmatrix} \bar{2} & \bar{2} \\ \bar{2} & \bar{1} \end{bmatrix}, \begin{bmatrix} \bar{2} & \bar{2} \\ \bar{2} & \bar{2} \end{bmatrix} \right\}.
$$

This remark is useful for the definition of an *e-hypervector space*.

Definition 6.7. Let $(F, +, \cdot)$ be an e-hyperfield. An ordered set $a = (a_1, a_2, ..., a_n)$ of n e-hypernumbers of F is called an *e-hypervector* and the e-hypernumbers a_i, $i \in \{1, 2, ..., n\}$ are called *components* of the e-hypervector a.

Two e-hypervectors are equals if they have equal corresponding components. The hypersums of two e-hypervectors a, b is defined as follows:

$$a + b = \{(c_1, c_2, ..., c_n) \mid c_i \in a_i + b_i, \ i \in \{1, 2, ..., n\}\}.$$

The scalar hypermultiplication of an e-hypervector a by an e-hypernumber λ is defined in a usual manner:

$$\lambda \circ a = \{(c_1, c_2, ..., c_n) \mid c_i \in \lambda \cdot a_i, \ i \in \{1, 2, ..., n\}\}.$$

The set F^n of all e-hypervectors with elements of F, endowed with the hypersum and the scalar hypermultiplication is called *n-dimensional e-hypervector space*. The set of $m \times n$ hypermatrices is an mn-dimensional e-hypervector space.

The next proposition can be easily verified.

Proposition 6.1. *Let F be an e-hyperfield and F^n be its n-dimensional e-hypervector space. Then the following assertions hold:*

(1) *the additive unit is the zero e-hypervector $0 = (0, 0, ..., 0)$;*
(2) $\lambda \circ (a + b) \cap (\lambda \circ a + \lambda \circ b) \neq \emptyset$, *for all $\lambda \in F$ and $a, b \in F^n$;*
(3) $(\lambda + \alpha) \circ a \cap (\lambda \circ a + \alpha \circ a) \neq \emptyset$, *for all $\lambda, \alpha \in F$ and $a \in F^n$;*
(4) $\lambda \circ (\alpha \circ a) \cap (\lambda \cdot \alpha) \circ a \neq \emptyset$, *for all $\lambda, \alpha \in F$ and $a \in F^n$;*
(5) $1 \circ a = a$, $\lambda \circ 0 = 0$, *for all $\lambda \in F$ and $a \in F^n$.*

Notice that by $(\lambda + \alpha) \cdot a_i$ we intend $\bigcup\limits_{t \in \lambda + \alpha} t \cdot a_i$, while $\lambda \cdot a_i + \alpha \cdot a_i$ means that

$$\bigcup\limits_{\substack{x \in \lambda \cdot a_i \\ y \in \alpha \cdot a_i}} (x + y).$$

Definition 6.8. An *e-hyperalgebra* over an e-hyperfield $(F, +, \cdot)$ is an n-dimensional e-hypervector space F^n, endowed with a multiplication of e-hypervectors $\odot$, such that $(F^n, +, \odot)$ is an e-hyperring and for all $\lambda \in F$ and all $x, y \in F^n$, we have

$$\lambda \circ (x \odot y) = (\lambda \circ x) \odot y = x \odot (\lambda \circ y).$$

The most important example of an e-hyperalgebra is the algebra of $n \times n$ square e-hypermatrices.

As it is well know, Lie's theory is at the foundation of all physical theories, including classical and quantum mechanics, particle physics, nuclear physics, superconductivity, chemistry, astrophysics, etc. Despite the mathematical and physical consistency, by no means Lie's theory can represent the totality of systems existing in the universe. We conclude the presentation of e-hyperstructures with the definition of an e-hyper-Lie-algebra.

Definition 6.9. Let $(L, +)$ be an e-hypervector space over an e-hyperfield $(F, +, \cdot)$. Consider any bracket or commutator hyperoperation:

$$[\, , \,] : L \times L \to \mathcal{P}^*(L) : \ (x, y) \mapsto [x, y].$$

Then L is an *e-hyper-Lie-algebra* over F if the following axioms hold:

(1) the bracket hyperoperation is bilinear, i.e.

$$\forall x, x_1, x_2, y, y_1, y_2 \in L, \forall \alpha_1, \alpha_2, \beta_1, \beta_2 \in F,$$
$$[\alpha_1 x_1 + \alpha_2 x_2, y] \cap (\alpha_1 [x_1, y] + \alpha_2 [x_2, y]) \neq \emptyset,$$
$$[x, \beta_1 y_1 + \beta_2 y_2] \cap (\beta_1 [x, y_1] + \beta_2 [x, y_2]) \neq \emptyset;$$

(2) $\forall x \in L, 0 \in [x, x]$;
(3) $\forall x, y, z \in L, 0 \in ([x, [y, z]] + [y, [z, x]] + [z, [x, y]])$.

The most important thing in studying e-hyper-Lie-algebras is to check if a subset is closed under the Lie bracket. This is so, because the product of hypermatrices normally has an enormous number of elements. However, for some interesting subclasses it is easy to check if they are closed or not.

Example 6.5.

(1) Consider the Lie bracket of the two traceless e-hypermatrices, over the e-hyperfield given in Example 6.4(1):

$$A = \begin{bmatrix} \bar{2} & \bar{2} \\ \bar{1} & \bar{1} \end{bmatrix}, \quad B = \begin{bmatrix} \bar{1} & \bar{0} \\ \bar{2} & \bar{2} \end{bmatrix}.$$

We obtain

$$[A, B] = \begin{bmatrix} \bar{2} & \bar{2} \\ \bar{1} & \bar{1} \end{bmatrix} \cdot \begin{bmatrix} \bar{1} & \bar{0} \\ \bar{2} & \bar{2} \end{bmatrix} - \begin{bmatrix} \bar{1} & \bar{0} \\ \bar{2} & \bar{2} \end{bmatrix} \cdot \begin{bmatrix} \bar{2} & \bar{2} \\ \bar{1} & \bar{1} \end{bmatrix}$$

$$= \begin{bmatrix} \bar{2} + \{\bar{1}, \bar{2}\} & \bar{0} + \{\bar{1}, \bar{2}\} \\ \bar{2} + \bar{2} & \bar{0} + \bar{2} \end{bmatrix} - \begin{bmatrix} \bar{2} + \bar{0} & \bar{2} + \bar{0} \\ \{\bar{1}, \bar{2}\} + \bar{2} & \{\bar{1}, \bar{2}\} + \bar{2} \end{bmatrix}$$

$$= \begin{bmatrix} \{\bar{0}, \bar{1}\} & \{\bar{1}, \bar{2}\} \\ \bar{1} & \bar{2} \end{bmatrix} - \begin{bmatrix} \bar{2} & \bar{2} \\ \{\bar{0}, \bar{1}\} & \{\bar{0}, \bar{1}\} \end{bmatrix}$$

$$= \begin{bmatrix} \{\bar{1}, \bar{2}\} & \{\bar{2}, \bar{0}\} \\ \{\bar{1}, \bar{0}\} & \{\bar{2}, \bar{1}\} \end{bmatrix}.$$

We notice that the Lie bracket of A and B consists of 16 e-hypermatrices and some of them are not traceless. For example,

$$\begin{bmatrix} \bar{1} & \bar{0} \\ \bar{1} & \bar{1} \end{bmatrix}, \quad \begin{bmatrix} \bar{2} & \bar{2} \\ \bar{0} & \bar{2} \end{bmatrix}.$$

Hence the set of traceless e-hypermatrices is not closed.

(2) Let F be a field and P be a set such that $1 \in P \subseteq F - \{0\}$. We define the following hyperoperation: $xP^*y = xPy$, for all $x, y \in F - \{0, 1\}$ and $xP^*y = xy$ otherwise.

For instance, if we take the field $\mathbb{Z}_7$ of integers modulo 7 and the set $P = \{\bar{1}, \bar{3}\}$, then the hyperoperation P^* is given by the following table:

P^*	$\bar{0}$	$\bar{1}$	$\bar{2}$	$\bar{3}$	$\bar{4}$	$\bar{5}$	$\bar{6}$
$\bar{0}$	$\bar{0}$	$\bar{0}$	$\bar{0}$	$\bar{0}$	$\bar{0}$	$\bar{0}$	$\bar{0}$
$\bar{1}$	$\bar{0}$	$\bar{1}$	$\bar{2}$	$\bar{3}$	$\bar{4}$	$\bar{5}$	$\bar{6}$
$\bar{2}$	$\bar{0}$	$\bar{2}$	$\{\bar{4},\bar{5}\}$	$\{\bar{6},\bar{4}\}$	$\{\bar{1},\bar{3}\}$	$\{\bar{3},\bar{2}\}$	$\{\bar{5},\bar{1}\}$
$\bar{3}$	$\bar{0}$	$\bar{3}$	$\{\bar{6},\bar{4}\}$	$\{\bar{2},\bar{6}\}$	$\{\bar{5},\bar{1}\}$	$\{\bar{1},\bar{3}\}$	$\{\bar{4},\bar{5}\}$
$\bar{4}$	$\bar{0}$	$\bar{4}$	$\{\bar{1},\bar{3}\}$	$\{\bar{5},\bar{1}\}$	$\{\bar{2},\bar{6}\}$	$\{\bar{6},\bar{4}\}$	$\{\bar{3},\bar{2}\}$
$\bar{5}$	$\bar{0}$	$\bar{5}$	$\{\bar{3},\bar{2}\}$	$\{\bar{1},\bar{3}\}$	$\{\bar{6},\bar{4}\}$	$\{\bar{4},\bar{5}\}$	$\{\bar{2},\bar{6}\}$
$\bar{6}$	$\bar{0}$	$\bar{6}$	$\{\bar{5},\bar{1}\}$	$\{\bar{4},\bar{5}\}$	$\{\bar{3},\bar{2}\}$	$\{\bar{2},\bar{6}\}$	$\{\bar{1},\bar{3}\}$

Then, $(\mathbb{Z}_7, +, P^*)$ is an e-hyperfield.

Now, consider the e-hyperfield based on $\mathbb{Z}_7$, where the multiplication is replaced by the P-hyperoperation given in the above table. Take the following e-hypermatrices:

$$A = \begin{bmatrix} \bar{1} & \bar{2} \\ \bar{0} & \bar{3} \end{bmatrix}, \quad B = \begin{bmatrix} \bar{4} & \bar{5} \\ \bar{0} & \bar{6} \end{bmatrix}.$$

Then the Lie bracket is:

$$[A, B] = \left[\begin{bmatrix} \bar{1} & \bar{2} \\ \bar{0} & \bar{3} \end{bmatrix}, \begin{bmatrix} \bar{4} & \bar{5} \\ \bar{0} & \bar{6} \end{bmatrix} \right]$$

$$= \begin{bmatrix} \bar{1}\cdot\bar{4}+\bar{2}\cdot\bar{0} & \bar{1}\cdot\bar{5}+\bar{2}\cdot\bar{6} \\ \bar{0}\cdot\bar{4}+\bar{3}\cdot\bar{0} & \bar{0}\cdot\bar{5}+\bar{3}\cdot\bar{6} \end{bmatrix} - \begin{bmatrix} \bar{4}\cdot\bar{1}+\bar{5}\cdot\bar{0} & \bar{4}\cdot\bar{2}+\bar{5}\cdot\bar{3} \\ \bar{0}\cdot\bar{1}+\bar{6}\cdot\bar{0}\} & \bar{0}\cdot\bar{2}+\bar{6}\cdot\bar{3} \end{bmatrix}$$

$$= \begin{bmatrix} \{\bar{4} & \{\bar{3},\bar{6}\} \\ \bar{0} & \{\bar{4},\bar{5}\} \end{bmatrix} - \begin{bmatrix} \bar{4} & \{\bar{2},\bar{4},\bar{6}\} \\ \bar{0} & \{\bar{4},\bar{5}\} \end{bmatrix}$$

$$= \begin{bmatrix} \bar{0} & \{\bar{1},\bar{6},\bar{4},\bar{2},\bar{0}\} \\ \bar{0} & \{\bar{0},\bar{6},\bar{1}\} \end{bmatrix}.$$

The Lie bracket of A and B has 15 elements among them 5 are strictly upper triangular e-hypermatrices. Therefore the set of upper triangular e-hypermatrices is closed under the Lie bracket hyperoperation.

Now, we connect the above e-hyperstructures to isotopies and genotopies. We give now an idea about these topics, which were constructed for physical needs.

Isotopies can be traced back to the early stages of set theory, where two Latin squares were said to be *isotopically related* when they can be made to coincide via permutations. Since Latin squares can be interpreted as the multiplication table of quasigroups, the isotopies propagated to quasigroups and then to Jordan algebras. Santilli used the term *isotopy* from its Greak meaning of *preserving the topology* and interpreted them as axiom-preserving. In fact, the new and old structures are indistinguishable at the abstract. Nowadays, the term "isotopies" denotes nonlinear, nonlocal and nonhamiltonian liftings of any given linear, local and Hamiltonian structure, which preserves linearity, locality and canonicity in generalized spaces over generalized fields.

The main novelty of the isotopies studied by Santilli with respect to the preceding ones is the lifting of the trivial n-dimensional unit $I = diag(1, 1, ..., 1)$ of a conventional theory into a nowhere singular, symmetric, real-valued, positive-defined and n-dimensional matrix:

$$\hat{I} = (\hat{I}_{i,j}) = (\hat{I}_{j,i}) = \hat{I}^{-1} = (\hat{I}_{i,j})^{-1} = (\hat{I}_{j,i})^{-1}, \ i,j \in \{1, 2, ..., n\},$$

whose elements have a smooth but otherwise arbitrary functional dependence on the local coordinates x, their derivates $\dot{x}$, $\ddot{x}$, ... with respect to an independent variable t and any needed additional local quantity,

$$\hat{I} \rightarrow \hat{I}(x, \dot{x}, \ddot{x}, ...).$$

The original theory is reconstructed in such a way to admit $\hat{I}$ as the new left and right unit. Thus, if $(F, +, \cdot)$ is a field of characteristic zero, then we can construct an *isofield* $\hat{F} = (\hat{F}, +, \circ)$, whose elements have the form $\hat{a} = a \cdot \hat{1}$, where $a \in F$ and $\hat{1}$ is a positive-defined element generally outside F. The new multiplication $\circ$, called *isomultiplication* is defined as follows:

$$\hat{a} \circ \hat{b} = \hat{a} \cdot \hat{1} \cdot \hat{b}, \text{ for all } \hat{a}, \hat{b}.$$

The element $\hat{1} = \hat{1}^{-1}$ is the left and right unit of $\hat{F}$. The structure $(\hat{F}, +, \circ)$ is a new field and it is called an *isotope* of F, while the lifting $F \rightarrow \hat{F}$ is called an *isotopy*. For instance, we obtain the isofields $(\mathbb{R}, +, \circ)$ of *isoreal numbers*,

$(\mathbb{C}, +, \circ)$ of *isocomplex numbers*, $(Q, +, \circ)$ of *isoquaternions*. Notice that $\hat{1}^{\hat{n}} = \underbrace{\hat{1} \circ ... \circ \hat{1}}_{n} = \hat{1}$, $\hat{1}/\hat{1} = \hat{1}$.

Genotopies were introduced by Santilli from the Greak meaning of *inducing topology* and interpreted them as liftings of a given theory verifying certain axioms into a form which verifies broader axioms admitting the original ones as particular cases. The main difference between isotopies and the genotopies is that the isomultiplication of two isonumbers $\hat{a}, \hat{b}$ has no ordering, while for the genotopies one must assume an ordering.

The multiplication of two quantities is ordered on the right and it is denoted by the symbol $>$, when the first quantity multiplies the second one on the right, while it is ordered on the left and denoted by the symbol $<$, when the second quantity multiplies the first one on the left.

The genotopies are based on the property that the restriction of the multiplication on the right in an ordered field permits the preservation of all axioms of a field. We obtain two fields $(F^>, +, >)$ and $(^<F, +, <)$, based on the multiplication on the right and on the left respectively. The genotopies emerge when the multiplication on the right is assumed to be different from that on the left. Hence we have two different generalized units, one for the multiplication on the right $\hat{I}^>$ and one for the multiplication on the left $^<\hat{I}$. For isotopies, we have the same isounit $\hat{I}$ for both isomultiplications. Isotopies are a particular case of genofields.

If $(F, +, \cdot)$ is a field of characteristic zero, then we can construct an *genofield on the right* $\hat{F}^> = (\hat{F}^>, +, \hat{>})$, whose elements have the form $\hat{a}^> = \hat{a} \cdot \hat{1}^>$ and are called *genonumbers on the right*, where $a \in F$ and $\hat{1}^>$ is a quantity generally outside F and $\hat{F}$. The new multiplication $\hat{>}$, called *genomultiplication on the right* is defined as follows:

$$\hat{a} \hat{>} \hat{b} = \hat{a} \cdot \hat{Q} \cdot \hat{b}, \text{ for all } \hat{a}, \hat{b}.$$

The element $\hat{Q}^{-1}$ is the left and right unit $\hat{1}^>$ of $\hat{F}^>$. In other words, for all $\hat{a}^> \in \hat{F}^>$, $\hat{1}^> \hat{>} \hat{a}^> = \hat{a}^> = \hat{a}^> \hat{>} \hat{1}^>$.

A *genofield on the left* $^<\hat{F} = (^<\hat{F}, +, \hat{<})$ is the image of $\hat{F}^> = (\hat{F}^>, +, \hat{>})$, under the replacement of the genomultiplication on the right $\hat{>}$ with the genomultiplication on the left:

$$^<\hat{a} \hat{<} {}^<\hat{b} = {}^< \hat{a} \cdot \hat{P} \cdot {}^<\hat{b},$$

with the corresponding *genounit on the left* $^<\hat{I} = \hat{P}^{-1}$, i.e. for all $^<\hat{a} \in{}^< \hat{F}$, we have $^<\hat{1} \hat{<} {}^<\hat{a} ={}^< \hat{a} ={}^< \hat{a} \hat{<} {}^<\hat{1}$.

For $P = Q$ we obtain isotopies.

The unit $\hat{I}$ of the isotopies and the units $^<\hat{I}$, $\hat{I}^>$ of the genotopies have a realization, e.g. via a given function or matrix. A first class of e-hyperstructures can be introduced as hyperstructures with hypermultiplication, for which e verifies the weak condition to be a left and a right unit and this class leads us to isotopies, while a second class of e-hyperstructures requires the further differentiation of the hypermultiplication on the right from that on the left and this leads us to genotopies. Hence isotopies and genotopies represent particular cases of e-hyperstructures. As a remark, we considered multiplicative hyperfields in order to define an e-hyperfield, since the sum is not lifted in isotopies.

In order to transfer Santilli's iso-theory theory into the hyperstructure case we introduce two general constructions on this direction.

Construction 6.4. (General enlargement). On a field $(F, +, \cdot)$ and on the isofield $\hat{F} = \hat{F}(\hat{a}, \hat{+}, \hat{\times})$ we replace in the results of the iso-product

$$\hat{a} \hat{\times} \hat{b} = \hat{a} \times \hat{T} \times \hat{b}, \text{ with } \hat{1} = \hat{T}^{-1}$$

of the element $\hat{T}$ by a set of elements $\hat{H}_{ab} = \{\hat{T}, \hat{x}_1, \hat{x}_2, \ldots\}$ where $\hat{x}_1, \hat{x}_2, \ldots \in \hat{F}$, containing $\hat{T}$, for all $\hat{a} \hat{\times} \hat{b}$ for which $\hat{a}, \hat{b} \notin \{\hat{0}, \hat{1}\}$ and $\hat{x}_1, \hat{x}_2, \ldots \in \hat{F} - \{\hat{0}, \hat{1}\}$. If one of $\hat{a}, \hat{b}$, or both, is equal to $\hat{0}$ or $\hat{1}$, then $\hat{H}_{ab} = \{\hat{T}\}$. Therefore the new iso-hyperoperation is

$$\hat{a} \hat{\times} \hat{b} = \hat{a} \times \hat{H}_{ab} \times \hat{b} = \hat{a} \times \{\hat{T}, \hat{x}_1, \hat{x}_2, \ldots\} \times \hat{b}, \text{ for all } \hat{a}, \hat{b} \in \hat{F}$$

$\hat{F} = \hat{F}(\hat{a}, \hat{+}, \hat{\times})$ becomes isoH_v-field. The elements of F are called *isoH_v-numbers or isonumbers*.

More important hyperoperations, of the above construction, are the ones where only for few ordered pairs $(\hat{a}, \hat{b})$ the result is enlarged, even more, the extra elements $\hat{x}_i$, are only few, preferable one. Thus, this special case is if there exists only one pair $(\hat{a}, \hat{b})$ for which

$$\hat{a} \hat{\times} \hat{b} = \hat{a} \times \{\hat{T}, \hat{x}\} \times \hat{b}, \text{ for all } \hat{a}, \hat{b} \in \hat{F}$$

and the rest are ordinary results, then we have a very thin isoH_v-field.

The assumption $\hat{H}_{ab} = \{\hat{T}\}$, $\hat{a}$ or $\hat{b}$, is equal to $\hat{0}$ or $\hat{1}$, with that $\hat{x}_i$, are not $\hat{0}$ or $\hat{1}$, give that the isoH_v-field has one scalar absorbing $\hat{0}$, one scalar $\hat{1}$, and for all $\hat{a} \in \hat{F}$ one inverse.

A *generalization of P-hyperoperations*, used in Santilli's isotheory, is the following [44]. Let $(G, \cdot)$ be abelian group and P a subset of G with $|P| > 1$. We define the hyperoperation $(\times_p)$ as follows:

$$x \times_p y = \begin{cases} x \cdot P \cdot y = \{x \cdot h \cdot y \mid h \in P\} & \text{if } x \neq e \text{ and } y \neq e \\ x \cdot y & \text{if } x = e \text{ or } y = e \end{cases}$$

we call this hyperoperation P_e-*hyperoperation*. The hyperstructure $(G, \times_p)$ is abelian H_v-group.

Construction 6.5. (The P-hyperoperation). Consider an isofield $\hat{F} = \hat{F}(\hat{a}, \hat{+}, \hat{\times})$ with $\hat{a} = a \times \hat{1}$, the isonumbers, where $a \in F$, and $\hat{1}$ is positive-defined outside F, with two operations $\hat{+}$ and $\hat{\times}$, where $\hat{+}$ is the sum with the conventional unit 0, and $\hat{\times}$ is the iso-product

$$\hat{a} \hat{\times} \hat{b} = \hat{a} \times \hat{T} \times \hat{b}, \text{ with } \hat{1} = \hat{T}^{-1}, \text{ for all } \hat{a}, \hat{b} \in \hat{F}.$$

Take a set $\hat{P} = \{\hat{T}, \hat{p}_1, ..., \hat{p}_s\}$, with $\hat{p}_1, ..., \hat{p}_s \in \hat{F} - \{\hat{0}, \hat{1}\}$, we define the isoP-$H_v$-field, $\hat{F} = \hat{F}(\hat{a}, \hat{+}, \hat{\times}_p)$, where the hyperoperation $\hat{\times}_P$ as follows:

$$\hat{a} \hat{\times}_P \hat{b} := \begin{cases} \hat{a} \times \hat{\hat{P}} \times \hat{b} = \{\hat{a} \times \hat{\hat{h}} \times \hat{b} \mid \hat{\hat{h}} \in \hat{\hat{P}}\} & \text{if } \hat{a} \neq \hat{1} \text{ and } \hat{b} \neq \hat{1} \\ \hat{a} \times \hat{\hat{T}} \times \hat{b} & \text{if } \hat{a} = \hat{1} \text{ or } \hat{b} = \hat{1}. \end{cases}$$

The elements of $\hat{F}$ are called *isoP-H_v-numbers*.

Remark 6.2. If $\hat{P} = \{\hat{T}, \hat{p}\}$, that is that $\hat{P}$ contains only one $\hat{p}$ except $\hat{T}$. The inverses in isoP-H_v-fields, are not necessarily unique.

Example 6.6. Non degenerate example on the above constructions:

In order to define a generalized P-hyperoperation on $\hat{\mathbb{Z}}_7 = \hat{\mathbb{Z}}_7(\underline{\hat{a}}, \hat{+}, \hat{\times})$, where we take $\hat{P} = \{\underline{\hat{1}}, \underline{\hat{6}}\}$, the weak associative multiplicative hyperoperation is described by the table:

$\hat{\times}$	$\underline{\hat{0}}$	$\underline{\hat{1}}$	$\underline{\hat{2}}$	$\underline{\hat{3}}$	$\underline{\hat{4}}$	$\underline{\hat{5}}$	$\underline{\hat{6}}$
$\underline{\hat{0}}$	$\underline{\hat{0}}$	$\underline{\hat{0}}$	$\underline{\hat{0}}$	$\underline{\hat{0}}$	$\underline{\hat{0}}$	$\underline{\hat{0}}$	$\underline{\hat{0}}$
$\underline{\hat{1}}$	$\underline{\hat{0}}$	$\underline{\hat{1}}$	$\underline{\hat{2}}$	$\underline{\hat{3}}$	$\underline{\hat{4}}$	$\underline{\hat{5}}$	$\underline{\hat{6}}$
$\underline{\hat{2}}$	$\underline{\hat{0}}$	$\underline{\hat{2}}$	$\{\underline{\hat{4}}, \underline{\hat{3}}\}$	$\{\underline{\hat{6}}, \underline{\hat{1}}\}$	$\{\underline{\hat{1}}, \underline{\hat{6}}\}$	$\{\underline{\hat{3}}, \underline{\hat{4}}\}$	$\{\underline{\hat{5}}, \underline{\hat{2}}\}$
$\underline{\hat{3}}$	$\underline{\hat{0}}$	$\underline{\hat{3}}$	$\{\underline{\hat{6}}, \underline{\hat{1}}\}$	$\{\underline{\hat{2}}, \underline{\hat{5}}\}$	$\{\underline{\hat{5}}, \underline{\hat{2}}\}$	$\{\underline{\hat{1}}, \underline{\hat{6}}\}$	$\{\underline{\hat{4}}, \underline{\hat{3}}\}$
$\underline{\hat{4}}$	$\underline{\hat{0}}$	$\underline{\hat{4}}$	$\{\underline{\hat{1}}, \underline{\hat{6}}\}$	$\{\underline{\hat{5}}, \underline{\hat{2}}\}$	$\{\underline{\hat{2}}, \underline{\hat{5}}\}$	$\{\underline{\hat{6}}, \underline{\hat{1}}\}$	$\{\underline{\hat{3}}, \underline{\hat{4}}\}$
$\underline{\hat{5}}$	$\underline{\hat{0}}$	$\underline{\hat{5}}$	$\{\underline{\hat{3}}, \underline{\hat{4}}\}$	$\{\underline{\hat{1}}, \underline{\hat{6}}\}$	$\{\underline{\hat{6}}, \underline{\hat{1}}\}$	$\{\underline{\hat{4}}, \underline{\hat{3}}\}$	$\{\underline{\hat{2}}, \underline{\hat{5}}\}$
$\underline{\hat{6}}$	$\underline{\hat{0}}$	$\underline{\hat{6}}$	$\{\underline{\hat{5}}, \underline{\hat{2}}\}$	$\{\underline{\hat{4}}, \underline{\hat{3}}\}$	$\{\underline{\hat{3}}, \underline{\hat{4}}\}$	$\{\underline{\hat{2}}, \underline{\hat{5}}\}$	$\{\underline{\hat{1}}, \underline{\hat{6}}\}$

The hyperstructure $\hat{\mathbb{Z}}_7 = \hat{\mathbb{Z}}_7(\underline{\hat{a}}, \hat{+}, \hat{\times})$ is commutative and associative on the product hyperoperation. Moreover the β^* classes on the iso-product $\hat{\times}$ are $\{\hat{1}, \hat{6}\}, \{\hat{5}, \hat{2}\}, \{\hat{3}, \hat{4}\}$.

6.3 The Lie-Santilli's admissibility

The structure of the laws in physics is largely based on symmetries. The objects in Lie theory are fundamental, interesting and innovating in both mathematics and physics. It has many applications to the spectroscopy of molecules, atoms, nuclei and hadrons. The central role of Lie algebra in particle physics is well known. A Lie-admissible algebra, introduced by Albert [3], is a (possibly non-associative) algebra that becomes a Lie algebra under the bracket $[a, b] = ab - ba$. Examples include associative algebras, Lie algebras and Okubo algebras. Lie admissible algebras arise in various topics, including geometry of invariant affine connections on Lie groups and classical and quantum mechanics.

For an algebra A over a field F, the commutator algebra A^- of A is the anti-commutative algebra with multiplication $[a, b] = ab - ba$ defined on the vector space A. If A^- is a Lie algebra, i.e., satisfies the Jacobi identity, then A^- is called Lie-admissible. Much of the structure theory of Lie-admissible algebras has been carried out initially under additional conditions such as the flexible identity or power-associativity.

Santilli obtained Lie admissible algebras (brackets) from a modified form of Hamilton's equations with external terms which represent a general non-self-adjoint Newtonian system in classical mechanics. In 1967, Santilli introduced the product

$$(A, B) = \lambda AB - \mu BA = \alpha(AB - BA) + \beta(AB + BA), \qquad (6.3)$$

where $\lambda = \alpha + \beta$, $\mu = \beta - \alpha$, which is jointly Lie admissible and Jordan admissible while admitting Lie algebras in their classification. Then, he introduced the following infinitesimal and finite generalizations of Heisenberg equations

$$i\frac{dA}{dt} = (A, H) = \lambda AH - \mu HA, \qquad (6.4)$$

where $A(t) = U(t)A(0)V(t)^\dagger = e^{Ht\mu i}A(0)e^{-i\lambda tH}$, $U = e^{Ht\mu i}$, $V = e^{i\lambda tH}UV^\dagger \neq I$ and H is the Hamiltonian. In 1978, Santilli introduced the following most general known realization of products that are jointly Lie admissible and Jordan admissible

$$\begin{aligned}(A, B) &= ARB - BSA = (ATB - BTA) + \{AWB + BWA\} \\ &= [A, B]^* + \{A, B\}^* \\ &= (ATH - HTA) + \{AWH + HWA\},\end{aligned} \qquad (6.5)$$

where $R = T + W$, $S = W - T$ and $R, S, R \pm S$ are non-singular operators. We refer the readers to [5, 92, 93] for details concerning Lie-Santilli algebras.

In [45, 44], Davvaz, Santilli and Vougiouklis studied multi-valued hyperstructures following the apparent existence in nature of a realization of two-
valued hyperstructures with hyperunits characterized by matter-antimatter
systems and their extentions where matter is represented with conventional
mathematics and antimatter is represented with isodual mathematics. On
the other hand, the main tools connecting the class of algebraic hyperstructures with the classical algebraic structures are the fundamental relations.
In continue, we study the notion of algebra, hyperalgebra and their connections by using the concept of fundamental relation. We introduce a special
class of Lie hyperalgebra. By this class of Lie hyperalgebra, we are able to
generalize the concept of Lie-Santilli theory to hyperstructure case.

Definition 6.10. Let $(L, +, \cdot)$ be a hypervector space over the hyperfield
$(F, +, \cdot)$. Consider the bracket (commutator) hyperoperation:

$$[\,,\,] : L \times L \to \mathcal{P}^*(L) : (x, y) \to [x, y]$$

then L is a Lie hyperalgebra over F if the following axioms are satisfied:

(L1) The bracket hyperoperation is bilinear, i.e. $[\lambda_1 x_1 + \lambda_2 x_2, y] = (\lambda_1[x_1, y] + \lambda_2[x_2, y])$, $[x, \lambda_1 y_1 + \lambda_2 y_2] = (\lambda_1[x, y_1] + \lambda_2[x, y_2])$, for all $x, x_1, x_2, y, y_1, y_2 \in L$, $\lambda_1, \lambda_2 \in F$;

(L2) $0 \in [x, x]$, for all $x \in L$;

(L3) $0 \in \Big([x, [y, z]] + [y, [z, x]] + [z, [x, y]]\Big)$, for all $x, y \in L$.

Definition 6.11. Let A be a hypervector space over a hyperfield F. Then,
A is called a *hyperalgebra* over the hyperfield F if there exists a mapping
$\cdot : A \times A \to \mathcal{P}^*(A)$ (images to be denoted by $x \cdot y$ for $x, y \in A$) such that
the following conditions hold:

(1) $(x + y) \cdot z = x \cdot z + y \cdot z$ and $x \cdot (y + z) = x \cdot y + x \cdot z$;

(2) $(cx) \cdot y = c(x \cdot y) = x \cdot (cy)$;

(3) $0 \cdot y = y \cdot 0 = 0$;

for all $x, y, z \in A$ and $c \in F$.

In the above definition, if all hyperoperations are ordinary operations,
then we have an *algebra*.

Example 6.7. Let F be a hyperfield and

$$A = \left\{ \begin{bmatrix} a & b \\ c & d \end{bmatrix} \mid a, b, c, d \in F \right\}.$$

We define the following hyperoperations on A:

$$\begin{bmatrix} a_1 & b_1 \\ c_1 & d_1 \end{bmatrix} \uplus \begin{bmatrix} a_2 & b_2 \\ c_2 & d_2 \end{bmatrix} = \left\{ \begin{bmatrix} a_3 & b_3 \\ c_3 & d_3 \end{bmatrix} \mid \right.$$
$$\left. a_3 \in a_1 + a_2, \; b_3 \in b_1 + b_2, \; c_3 \in c_1 + c_2, \; d_3 \in d_1 + d_2 \right\},$$

$$\begin{bmatrix} a_1 & b_1 \\ c_1 & d_1 \end{bmatrix} \otimes \begin{bmatrix} a_2 & b_2 \\ c_2 & d_2 \end{bmatrix} = \left\{ \begin{bmatrix} a_3 & b_3 \\ c_3 & d_3 \end{bmatrix} \mid \right.$$
$$a_3 \in a_1 \cdot a_2 + b_1 \cdot d_2, \; b_3 \in a_1 \cdot b_2 + b_1 \cdot d_2, \; c_3 \in c_1 \cdot a_2 + d_1 \cdot c_2,$$
$$\left. d_3 \in c_1 \cdot b_2 + d_1 \cdot d_2 \right\},$$

$$a \bullet \begin{bmatrix} a_1 & b_1 \\ c_1 & d_1 \end{bmatrix} = \left\{ \begin{bmatrix} a_2 & b_2 \\ c_2 & d_2 \end{bmatrix} \mid \; a_2 \in aa_1, \; b_2 \in ab_1, \; c_2 \in ac_2, \; d_3 \in ad_2 \right\}.$$

Then A together with the above hyperoperations is a hyperalgebra over F.

Example 6.8. We can generalize Example 6.7 to $n \times n$ matrices.

A non-empty subset A' of a hyperalgebra A is called a *sub hyperalgebra* if it is a subhyperspace of A and for all $x, y \in A'$ we have $xy \in A'$.

In connection with the explicit forms of the hyperproduct let us consider an associative hyperalgebra A, with hyperproduct $a \cdot b$, over a hyperfield F. It is possible to construct a new hyperalgebra, denoted by A^-, by means of the anti-commutative hyperproduct

$$[a, b] = a \cdot b - b \cdot a = \bigcup_{\substack{x \in a \cdot b \\ y \in b \cdot a}} x - y. \tag{6.6}$$

Lemma 6.1. *For any non-empty subset S of A, we have $0 \in S - S$.*

Proof. It is straightforward. ∎

Proposition 6.2. A^- *is a Lie hyperalgebra.*

Proof. For all $x, x_1, x_2, y, y_1, y_2 \in L$, $\lambda_1, \lambda_2 \in F$, we have

$$\begin{aligned} [\lambda_1 x_1 + \lambda_2 x_2, y] &= (\lambda_1 x_1 + \lambda_2 x_2) \cdot y - y \cdot (\lambda_1 x_1 + \lambda_2 x_2) \\ &= (\lambda_1 x_1) \cdot y + (\lambda_2 x_2) \cdot y - y \cdot (\lambda_1 x_1) - y \cdot (\lambda_2 x_2) \\ &= \Big((\lambda_1 x_1) \cdot y - y \cdot (\lambda_1 x_1) \Big) + \Big((\lambda_2 x_2) \cdot y - y \cdot (\lambda_2 x_2) \Big) \\ &= [\lambda_1 x_1, y] + [\lambda_2 x_2, y], \\ &= \lambda_1 [x_1, y] + \lambda_2 [x_2, y], \end{aligned}$$

and similarly we obtain $[x, \lambda_1 y_1 + \lambda_2 y_2] = (\lambda_1 [x, y_1] + \lambda_2 [x, y_2])$.

Now, we prove (L2). Since $x \cdot x$ is non-empty, there exists $a_0 \in x \cdot x$. hence, $-a_0 \in -x \cdot x$. Thus, $0 \in a_0 - a_0 \subseteq \bigcup_{a \in x \cdot x} a - a = x \cdot x - x \cdot x = [x, x]$.

It remains to show that (L3) is also satisfied. For,

$$\begin{aligned}
[x, [y, z]] &= x \cdot [y, z] - [y, z] \cdot x \\
&= x \cdot (y \cdot z - z \cdot y) - (y \cdot z - z \cdot y) \cdot x \\
&= x \cdot y \cdot z - x \cdot z \cdot y - y \cdot z \cdot x - z \cdot y \cdot x.
\end{aligned}$$

Hence,

$$\begin{aligned}
\Big([x, [y, z]] + [y, [z, x]] + [z, [x, y]] \Big) &= x \cdot y \cdot z - x \cdot z \cdot y - y \cdot z \cdot x - z \cdot y \cdot x \\
&\quad + (y \cdot z \cdot x - y \cdot x \cdot z - z \cdot x \cdot y + x \cdot z \cdot y) \\
&\quad + (z \cdot x \cdot y - z \cdot y \cdot x - x \cdot y \cdot z + y \cdot x \cdot z) \\
&= (x \cdot y \cdot z - x \cdot y \cdot z) + (x \cdot z \cdot y - x \cdot z \cdot y) \\
&\quad + (z \cdot y \cdot x - z \cdot y \cdot x) + (y \cdot z \cdot x - y \cdot z \cdot x) \\
&\quad + (y \cdot x \cdot z - y \cdot x \cdot z) + (z \cdot x \cdot y - z \cdot x \cdot y).
\end{aligned}$$

By Lemma 6.1, 0 is belong to the right hand of the above equality, so $0 \in \Big([x, [y, z]] + [y, [z, x]] + [z, [x, y]] \Big)$. ∎

Definition 6.12. Corresponding to any hyperalgebra A with hyperproduct $a \cdot b$ it is possible to define an anticommutative hyperalgebra A^- which is the same hypervector space as A with the new hyperproduct

$$[a, b]_{A^-} = a \cdot b - b \cdot a. \tag{6.7}$$

A hyperalgebra A is called *Lie-admissible* if the hyperalgebra A^- is a Lie hyperalgebra.

If A is an associative hyperalgebra, then the hyperproduct (6.7) coincide with (6.6) and A^- is a Lie hyperalgebra in its more usual form. Thus, the associative hyperalgebras constitute a basic class of Lie-admissible hyperalgebras.

A *Jordan algebra* is a (non-associative) algebra over a field whose multiplication satisfies the following axioms:

(1) $xy = yx$ (commutative law);
(2) $(xy)(xx) = x(y(xx))$ (Jordan identity).

Definition 6.13. A *Jordan hyperalgebra* is a (non-associative) hyperalgebra over a hyperfield whose multiplication satisfies the following axioms:

(J1) $x \cdot y = y \cdot x$ (commutative law);
(J2) $(x \cdot y) \cdot (x \cdot x) = x \cdot (y \cdot (x \cdot x))$ (Jordan identity).

Let A be an associative hyperalgebra over a hyperfield F. It is possible to construct a new hyperalgebra, denoted by A^+, by means of the commutative hyperproduct

$$\{a,b\} = a \cdot b + b \cdot a = \bigcup_{\substack{x \in a \cdot b \\ y \in b \cdot a}} x + y. \tag{6.8}$$

Proposition 6.3. A^+ *is a Jordan hyperalgebra.*

Proof. It is straightforward. ∎

Definition 6.14. Corresponding to any hyperalgebra A with hyperproduct $a \cdot b$ it is possible to define, as for A^-, a commutative hyperalgebra A^+ which is the same hypervector space as A but with the new hyperproduct

$$\{a,b\}_{A^+} = a \cdot b + b \cdot a. \tag{6.9}$$

In this connection the most interesting case occure when A^+ is a (commutative) Jordan hyperalgebra.

A hyperalgebra A is said to be *Jordan admissible* if A^+ is a (commutative) Jordan hyperalgebra.

If A is an associative hyperalgebra, then the hyperproduct (6.9) reduces to (6.8) and A^+ is a special Jordan hyperalgebra. Thus, associative hyperalgebras constitute a basis class of Jordan-admissible hyperalgebras.

Definition 6.15. The fundamental relation ε^* is defined in a hyperalgebra as the smallest equivalence relation such that the quotient is an algebra.

By using strongly regular relations, we can connect hyperalgebras to algebras. More exactly, starting with a hyperalgebra and using a strongly regular relation, we can construct an algebra structure on the quotient set. An equivalence relation ρ on a hyperalgebra A is called right (resp. left) strongly regular if and only if $x\rho y$ implies that $(x+z)\overline{\overline{\rho}}(y+z)$ and $(x\alpha z)\overline{\overline{\rho}}(y \cdot z)$ for every $z \in A$ (resp. $(z+x)\overline{\overline{\rho}}(z+y)$ and $(z \cdot x)\overline{\overline{\rho}}(z \cdot y)$), and ρ is strongly regular if it is both left and right strongly regular.

Theorem 6.6. *Let A be a hyperalgebra over the hyperfield F. Denote by ϑ the set of all finite polynomials of elements of A over F. We define the relation ε on A as follows:*

$$x\varepsilon y \quad \text{if and only if } \{x,y\} \subseteq u, \text{ where } u \in \vartheta.$$

Then, the ε^* is the transitive closure of ε and is called the fundamental equivalence relation on A.

Proof. The proof is similar to the proof of Theorem 6.1.   ∎

Remark 6.3. Note that the relation ε^* is a strongly regular relation.

Remark 6.4. In A^-/ε^*, the binary operations and external operation are defined in the usual manner:

$$\varepsilon^*(x) \oplus \varepsilon^*(y) = \varepsilon^*(z), \text{ for all } z \in \varepsilon^*(x) + \varepsilon^*(y),$$
$$\varepsilon^*(x) \odot \varepsilon^*(y) = \varepsilon^*(z), \text{ for all } z \in \varepsilon^*(x) \cdot \varepsilon^*(y),$$
$$\gamma^*(r) \circ \varepsilon^*(x) = \varepsilon^*(z), \text{ for all } z \in \gamma^*(r)\varepsilon^*(x).$$

Theorem 6.7. *Let A be an associative hyperalgebra over a hyperfield F. Then, A^-/ε^* is a Lie-admissible algebra with the following product:*

$$\left\langle \varepsilon^*(x), \varepsilon^*(y) \right\rangle = \varepsilon^*(x) \odot \varepsilon^*(y) \ominus \varepsilon^*(y) \odot \varepsilon^*(x). \tag{6.10}$$

Proof. By Definition 6.15 and Theorem 6.6, A/ε^* is an ordinary associative algebra. So, it is enough to show that it is a Lie algebra with the hyperproduce (6.10). By Proposition 6.2, A^- is a Lie hyperalgebra with the hyperproduct $[a, b] = a \cdot b - b \cdot a$.

(1) By (L1), for all $x_1, x_2, y \in A$, $\lambda_1, \lambda_2 \in F$, we have $[\lambda_1 x_1 + \lambda_2 x_2, y] = \lambda_1[x_1, y] + \lambda_2[x_2, y]$. Hence,

$$(\lambda_1 x_1 + \lambda_2 x_2) \cdot y - y \cdot (\lambda_1 x_1 + \lambda_2 x_2) = \lambda_1(x_1 \cdot y - y \cdot x_1) + \lambda_2(x_2 \cdot y - y \cdot x_2), \tag{6.11}$$

and so

$$\varepsilon^*\Big((\lambda_1 x_1 + \lambda_2 x_2) \cdot y - y \cdot (\lambda_1 x_1 + \lambda_2 x_2)\Big)$$
$$= \varepsilon^*\Big(\lambda_1(x_1 \cdot y - y \cdot x_1) + \lambda_2(x_2 \cdot y - y \cdot x_2)\Big). \tag{6.12}$$

This implies that

$$\Big(\gamma^*(\lambda_1) \circ \varepsilon^*(x_1) \oplus \gamma^*(\lambda_2) \circ \varepsilon^*(x_2)\Big) \odot \varepsilon^*(y)$$
$$\ominus \varepsilon^*(y) \odot \Big(\gamma^*(\lambda_1) \circ \varepsilon^*(x_1) \oplus \gamma^*(\lambda_2) \circ \varepsilon^*(x_2)\Big)$$
$$= \gamma^*(\lambda_1) \circ (\varepsilon^*(x_1) \odot \varepsilon^*(y) \ominus \varepsilon^*(y) \odot \varepsilon^*(x_1)) \tag{6.13}$$
$$+ \gamma^*(\lambda_2) \circ (\varepsilon^*(x_2) \odot \varepsilon^*(y) \ominus \varepsilon^*(y) \odot \varepsilon^*(x_2)).$$

Therefore,

$$\left\langle \gamma^*(\lambda_1) \circ \varepsilon^*(x_1) \oplus \gamma^*(\lambda_2) \circ \varepsilon^*(x_2), \varepsilon^*(y) \right\rangle$$
$$= \gamma^*(\lambda_1) \circ \left\langle \varepsilon^*(x_1), \varepsilon^*(y) \right\rangle \oplus \gamma^*(\lambda_2) \circ \left\langle \varepsilon^*(x_2), \varepsilon^*(y) \right\rangle. \tag{6.14}$$

Similarly, for all $x, y_1, y_2 \in A$, $\lambda_1, \lambda_2 \in F$, we obtain

$$\left\langle \varepsilon^*(x), \gamma^*(\lambda_1) \circ \varepsilon^*(y_1) \oplus \gamma^*(\lambda_2) \circ \varepsilon^*(y_2) \right\rangle$$
$$= \gamma^*(\lambda_1) \circ \left\langle \varepsilon^*(x), \varepsilon^*(y_1) \right\rangle \oplus \gamma^*(\lambda_2) \circ \left\langle \varepsilon^*(x), \varepsilon^*(y_2) \right\rangle. \tag{6.15}$$

(2) By (L2), $0 \in x \cdot x - x \cdot x$, so

$$\varepsilon^*(0) = \varepsilon^*(x \cdot x - x \cdot x) = \varepsilon^*(x) \odot \varepsilon^*(x) \ominus \varepsilon^*(x) \odot \varepsilon^*(x)$$
$$= \Big\langle \varepsilon^*(x), \varepsilon^*(x) \Big\rangle.$$

(3) By (L3), we have $0 \in \Big([x, [y, z]] + [y, [z, x]] + [z, [x, y]] \Big)$, for all $x, y \in A$. Thus,

$$\varepsilon^*(0) = \Big\langle \varepsilon^*(x), \Big\langle \varepsilon^*(y), \varepsilon^*(z) \Big\rangle \Big\rangle + \Big\langle \varepsilon^*(y), \varepsilon^*(z), \varepsilon^*(x) \Big\rangle \Big\rangle$$
$$+ \Big\langle \varepsilon^*(z), \Big\langle \varepsilon^*(x), \varepsilon^*(y) \Big\rangle \Big\rangle. \tag{6.16}$$

∎

Theorem 6.8. *Let A be an associative hyperalgebra over a hyperfield F. Then, A^+/ε^* is a Jordan-admissible algebra with the following product:*

$$\Big[\varepsilon^*(x), \varepsilon^*(y) \Big] = \varepsilon^*(x) \odot \varepsilon^*(y) \oplus \varepsilon^*(y) \odot \varepsilon^*(x). \tag{6.17}$$

Proof. By Definition 6.15 and Theorem 6.6, A/ε^* is an ordinary associative algebra. So, it is enough to show that it is a Jordan algebra with the hyperproduct (6.17). By Proposition 6.3, A^+ is a Jordan hyperalgebra with the hyperproduct $\{a, b\} = a \cdot b + b \cdot a$.

(1) By (J1), for all $x, y \in A$, $\{x, y\} = \{y, x\}$. So, $\varepsilon^*(x \cdot y + y \cdot x) = \varepsilon^*(y \cdot x + x \cdot y)$ which implies that $\varepsilon^*(x) \odot \varepsilon^*(y) \oplus \varepsilon^*(y) \odot \varepsilon^*(x) = \varepsilon^*(y) \odot \varepsilon^*(x) \oplus \varepsilon^*(x) \odot \varepsilon^*(y)$. Thus, $\Big[\varepsilon^*(x), \varepsilon^*(y) \Big] = \Big[\varepsilon^*(y), \varepsilon^*(x) \Big]$.

(2) By (J2), for all $x, y \in A$, $\Big\{ \{x, y\}, \{x, x\} \Big\} = \Big\{ x, \{y, \{x, x\}\} \Big\}$. Hence,

$$\varepsilon^* \Big(\Big\{ \{x, y\}, \{x, x\} \Big\} \Big) = \varepsilon^* \Big(\Big\{ x, \{y, \{x, x\}\} \Big\} \Big), \tag{6.18}$$

which implies that

$$\Big[\lceil \varepsilon^*(x), \varepsilon^*(y) \rceil, \lceil \varepsilon^*(x), \varepsilon^*(x) \rceil \Big] = \Big[\varepsilon^*(x), \lceil \varepsilon^*(y), \lceil \varepsilon^*(x), \varepsilon^*(x) \rceil \rceil \Big]. \tag{6.19}$$

This completes the proof.

∎

In the same way it is possible to introduce always in terms of the associative product the following bilinear form

$$(a, b) = \lambda a \cdot b + (1 - \lambda) b \cdot a = \lambda [a, b] + b \cdot a, \tag{6.20}$$

where λ is a free element belonging to the hyperfield F, which characterizes the λ-mutations $A(\lambda)$ of A. Clearly, $A(1)$ is isomorphic to A.

In this connection, a more general bilinear form in terms of the associative hyperproduct is given by

$$(a, b) = \lambda a \cdot b + \mu b \cdot a = \alpha[a, b] + \beta\{a, b\}, \qquad (6.21)$$

where $\lambda = \alpha + \beta$ and $\mu = \beta - \alpha$ are free elements belonging to the hyperfield F, which constitutes the basic hyperproduct of the (λ, μ)-mutations $A(\lambda, \mu)$ of A. Clearly, $A(1, 0)$ is isomorphic to A and $A(1, -1)$ is isomorphic to A^-; $A(1, 1)$ is isomorphic to A^+ and $A(\lambda, 1 - \lambda)$ is isomorphic to $A(\lambda)$.

Let A be an associative hyperalgebra. We can define another hyperoperation by means an element T which is denoted by $[\cdot, \cdot]_T$ and is defined by

$$\begin{aligned} &[\cdot, \cdot]_T : A \times A \to \mathcal{P}^*(A), \\ &[\cdot, \cdot]_T : (x, y) \mapsto [x, y]_T = x \cdot T \cdot y - y \cdot T \cdot x. \end{aligned} \qquad (6.22)$$

Proposition 6.4. *The hyperoperation $[\cdot, \cdot]_T$ satisfies the following conditions:*

(1) $0 \in [x, x]_T$, *for all* $x \in A$;

(2) $0 \in \Big([x, [y, z]_T]_T + [y, [z, x]_T]_T + [z, [x, y]_T]_T\Big)$, *for all* $x, y \in T$.

Proof.   (1) Since $x \cdot T \cdot x$ is non-empty, there exists $a_0 \in x \cdot T \cdot x$. hence, $-a_0 \in -x \cdot T \cdot x$. Thus, $0 \in a_0 - a_0 \subseteq \bigcup_{a \in x \cdot T \cdot x} a - a = x \cdot T \cdot x - x \cdot T \cdot x = [x, x]_T$.

(2) We have

$$\begin{aligned} [x, [y, z]_T]_T &= x \cdot T \cdot [y, z]_T - [y, z]_T \cdot T \cdot x \\ &= x \cdot T \cdot (y \cdot T \cdot z - z \cdot T \cdot y) - (y \cdot T \cdot z - z \cdot T \cdot y) \cdot T \cdot x \\ &= x \cdot T \cdot y \cdot T \cdot z - x \cdot T \cdot z \cdot T \cdot y - y \cdot T \cdot z \cdot T \cdot x - z \cdot T \cdot y \cdot T \cdot x. \end{aligned}$$

Hence,

$$\begin{aligned} \Big(&[x, [y, z]_T]_T + [y, [z, x]_T]_T + [z, [x, y]_T]_T\Big) \\ &= x \cdot T \cdot y \cdot T \cdot z - x \cdot T \cdot z \cdot T \cdot y - y \cdot T \cdot z \cdot T \cdot x - z \cdot T \cdot y \cdot T \cdot x \\ &\quad + (y \cdot T \cdot z \cdot T \cdot x - y \cdot T \cdot x \cdot T \cdot z - z \cdot T \cdot x \cdot T \cdot y + x \cdot T \cdot z \cdot T \cdot y) \\ &\quad + (z \cdot T \cdot x \cdot T \cdot y - z \cdot T \cdot y \cdot T \cdot x - x \cdot T \cdot y \cdot T \cdot z + y \cdot T \cdot x \cdot T \cdot z) \\ &= (x \cdot T \cdot y \cdot T \cdot z - x \cdot T \cdot y \cdot T \cdot z) + (x \cdot T \cdot z \cdot T \cdot y - x \cdot T \cdot z \cdot T \cdot y) \\ &\quad + (z \cdot T \cdot y \cdot T \cdot x - z \cdot T \cdot y \cdot T \cdot x) + (y \cdot T \cdot z \cdot T \cdot x - y \cdot T \cdot z \cdot T \cdot x) \\ &\quad + (y \cdot T \cdot x \cdot T \cdot z - y \cdot T \cdot x \cdot T \cdot z) + (z \cdot T \cdot x \cdot T \cdot y - z \cdot T \cdot x \cdot T \cdot y). \end{aligned}$$

By Lemma 6.1, 0 is belong to the right hand of the abobe equality, so $0 \in \Big([x, [y, z]_T]_T + [y, [z, x]_T]_T + [z, [x, y]_T]_T\Big)$. ∎

Remark 6.5. If A is an algebra and $[\cdot, \cdot]_T : A \times A \to A$, then we have Lie-Santilli bracket.

Definition 6.16. Corresponding to any hyperalgebra A with hyperproduct $a \cdot b$ it is possible to define a A^* which is the same hypervector space as A with the new hyperproduct

$$[a, b]_{A^*} = a \cdot T \cdot b - b \cdot T \cdot a. \tag{6.23}$$

A hyperalgebra A is called *Lie-Santilli-admissible* if the hyperalgebra A^* is a Lie hyperalgebra.

Corollary 6.2. *If A is an associative hyperalgebra, then the hyperproduct (6.23) coincide with (6.22) and A^* is a Lie hyperalgebra.*

Theorem 6.9. *Let A be an associative hyperalgebra over a hyperfield F. Then, A^*/ε^* is a Lie-Santilli-admissible algebra with the following product:*

$$\left\langle \varepsilon^*(x), \varepsilon^*(y) \right\rangle_T = \varepsilon^*(x) \odot \varepsilon^*(T) \odot \varepsilon^*(y) \ominus \varepsilon^*(y) \odot \varepsilon^*(T) \odot \varepsilon^*(x). \tag{6.24}$$

Proof. By Definition 6.15 and Theorem 6.6, A/ε^* is an ordinary associative algebra. So, it is enough to show that it is a Lie algebra with the hyperproduce (6.6). By Corollary 6.2, A^* is a Lie hyperalgebra with the hyperproduct $[a, b]_T = a \cdot T \cdot b - b \cdot T \cdot a$.

(1) By (L1), for all $x_1, x_2, y \in$, $\lambda_1, \lambda_2 \in F$, we have $[\lambda_1 x_1 + \lambda_2 x_2, y]_T = \lambda_1 [x_1, y]_T + \lambda_2 [x_2, y]_T$. Hence,

$$(\lambda_1 x_1 + \lambda_2 x_2) \cdot T \cdot y - y \cdot T \cdot (\lambda_1 x_1 + \lambda_2 x_2)$$
$$= \lambda_1 (x_1 \cdot T \cdot y - y \cdot T \cdot x_1) + \lambda_2 (x_2 \cdot T \cdot y - y \cdot T \cdot x_2), \tag{6.25}$$

and so

$$\varepsilon^* \Big((\lambda_1 x_1 + \lambda_2 x_2) \cdot T \cdot y - y \cdot T \cdot (\lambda_1 x_1 + \lambda_2 x_2) \Big)$$
$$= \varepsilon^* \Big(\lambda_1 (x_1 \cdot T \cdot y - y \cdot T \cdot x_1) + \lambda_2 (x_2 \cdot T \cdot y - y \cdot T \cdot x_2) \Big). \tag{6.26}$$

This implies that

$$\Big(\gamma^*(\lambda_1) \circ \varepsilon^*(x_1) \oplus \gamma^*(\lambda_2) \circ \varepsilon^*(x_2) \Big) \odot \varepsilon^*(T) \odot \varepsilon^*(y)$$
$$\ominus \varepsilon^*(y) \odot \varepsilon^*(T) \odot \Big(\gamma^*(\lambda_1) \circ \varepsilon^*(x_1) \oplus \gamma^*(\lambda_2) \circ \varepsilon^*(x_2) \Big)$$
$$= \gamma^*(\lambda_1) \circ \Big(\varepsilon^*(x_1) \odot \varepsilon^*(T) \odot \varepsilon^*(y) \ominus \varepsilon^*(y) \odot \varepsilon^*(T) \odot \varepsilon^*(x_1) \Big)$$
$$\oplus \gamma^*(\lambda_2) \circ \Big(\varepsilon^*(x_2) \odot \varepsilon^*(T) \odot \varepsilon^*(y) \ominus \varepsilon^*(y) \odot \varepsilon^*(T) \odot \varepsilon^*(x_2) \Big). \tag{6.27}$$

Therefore,

$$\left\langle \gamma^*(\lambda_1) \circ \varepsilon^*(x_1) \oplus \gamma^*(\lambda_2) \circ \varepsilon^*(x_2), \varepsilon^*(y) \right\rangle_T$$
$$= \gamma^*(\lambda_1) \circ \left\langle \varepsilon^*(x_1), \varepsilon^*(y) \right\rangle_T \oplus \gamma^*(\lambda_2) \circ \left\langle \varepsilon^*(x_2), \varepsilon^*(y) \right\rangle_T. \tag{6.28}$$

Similarly, for all $x, y_1, y_2 \in A$, $\lambda_1, \lambda_2 \in F$, we obtain

$$\left\langle \varepsilon^*(x), \gamma^*(\lambda_1) \circ \varepsilon^*(y_1) \oplus \gamma^*(\lambda_2) \circ \varepsilon^*(y_2) \right\rangle_T$$

$$= \gamma^*(\lambda_1) \circ \left\langle \varepsilon^*(x), \varepsilon^*(y_1) \right\rangle_T \oplus \gamma^*(\lambda_2) \circ \left\langle \varepsilon^*(x), \varepsilon^*(y_2) \right\rangle_T. \tag{6.29}$$

(2) By (L2), $0 \in x \cdot T \cdot x - x \cdot T \cdot x$, so

$$\varepsilon^*(0) = \varepsilon^*(x \cdot T \cdot x - x \cdot T \cdot x)$$
$$= \varepsilon^*(x) \odot \varepsilon^*(T) \odot \varepsilon^*(x) \ominus \varepsilon^*(x) \odot \varepsilon^*(T) \odot \varepsilon^*(x)$$
$$= \left\langle \varepsilon^*(x), \varepsilon^*(x) \right\rangle_T.$$

(3) By (L3), we have $0 \in \left([x, [y, z]] + [y, [z, x]] + [z, [x, y]] \right)$, for all $x, y \in A$. Thus,

$$\varepsilon^*(0) = \left\langle \varepsilon^*(x), \left\langle \varepsilon^*(y), \varepsilon^*(z) \right\rangle_T \right\rangle_T + \left\langle \varepsilon^*(y), \left\langle \varepsilon^*(z), \varepsilon^*(x) \right\rangle_T \right\rangle_T$$

$$+ \left\langle \varepsilon^*(z), \left\langle \varepsilon^*(x), \varepsilon^*(y) \right\rangle_T \right\rangle_T. \tag{6.30}$$

∎

In some cases we can start from a hyperalgebra which is not a Lie hyperalgebra with respect to the Lie bracket. However it can be Lie-Santilli hyperalgebra with respect to Lie-Santilli bracket.

Example 6.9. Let F be a hyperfield and

$$A = \left\{ (a_{ij})_{m \times n} \mid a_{ij} \in F \right\}.$$

Similar to Examples 6.7 and 6.8, we can define the hyperoperation ⊎ and the external hyperproduct •. Note that it is impossible to define the hyperoperation ⊗ between two elements of A and so A is not a Lie hyperalgebra but it is Lie-Santilli hyperalgebra with respect to

$$\left[(a_{ij})_{m \times n}, (b_{ij})_{m \times n} \right]_{(T_{ij})_{m \times n}}$$
$$= (a_{ij})_{m \times n} \otimes (T_{ij})^t_{m \times n} \otimes (a_{ij})_{m \times n} \oslash (b_{ij})_{m \times n} \otimes (T_{ij})_{m \times n} \otimes (a_{ij})_{m \times n}, \tag{6.31}$$

where $(T_{ij})^t_{m \times n}$ denotes the transpose of the matrix $(T_{ij})_{m \times n}$.

6.4 H_v-matrix representations

Hyperstructures and H_v-structures are used in representation. H_v-matrix representations are introduced by Vougiouklis in [120], also see [113, 115, 125].

Definition 6.17. H_v-matrix (or h/v-matrix) is a matrix with entries from an H_v-ring or h/v-ring or h/v-field. The hyperproduct of two H_v-matrices $= (a_{ij})$ and $= (b_{ij})$, of type $m \times n$ and $n \times r$ respectively, is defined in the usual manner, and it is a set of $m \times r$ H_v-matrices. The sum of products of elements of the H_v-ring is considered to be the n-ary circle hyperoperation on the hyperaddition. The hyperproduct of H_v-matrices is not weak associative.

The problem of the H_v-matrix (or h/v-group) representations is the following:

Definition 6.18. Let $(H, \cdot)$ be H_v-group (or h/v-group). Find H_v-ring (or h/v-ring) $(R, +, \cdot)$, a set $M_R = \{(a_{ij}) \mid a_{ij} \in R\}$, and a map $T : H \to M_R :$ $h \mapsto T(h)$ such that

$$T(h_1 h_2) \cap T(h_1) T(h_2) \neq \emptyset, \text{ forall } h_1, h_2 \in H.$$

T is H_v-*matrix (or h/v-matrix) representation.* If

$$T(h_1 h_2) \subseteq T(h_1) T(h_2), \text{ forall } h_1, h_2 \in H,$$

then T is an *inclusion representation.* If

$$T(h_1 h_2) = T(h_1) T(h_2), \text{ for all } h_1, h_2 \in H,$$

then T is *good representation* and an induced one T^* for the hypergroup algebra is obtained.

The main theorem of the theory of representations is the following.

Theorem 6.10. *A necessary condition in order to have an inclusion representation T of an h/v-group $(H, \cdot)$ by $n \times n$ h/v-matrices over the h/v-ring $(R, +, \cdot)$ is the following:*
For all classes $\beta^*(x)$, $x \in H$ there must exist elements $a_{ij} \in H, i, j \in \{1, ..., n\}$ such that

$$T(\beta^*(a)) \subseteq \{A = (a'_{ij}) \mid a'_{ij} \in \gamma^*(a_{ij}), i, j \in \{1, ..., n\}\}.$$

Thus, inclusion representation $T : H \to M_R : a \mapsto T(a) = (a_{ij})$ induces an homomorphic T^* of H/β^* over R/γ^* by setting $T^*(\beta^*(a)) = [\gamma * (a_{ij})]$, for all $\beta^*(a)H/\beta^*$, where the $\gamma^*(a_{ij})R/\gamma^*$ is the ij entry of $T^*(\beta * (a))$. T^* is called fundamental induced representation of T.

Let T be a representation of an h/v-group H by h/v-matrices and $tr_\phi(T(x)) = \gamma^*(Tx_{ii})$ be the fundamental trace, then is called *fundamental character*, the mapping

$$X_T : H \to R/\gamma^* : x \mapsto X_T(x) = tr_\phi(T(x)) = trT^*(x).$$

In representations of H_v-groups there are two difficulties: First to find an H_v-ring or an H_v-field and second, an appropriate set of H_v-matrices.

We now give a definition of a set of h/v-matrices which is closed only under the product of matrices, not under the sum.

Definition 6.19. We call an h/v-matrix $A = (a_{ij}) \in M_{n \times n}$, U_0-matrix if it is upper triangular where the condition $[a_{11}] \cdot [a_{22}] \cdots [a_{nn}] \neq [0]$, is valid, where $[x]$ is the fundamental class of x. We notice that this set is closed under the product only, not in addition. Thus, it is interesting only when, in the h/v-matrix representations, the product is used. Moreover, we define the product of an element with a class to be the corresponding class.

Theorem 6.11. *In the case of 3×3 U_0-matrices the unit U_0-matrices are*

$$I = [1]E_{11} + [1]E_{22} + [1]E_{33}$$

and the inverses of the U_0-matrix $A = (a_{ij}) \in M_{3 \times 3}$, are given as follows:

$$\begin{bmatrix} a_{11}^{-1} & -(a_{11})^{-1}a_{12}[a_{22}]^{-1} & (a_{11})^{-1}a_{12}(a_{22})^{-1}a_{23}[a_{33}]^{-1} - (a_{11})^{-1}a_{13}[a_{33}]^{-1} \\ 0 & [a_{22}^{-1}] & -(a_{22})^{-1}a_{23}[a_{33}]^{-1} \\ 0 & 0 & [a_{33}]^{-1} \end{bmatrix}$$

Proof. Let us denote by E_{ij} the matrix with 1 in the ij-entry and zero in the rest entries. First, it is clear that all the h/v-matrices of the set $I = [1]E_{11} + [1]E_{22} + [1]E_{33}$, are unit h/v-matrices in the sense that $(I \cdot A \cup A \cdot I) \subseteq [A] = ([a_{ij}])$.

In order to find the inverse of the given h/v-matrix

$$A = (a_{ij}) = a_{11}E_{11} + a_{22}E_{22} + a_{33}E_{33} + a_{12}E_{12} + a_{13}E_{13} + a_{23}E_{23}$$

we have to find the set of h/v-matrices

$$X = ([x_{ij}]) = [x_{11}]E_{11} + [x_{22}]E_{22} + [x_{33}]E_{33} + [x_{12}]E_{12} + [x_{13}]E_{13} + [x_{23}]E_{23}$$

such that $A \cdot X = X \cdot A = I$, therefore we must have, on the one side and taking into accound that the product of one element with a fundamental class is considered the entire class,

$$a_{11}[x_{11}] = 1, a_{22}[x_{22}] = 1, a_{33}[x_{33}] = 1,$$
$$a_{11}[x_{12}] + a_{12}[x_{22}] = 0, a_{11}[x_{13}] + a_{12}[x_{23}] + a_{13}[x_{33}] = 0,$$
$$a_{22}[x_{23}] + a_{23}[x_{33}] = 0.$$

Therefore, we have

$$[x_{11}] = [a_{11}]^{-1}, [x_{22}] = [a_{22}]^{-1}, [x_{33}] = [a_{33}]^{-1},$$

and then,

$$[x_{12}] = -(a_{11})^{-1}a_{12}[a_{22}]^{-1}, [x_{23}] = -(a_{22})^{-1}a_{23}[a_{33}]^{-1},$$
$$[x_{13}] = (a_{11})^{-1}a_{12}(a_{22})^{-1}a_{23}[a_{33}]^{-1} - (a_{11})^{-1}a_{13}[a_{33}]^{-1}. \quad \blacksquare$$

Example 6.10. Consider the h/v-field $(\mathbb{Z}_{10}, +, \otimes)$ where only $3 \otimes 8 = \{4, 9\}$ is a hyperproduct. Let us take the h/v-matrix

$$A = 3E_{11} + E_{22} + 2E_{33} + 6E_{12} + 2E_{13} + 9E_{23}.$$

Then from the above formulas we obtain that the set of inverse h/v-matrices is

$$A^{-1} = [2]E_{11} + [1]E_{22} + [3]E_{33} + [3]E12 + [2]E_{13} + [3]E_{23}.$$

So, for example, if we take the h/v-matrix

$$A^{-1} = 7E_{11} + 6E_{22} + 8E_{33} + 8E_{12} + 2E_{13} + 3E_{23},$$

then we obtain that

$$A \cdot A^{-1} = E_{11} + E_{22} + E_{33} + \{0, 5\}E_{12} + 5E_{23}.$$

Therefore, it contains a unit h/v-matrix.

Using several classes of H_v-structures, Vougiouklis studied several representations.

Definition 6.20. Let $M = M_{m \times n}$ be a module of $m \times n$ matrices over a ring R and $P = \{P_i \mid i \in I\} \subseteq M$. We define, a kind of, a P-hyperoperation $\underline{P}$ on M as follows

$$\underline{P} : M \times M \to \mathcal{P}^*(M) : (A, B) \to A\underline{P}B = \{AP_i^t B : i \in I\} \subseteq M,$$

where P^t denotes the transpose of the matrix P.

The hyperoperation $\underline{P}$, which is a bilinear map, is a generalization of Rees' operation where, instead of one sandwich matrix, a set of sandwich matrices is used. The hyperoperation P is strong associative and the inclusion distributivity with respect to addition of matrices

$$A\underline{P}(B + C) \subseteq A\underline{P}B + A\underline{P}C, \text{ for all } A, B, C \in M$$

is valid. Therefore, $(M, +, \underline{P})$ defines a multiplicative hyperring on non-square matrices.

Definition 6.21. Let $M = M_{m \times n}$ be a module of $m \times n$ matrices over R and let us take sets

$$S = \{s_k \mid k \in K\} \subseteq R,$$
$$Q = \{Q_j \mid j \in J\} \subseteq M,$$
$$P = \{P_i \mid i \in I\} \subseteq M.$$

Define three hyperoperations as follows:

$$\underline{S} : R \times M \to \mathcal{P}^*(M) : (r, A) \to r\underline{S}A = \{(rs_k)A : k \in K\} \subseteq M,$$
$$\underline{Q}_+ : M \times M \to \mathcal{P}^*(M) : (A, B) \to A\underline{Q}_+B = \{A + Q_j + B : j \in J\} \subseteq M,$$
$$\underline{P} : M \times M \to \mathcal{P}^*(M) : (A, B) \to A\underline{P}B = \{AP_i^tB : i \in I\} \subseteq M.$$

Then $(M, \underline{S}, \underline{Q}_+, \underline{P})$ is a hyperalgebra over R called *general matrix P-hyperalgebra*.

Remark 6.6. In a similar way a generalization of this hyperalgebra can be defined if one consider an H_v-ring or an H_v-field instead of a ring and using H_v-matrices instead of matrices.

Definition 6.22. Let $A = (a_{ij}), B = (b_{ij}) \in M_{m \times n}$, we call (A, B) a *unitize pair of matrices* if $A^tB = I_n$, where I_n denotes the $n \times n$ unit matrix.

We prove the following theorem which can be applied in the classical theory.

Theorem 6.12. *In the above notation if $m < n$, then there is no unitize pair.*

Proof. Suppose that $A^tB = (c_{ij})$, that is $c_{ij} = \sum\limits_{k=1}^{n} a_{ik}b_{kj}$, and we denote by A_m the block of the matrix A such that $A_m = (a_{ij}) \in M_{m \times m}$, i.e. we consider the matrix of the first m columns. Then we suppose that we have $(A_m)^tB_m = I_m$, therefore we must have $det(A_m) \neq 0$. Now, since $n > m$, we can consider the homogeneous system with respect to the 'unknowns' $b_{1n}, b_{2n}, ..., b_{mn}$:

$$c_{in} = \sum\limits_{k=1}^{n} a_{ik}b_{kn} = 0 \text{ for } i = 1, 2, ..., m.$$

From which, since $det(A_m) \neq 0$, we obtain that $b_{1n} = b_{2n} = ... = b_{mn} = 0$. Using this fact on the last equation, on the same unknowns, $c_{nn} = \sum\limits_{k=1}^{n} a_{nk}b_{kn} = 1$ we have 0=1, absurd. ∎

We recall the following definition and notations from Section 4.4.

Let $A = (a_{ij}) \in M_{m \times n}$ be matrix and $s, t \in N$ be naturals such that $1 \leq s \leq m, 1 \leq t \leq n$. Then we define the characteristic-like map $\underline{cst}$ from $M_{m \times n}$ to $M_{s \times t}$ by corresponding to A the matrix $A\underline{cst} = (a_{ij})$ where $1 \leq i \leq s, 1 \leq j \leq t$. This map is called *cut-projection* of type $\underline{st}$. In other words $A\underline{cst}$ is a matrix obtained from A by cutting the lines, with index greater than s, and columns, with index greater than t.

We can use cut-projections on several types of matrices to define sums and products, however, in this case we have ordinary operations, not multivalued.

In the same attitude we define hyperoperations on any type of matrices.

Let $A = (a_{ij}) \in M_{m \times n}$ be matrix and $s, t \in N$, such that $1 \leq s \leq m, 1 \leq t \leq n$. We define the mod-like map $\underline{st}$ from $M_{m \times n}$ to $M_{s \times t}$ by corresponding to A the matrix $A\underline{st} = (\underline{a}_{ij})$ which has as entries the sets

$$\underline{a}_{ij} = \{a_{i+\kappa s, j+\lambda t} | 1 \leq i \leq s, 1 \leq j \leq t, \text{ and } \kappa, \lambda \in N, i + \kappa s \leq m, j + \lambda t \leq n\}.$$

We call this multivalued map *helix-projection* of type $\underline{st}$. Thus $A\underline{st}$ is a set of $s \times t$-matrices $X = (x_{ij})$ such that $x_{ij} \in \underline{a}_{ij}$, for all i, j. Obviously $A\underline{mn} = A$. We may define helix-projections on 'matrices' of which their entries are sets.

Let $A = (a_{ij}) \in M_{m \times n}$ be matrix and $s, t \in N$, such that $1 \leq s \leq m, 1 \leq t \leq n$. Then it is clear that

$$(A\underline{sn})\underline{st} = (A\underline{mt})\underline{st} = A\underline{st}.$$

Let $A = (a_{ij}) \in M_{m \times n}$ be matrix and $s, t \in N$, such that $1 \leq s \leq m, 1 \leq t \leq n$. Then if $A\underline{st}$ is not a set of matrices but one single matrix then A is called *cut-helix matrix* of type $s \times t$. In other words, A is a helix matrix of type $s \times t$, if $A\underline{cst} = A\underline{st}$.

Definition 6.23. Let $A = (a_{ij}) \in M_{m \times n}$ and $B = (b_{ij}) \in M_{u \times v}$ be matrices and $s = min(m, u)$, $t = min(n, u)$. We define a hyperoperation, called *helix-addition* or *helix-sum*, as follows:

$$\oplus : M_{m \times n} \times M_{u \times v} \to \mathcal{P}^*(M_{s \times t}) :$$

$$(A, B) \to A \oplus B = A\underline{st} + B\underline{st} = (\underline{a}_{ij}) + (\underline{b}_{ij}) \subset M_{s \times t},$$

where

$$(\underline{a}_{ij}) + (\underline{b}_{ij}) = \{(c_{ij}) = (a_{ij} + b_{ij}) | a_{ij} \in \underline{a}_{ij} \text{ and } b_{ij} \in \underline{b}_{ij}\}.$$

Let $A = (a_{ij}) \in M_{m \times n}$ and $B = (b_{ij}) \in M_{u \times v}$ be two matrices and $s = min(m, u)$. We define a hyperoperation, called *helix-multiplication or helix-product*, as follows:

$$\otimes : M_{m \times n} \times M_{u \times v} \to \mathcal{P}^*(M_{m \times v}) :$$

$$(A, B) \to A \otimes B = A\underline{ms} \cdot B\underline{sv} = (\underline{a}_{ij}) \cdot (\underline{b}_{ij}) \subset M_{m \times v},$$

where

$$(\underline{a}_{ij}) \cdot (\underline{b}_{ij}) = \left\{ (c_{ij}) = (\textstyle\sum a_{it}b_{tj}) | a_{ij} \in \underline{a}_{ij} \text{ and } b_{ij} \in \underline{b}_{ij} \right\}.$$

For the helix-multiplication we remark that we have $A \otimes B = A\underline{ms} \cdot B\underline{sv}$ so we have either $A\underline{ms} = A$ or $B\underline{sv} = B$, that means that the helix-projection was applied only in one matrix and only in the rows or in the columns.

The commutativity is valid in the helix-addition. If the appropriate matrices in the helix-sum and in the helix-product are cut-helix, then the result is singleton.

Remark 6.7. From the fact that the helix-product on non square matrices is defined, the definition of a Lie-bracket is immediate, therefore the *helix-Lie Algebra* is defined, as well. This algebra is an H_v-Lie Algebra where the fundamental relation ϵ^* gives, by a quotient, a Lie algebra, from which a classification is obtained.

In the following we restrict ourselves on the matrices $M_{m \times n}$ where $m < n$. Obviously we have analogous results in the case where $m > n$ and for $m = n$ we have the classical theory.

In order to simplify the notation, since we have results on $\mod m$, we use:

For given $\kappa \in \mathbb{N} \setminus \{0\}$, we denote by $\underline{\kappa}$ the remainder resulting from its division by m if the remainder is non zero, and $\underline{\kappa} = m$ if the remainder is zero. Thus a matrix $A = (a_{\kappa\lambda}) \in M_{m \times n}$, $m < n$ is a cut-helix if we have $a_{\kappa\lambda} = a_{\underline{\kappa}\underline{\lambda}}$, for all $\kappa, \lambda \in \mathbb{N} \setminus \{0\}$. Moreover, let us denote by $I_c = (c_{\kappa\lambda})$ the *cut-helix unit matrix* which the cut matrix is the unit matrix I_m. Therefore, since $I_m = (\delta_{\kappa\lambda})$, where $\delta_{\kappa\lambda}$ is the Kroneckers delta, we obtain that, for all κ, λ, we have $c_{\kappa\lambda} = \delta_{\underline{\kappa}\underline{\lambda}}$.

Proposition 6.5. *For $m < n$ in $(M_{m \times n}, \otimes)$ the cut-helix unit matrix $I_c = (c_{\kappa\lambda})$, where $c_{\kappa\lambda} = \delta_{\underline{\kappa}\underline{\lambda}}$, is a left scalar unit and a right unit. It is the only one left scalar unit.*

Proof. If $A, B \in M_{m \times n}$, then in the helix-multiplication, since $m < n$, we take helix projection of A, therefore the result $A \otimes B$ is singleton if A is a cut-helix matrix of type $m \times m$. Moreover in order to have $A \otimes B = A_{\underline{mm}} \cdot B = B$, the $A_{\underline{mm}}$ must be the unit matrix. Consequently $I_c = (c_{\kappa\lambda})$, where $c_{\kappa\lambda} = \delta_{\kappa\underline{\lambda}}$, for all $\kappa, \lambda \in \mathbb{N} \setminus \{0\}$, is necessarily the left scalar unit element.

Now we remark that it is not possible to have the same case for the right matrix B, therefore we have only to prove that I_c is a right unit but it is not a scalar, consequently it is not unique.

Let $A = (a_{uv}) \in M_{m \times n}$ and consider the hyperproduct $A \otimes I_c$. In the entry $\kappa\lambda$ of this hyperproduct there are sets, for all $1 \le \kappa \le m$, $1 \le \lambda \le n$, of the form

$$\sum \underline{a}_{\kappa s} c_{s\lambda} = \sum \underline{a}_{\kappa s} \delta_{s\lambda} = \underline{a}_{\kappa\underline{\lambda}} \ni a_{\kappa\lambda}.$$

Therefore, $A \otimes I_c \ni A$, for all $A \in M_{m \times n}$. ∎

In the following examples we denote E_{ij} any type of matrices which have the ij-entry 1 and in all the other entries we have 0.

Construction 6.13. Consider the 2×3 matrices of the following form, for $\kappa \in \mathbb{N}$,

$$A_\kappa = E_{11} + \kappa E_{21} + E_{22} + E_{23}, \quad B_\kappa = \kappa E_{21} + E_{22} + E_{23}.$$

Then we obtain $A_\kappa \otimes A_\lambda = \{A_{\kappa+\lambda}, A_{\lambda+1}, B_{\kappa+\lambda}, B_{\lambda+1}\}$.
Similarly, we have $B_\kappa \otimes A_\lambda = \{B_{\kappa+\lambda}, B_{\lambda+1}\}, A_\kappa \otimes B_\lambda = B_\lambda = B_\kappa \otimes B_\lambda$.

Thus the set $\{A_\kappa, B_\lambda | \kappa, \lambda \in \mathbb{N}\}$ becomes an H_v-semigroup which is not COW because for $\kappa \ne \lambda$ we have $B_\kappa \otimes B_\lambda = B_\lambda \ne B_\kappa = B_\lambda \otimes B$, however

$$(A_\kappa \otimes A_\lambda) \cap (A_\lambda \otimes A_\kappa) = \{A_{\kappa+\lambda}, B_{\kappa+\lambda}\} \ne \varnothing, \text{ for all } \kappa, \lambda \in \mathbb{N}.$$

All elements B_λ are right absorbing and B_1 is a left scalar element, because $B_1 \otimes A_\lambda = B_{\lambda+1}$ and $B_1 \otimes B_\lambda = B_\lambda$. The element A_0 is a unit.

Construction 6.14. Consider the 2×3 matrices of the following form, for $\kappa \in \mathbb{N}$,

$$A_{\kappa\lambda} = E_{11} + E_{13} + \kappa E_{21} + E_{22} + \lambda E_{23}.$$

Then we obtain $A_{\kappa\lambda} \otimes A_{st} = \{A_{\kappa+s,\kappa+t}, A_{\kappa+s,\lambda+t}, A_{\lambda+s,\kappa+t}, A_{\lambda+s,\lambda+t}\}$.
Moreover $A_{st} \otimes A_{\kappa\lambda} = \{A_{\kappa+s,\lambda+s}, A_{\kappa+s,\lambda+t}, A_{\kappa+t,\lambda+s}, A_{\kappa+t,\lambda+t}\}$, so $A_{\kappa\lambda} \otimes A_{st} \cap A_{st} \otimes A_{\kappa\lambda} = \{A_{\kappa+s,\lambda+t}\}$, thus $(\otimes)$ is COW. The helix multiplication $(\otimes)$ is associative.

Consider all $m \times n$ matrices with $m \leq n$ and we write these matrices as block matrices of the form $A = (\underline{A}|\underline{A}')$, where $\underline{A}$ be a square $m \times m$ matrix and $\underline{A}'$ be of type $m \times (n - m)$. Denote $M_{\underline{m} \times \underline{n}}$ the set of all $m \times n$ matrices (with $m \leq n$) such that in every A the square matrix $\underline{A}$ is invertible. Take any $P \subset M_{\underline{m} \times \underline{n}}$ and define a P-hyperoperation as follows:

$$A \circ B = A P^t B, \text{ for all } A, B \in M_{\underline{m} \times \underline{n}},$$

where P^t is the set of all transpose matrices from the set P. Then the $(M_{\underline{m} \times \underline{n}}, +, \circ)$ becomes a multiplicative hyperring where all matrices of type $A_e = (\underline{A}^{-1}|\underline{0})$, for $A \in$ P are left units. Indeed

$$A_e \circ B = A_e P^t B \ni A_e A B = I_{m \times m} B = B, \text{ for all } B \in M_{\underline{m} \times \underline{n}}.$$

During last decades the hyperstructures have a variety of applications in other branches of mathematics and in many other sciences. These applications range from biomathematics -conchology, inheritance- and hadronic physics to mention but a few. The hyperstructures theory is closely related to fuzzy theory; consequently, hyperstructures can now be widely applicable in industry and production, too.

A general construction based on the partial ordering of the H_v-structures:

Theorem 6.15. (The Main e-Construction). *Given a group $(G, \cdot)$, where e is the unit, then we define in G, a large number of hyperoperations ($\otimes$) by extending $(\cdot)$ as follows:*

$$x \otimes y = \{xy, g_1, g_2, ...\}, \text{ for all } x, y \in G \setminus \{e\}, \text{ and } g_1, g_2, ... \in G \setminus \{e\}.$$

$g_1, g_2, ...$ are not necessarily the same for each pair (x,y). Then $(G, \otimes)$ becomes an H_v-group, actually is an H_b-group which contains the $(G, \cdot)$. The H_v-group $(G, \otimes)$ is an e-hypergroup. Moreover, if for each x,y such that $xy = e$, so we have $x \otimes y = xy$, then $(G, \otimes)$ becomes a strong e-hypergroup.

Proof. The proof is immediate since we enlarge the results of the group by putting elements from G and applying the Little Theorem. Moreover one can see that the unit e is a unique scalar and for each x in G, there exists a unique inverse x^{-1}, such that $1 \in x \cdot x^{-1} \cap x^{-1} \cdot x$ and if this condition is valid then we have $1 = x \cdot x^{-1} = x^{-1} \cdot x$. So the hyperstructure $(G, \otimes)$ is a strong e-hypergroup. ■

Chapter 7

Outline of applications and modeling

7.1 Chemical examples

Chemistry is the study of matter and of the changes matter undergoes. A chemical equation describes the products of a reaction that from the starting molecules or atoms. Chemistry seeks to predict the products that result from the reaction of specific quantities of atoms or molecules. Chemists accomplish this task by writing and balancing chemical equations. Symmetry is very important in chemistry researches and group theory is the tool that is used to determine symmetry. Classical algebraic structures (group theory) is a mathematical method by which aspects of a molecules symmetry can be determined. Algebraic hyperstructures are generalizations of classical algebraic structures. A motivation for the study of hyperstructures comes from chemical reactions. In [33], Davvaz and Dehghan-Nezhad provided examples of hyperstructures associated with chain reactions. In [36], Davvaz et al. introduced examples of weak hyperstructures associated with dismutation reactions. In [38], Davvaz et al. investigated the examples of hyperstructures and weak hyperstructures associated with redox reactions. In [2], Al Tahan and Davvaz presented three different examples of hyperstructures associated to electrochemical cells. In this section we review these examples. For more details we refer to the main references [2, 15, 31, 33–39].

7.1.1 *Chain reactions*

Chain reaction, in chemistry and physics, process yielding products that initiate further processes of the same kind, a self-sustaining sequence. Examples from chemistry are burning a fuel gas, the development of rancidity in fats, "knock" in internal-combustion engines, and the polymerization of ethylene to polyethylene. The best-known examples in physics are nuclear fissions brought about by neutrons. Chain reactions are in general very rapid but are also highly sensitive to reaction conditions, probably because the substances that sustain the reaction are easily affected by substances other than the reactants themselves. An atom or group of atoms possessing an odd (unpaired) electron is called radical. Radical species can be electrically neutral, in which case they are sometimes referred to as free radicals. Pairs of electrically neutral "free" radicals are formed via homolytic bond breakage. This can be achieved by heating in non-polar solvents or the vapor phase. At elevated temperature or under the influence ultraviolet light at room temperature, all molecular species will dissociate into radicals. Homolsis or homolytic bond fragmentation occurs when (in the language of Lewis theory) a two electron covalent bond breaks and one electron goes to each of the partner species.

For example, chlorine, Cl_2, forms chlorine radicals $(Cl^\bullet)$ and peroxides form oxygen radicals.

$$X{-}X \longrightarrow 2X^\bullet$$
$$Cl{-}Cl \longrightarrow 2Cl^\bullet$$
$$R{-}O{-}O{-}R \longrightarrow R{-}O^\bullet$$

Radical bond forming reactions (radical couplings) are rather rare processes. The reason is because radicals are normally present at low concentrations in a reaction medium, and it is statistically more likely they will abstract a hydrogen, or undergo another type of a substitution process, rather than reacting with each other by coupling. And as radicals are uncharged, there is little long range columbic attraction between two radical centers. Radical substitution reactions tend to proceed as chain reaction processes, often with many thousands of identical propagation steps. The propensity for chain reactivity gives radical chemistry a distinct feel compared with polar Lewis acid/base chemistry where chain reactions are less common. Methane can be chlorinated with chlorine to give chloromethane and hydrogen chloride. The reaction proceeds as a chain, radical, substitution mechanism. The process is a little more involved, and three steps are involved: initiation, propagation and termination:

(1) $Cl_2 \longrightarrow 2Cl^\bullet$
     (1) is called Chain-initiating step.
(2) $Cl^\bullet + CH_4 \longrightarrow HCl + CH_3^\bullet$

(3) $CH_3^• + Cl_2 \longrightarrow CH_3Cl + Cl^•$

then (2), (3), (2), (3), etc, until finally:

(2) and (3) are called Chain-propagating steps.

(4) $Cl^• + Cl^• \longrightarrow Cl_2$ or

(5) $CH_3^• + CH_3^• \longrightarrow CH_3CH_3$ or

(6) $CH_3^• + Cl^• \longrightarrow CH_3Cl$.

(4), (5) and (6) are called Chain-terminating steps.

First in the chain of reactions is a chain-initiating step, in which energy is absorbed and a reactive particle generated; in the present reaction it is the cleavage of chlorine into atoms (Step 1). There are one or more chain-propagating steps, each of which consumes a reactive particle and generates another; there they are the reaction of chlorine atoms with methane (Step 2), and of methyl radicals with chlorine (Step 3).

A chlorine radical abstracts a hydrogen from methane to give hydrogen chloride and a methyl radical. The methyl radical then abstracts a chlorine atom (a chlorine radical) from Cl_2 to give methyl chloride and a chlorine radical... which abstracts a hydrogen from methane... and the cycle continues... Finally there are chain-terminating steps, in which reactive particles are consumed but not generated; in the chlorination of methane these would involve the union of two of the reactive particles, or the capture of one of them by the walls of the reaction vessel.

The halogens are all typical non-metals. Although their physical forms differ-fluorine and chlorine are gases, bromine is a liquid and iodine is a solid at room temperature, each consists of diatomic molecules; F_2, Cl_2, Br_2 and I_2. The halogens all react with hydrogen to form gaseous compounds, with the formulas HF, HCl, HBr, and HI all of which are very soluble in water. The halogens all react with metals to give halides.

$$: \ddot{F} \; \text{-} \; \ddot{F} :, \qquad : \ddot{C}l \text{-} \underset{..}{C}l :, \qquad : \ddot{B}r \; \text{-} \; \ddot{B}r :, \qquad : \ddot{I} \; \text{-} \underset{..}{I} :$$

The reader will find in [83] a deep discussion of chain reactions and halogens.

In during chain reaction

$$A_2 + B_2 \overset{\text{Heat or Light}}{\longleftrightarrow} 2AB$$

there exist all molecules A_2, B_2, AB and whose fragment parts $A^•, B^•$ in experiment. Elements of this collection can by combine with each other. All combinational probability for the set $H = \{A^•, B^•, A_2, B_2, AB\}$ to do without energy can be displayed as in Table 7.1.

Then, $(H, \oplus)$ is an H_v-group [33]. Moreover, $X = \{A^•, A_2\}$ and $Y = \{B^•, B_2\}$ are only H_v-subgroups of $(H, \oplus)$ [33].

Table 7.1   Chain reactions.

$\oplus$	$A^\bullet$	$B^\bullet$	A_2	B_2	AB
$A^\bullet$	$A^\bullet, A_2$	$A^\bullet, B^\bullet, AB$	$A^\bullet, A_2$	$A^\bullet, B_2, B^\bullet, AB$	$A^\bullet, AB, A_2, B^\bullet$
$B^\bullet$	$A^\bullet, B^\bullet, AB$	$B^\bullet, B_2$	$A^\bullet, B^\bullet, AB, A_2$	$B^\bullet, B_2$	$A^\bullet, B^\bullet, AB, B_2$
A_2	$A^\bullet, A_2$	$A^\bullet, B^\bullet, AB, A_2$	$A^\bullet, A_2$	$A^\bullet, B^\bullet, A_2, B_2, AB$	$A^\bullet, B^\bullet, A_2, AB$
B_2	$A^\bullet, B^\bullet, B_2, AB$	$B^\bullet, B_2$	$A^\bullet, B^\bullet, A_2, B_2, AB$	$B^\bullet, B_2$	$A^\bullet, B^\bullet, B_2, AB$
AB	$A^\bullet, AB, A_2, B^\bullet$	$A^\bullet, B^\bullet, AB, B_2$	$A^\bullet, B^\bullet, A_2, AB$	$A^\bullet, B^\bullet, B_2, AB$	$A^\bullet, B^\bullet, A_2, B_2, AB$

Table 7.2   For H and I.

$\oplus$	$H^\bullet$	$I^\bullet$	H_2	I_2	HI
H°	$H^\bullet, H_2$	$H^\bullet, I^\bullet, HI$	$H^\bullet H_2$	$H^\bullet, I_2, I^\bullet, HI$	$H^\bullet, HI, H_2, I^\bullet$
$I^\bullet$	$H^\bullet, I^\bullet, HI$	$I^\bullet, I_2$	$H^\bullet, I^\bullet, HI, H_2$	$I^\bullet, I_2$	$H^\bullet, I^\bullet, HI, I_2$
H_2	$H^\bullet, H_2$	$H^\bullet, I^\bullet, HI, I_2$	$H^\bullet, H_2$	$H^\bullet, I^\bullet, H_2, I_2, HI$	$H^\bullet, I^\bullet, H_2, HI$
I_2	$H^\bullet, I^\bullet, I_2, HI$	$H^\bullet, I_2$	$H^\bullet, I^\bullet, H_2, I_2, HI$	$H^\bullet, I_2$	$H^\bullet, I^\bullet, I_2, HI$
HI	$H^\bullet, HI, H_2, I^\bullet$	$H^\bullet, I^\bullet, HI, I_2$	$H^\bullet, I^\bullet, H_2, HI$	$H^\bullet, I^\bullet, H_2, HI$	$H^\bullet, I^\bullet, H_2, I_2, HI$

If we consider $A = H$ and $B \in \{F, CL, Br, I\}$ (for example $B = I$), the complete reactions table becomes Table 7.2.

7.1.2   Dismutation reactions

In a redox reactions or oxidation-reduction reaction, electrons are transferred from one reactant to another. Oxidation refers to the loss of electrons, while reduction refers to the gain of electrons. A substance that has strong affinity for electrons and tends to extract them from other species is called an oxidizing agent or an oxidant. A reducing agent, or reductant, is a reagent that readily donates electrons to another species [101]. A half reaction is a reduction or an oxidation reaction. Two half-reactions are needed to form a whole reaction. Redox reactions have a number of similarities to acid-base reactions. Like acid-base reactions, redox reactions are a matched set; you don't have an oxidation reaction without a reduction reaction happening at the same time. When the change in free energy (ΔG) is negative, a process or chemical reaction proceeds spontaneously in the forward direction. When ΔG is positive, the process proceeds spontaneously in reverse. In electrochemical reactions $\Delta G = -nFE$, where n, F and E are number of electrons transferred in the reaction, Faraday constant and cell potential, respectively [101].

The change in the oxidation state of a species lets you know if it has undergone oxidation or reduction. Oxidation is the process in which an atom undergoes an algebraic increase in oxidation number, and reduction is the process in which an atom undergoes an algebraic decrease in oxidation number. On this basis, oxidation-reduction is involved in the reaction;

$$O_2 + C \longrightarrow CO_2.$$

In the reaction, oxidation number of the C atom increases from zero to $+4$ whereas, the oxidation number of O atom decreases from zero to -2. Furthermore, the total increase in the oxidation number equals to the total decrease in oxidation number [84].

Disproportionation or dismutation is used to describe two particular types of chemical reaction:

(1) A chemical reaction of the type $2A \longrightarrow A' + A''$, where A, A' and A'' are different chemical species [108]. Most but not all are redox reactions. For example $2H_2O \longrightarrow H_3O^+ + OH^-$ is a *disproportionation*, but is not a redox reaction.

(2) A chemical reaction in which two or more atoms of the same element originally having the same oxidation state react with other chemical(s) or themselves to give different oxidation numbers. In another word, disproportionation is a reaction in which a species is simultaneously reduced and oxidized to form two different oxidation numbers. The reverse of disproportionation is called comproportionation. *Comproportionation* is a chemical reaction where two reactants, each containing the same element but with a different oxidation number, will form a product with an oxidation number intermediate of the two reactants. For example, an element tin in the oxidation states 0 and $+4$ can comproportionate to the state $+2$. The standard reduction potentials of all half reactions are: $E^\circ{}_{Sn^{4+}/Sn^{2+}} = 0.154\ V, E^\circ{}_{Sn^{2+}/Sn} = -0.136\ V, E^\circ{}_{Sn^{4+}/Sn} = 0.009\ V$. Therefore, the comproportionation reaction is spontaneous.

$$Sn + Sn^{4+} \longrightarrow 2Sn^{2+}.$$

All combinational probability for the set $S = \{Sn,\ Sn^{2+},\ Sn^{4+}\}$ to do without energy can be displayed as follows. The major products are written in Table 7.3.

Then, $(S, \oplus)$ is weak associative. Also, we can conclude that $(\{Sn,\ Sn^{2+}\}, \oplus)$ is a hypergroup and $(\{Sn^{2+},\ Sn^{4+}\}, \oplus)$ is an H_v-semigroup [36].

Chlorine gas reacts with dilute hydroxide to form chloride, chlorate and water. The ionic equation for this reaction is as follows [67]:

$$3Cl_2 + 6OH^- \longrightarrow 5Cl^- + ClO_3{}^- + 3H_2O$$

As a reactant, the oxidation number of the elemental chlorine, chloride and chlorate are 0, -1 and +5, respectively. Therefore, chlorine has been oxidized to chlorate whereas; it has been reduced to chloride [67].

Table 7.3   Dismutation reactions Sn.

$\oplus$	Sn	Sn^{2+}	Sn^{4+}
Sn	Sn	Sn, Sn^{2+}	Sn^{2+}
$Sn2^+$	Sn, Sn^{2+}	Sn^{2+}	Sn^{2+}, Sn^{4+}
Sn^{4+}	Sn^{2+}	Sn^{2+}, Sn^{4+}	Sn^{4+}

Table 7.4   Dismutation reactions In.

$\oplus$	In	In^+	In^{3+}
In	In	In, In^+	In, In^{3+}
In^+	In, In^+	In, In^{3+}	In^+, In^{3+}
In^{3+}	In, In^{3+}	In^+, In^{3+}	In^{3+}

Indium has three oxidation states $0, +1$ and $+3$. The standard reduction potentials of all half reactions are: $E°_{In^{3+}/In^+} = -0.434\ V, E°_{In^+/In} = -0.147\ V, E°_{In^{3+}/In} = -0.338\ V$. According to the standard reduction potentials, disproportionation reaction of In^+ is spontaneous. All combinational probability for the set $S = \{In,\ In^+,\ In^{3+}\}$ to do without energy can be displayed as Table 7.4.

Then, $(S, \oplus)$ is weak associative. Clearly, $\oplus$ is commutative. Also, the reproduction axiom holds. Therefore, $(S, \oplus)$ is a commutative H_v-group [36].

Vanadium forms a number of different ions including $V, V^{2+}, V^{3+}, VO^{2+}$ and VO_2^+. The oxidation states of these species are $0, +2, +3, +4$ and $+5$, respectively. The standard reduction potentials of all corresponding half reactions are:

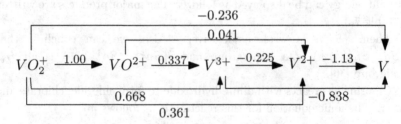

All combinational probability for the set $S = \{V, V^{2+}, V^{3+}, VO^{2+}, VO_2^+\}$ to do without energy in acidic media can be displayed as Table 7.5. When the reactants are added in appropriate stoichiometric ratios. For example

Table 7.5   Vanadium.

$\oplus$	V	V^{2+}	V^{3+}	VO^{2+}	$VO_2{}^+$
V	V	V,V^{2+}	V^{2+}	V^{2+},V^{3+}	V^{3+}
V^{2+}	V,V^{2+}	V^{2+}	V^{2+},V^{3+}	V^{3+}	V^{3+},VO^{2+}
V^{3+}	V^{2+}	V^{2+},V^{3+}	V^{3+}	V^{3+},VO^{2+}	VO^{2+}
VO^{2+}	V^{2+},V^{3+}	V^{3+}	V^{3+},VO^{2+}	VO^{2+}	$VO^{2+},VO_2{}^+$
$VO_2{}^+$	V^{3+}	V^{3+},VO^{2+}	VO^{2+}	$VO^{2+},VO_2{}^+$	$VO_2{}^+$

Table 7.6   The major products between all forms of vanadium.

$\oplus$	V	V^{2+}	V^{3+}	VO^{2+}	$VO_2{}^+$
V	V	V,V^{2+}	V,V^{2+}	V^{2+}	V^{2+},V^{3+}
V^{2+}	V,V^{2+}	V^{2+}	V^{2+},V^{3+}	V^{3+}	V^{3+},VO^{2+}
V^{3+}	V,V^{2+}	V^{2+},V^{3+}	V^{3+}	V^{3+},VO^{2+}	VO^{2+}
VO^{2+}	V^{2+}	V^{3+}	V^{3+},VO^{2+}	VO^{2+}	$VO^{2+},VO_2{}^+$
$VO_2{}^+$	V^{2+},V^{3+}	V^{3+},VO^{2+}	VO^{2+}	$VO^{2+},VO_2{}^+$	$VO_2{}^+$

vanadium (V) reacts with $VO_2{}^+$ as follows:

$$2V + 3VO_2{}^+ + 12H^+ \longrightarrow 5V^{3+} + 6H_2O$$

Then, $(S, \oplus)$ is a hyperstructure. The hyperstructures $(\{V,\ V^{2+}\}, \oplus)$, $(\{V^{2+},\ V^{3+}\}, \oplus)$, $(\{V^{3+},\ VO^{2+}\}, \oplus)$ and $(\{VO^{2+},\ VO_2^+\}, \oplus)$ are hypergroups [36]. Moreover, we have:

$$(\{V, V^{2+}\}, \oplus) \cong (\{V^{2+}, V^{3+}\}, \oplus) \cong (\{V^{3+}, VO^{2+}\}, \oplus) \cong (\{VO^{2+}, VO_2{}^+\}, \oplus).$$

The major products between all forms of vanadium are showed in Table 7.6. It is assumed the reactants are added together in 1 : 1 mole ratios.

Then, $(S, \oplus)$ is a hyperstructure. The hyperstructures $(\{V,\ V^{2+}\}, \oplus)$, $(\{V^{2+},\ V^{3+}\}, \oplus)$, $(\{V^{3+},\ VO^{2+}\}, \oplus)$ and $(\{VO^{2+},\ VO_2^+\}, \oplus)$ are hypergroups. Moreover, we have:

$$(\{V, V^{2+}\}, \oplus) \cong (\{V^{2+}, V^{3+}\}, \oplus) \cong (\{V^{3+}, VO^{2+}\}, \oplus) \cong (\{VO^{2+}, VO_2{}^+\}, \oplus).$$

7.1.3   Redox reactions

Redox (reduction-oxidation) reactions include all chemical reactions in which atoms have their oxidation state changed. This can be either a simple redox process, such as the oxidation of carbon to yield carbon dioxide

(CO_2) or the reduction of carbon by hydrogen to yield methane (CH_4), or a complex process such as the oxidation of glucose $(C_6H_{12}O_6)$ in the human body through a series of complex electron transfer processes. Oxidation is the loss of electrons or an increase in oxidation state, and reduction is the gain of electrons or a decrease in oxidation state by an analyte (molecule, atom or ion). There can not be an oxidation reaction without a reduction reaction happening simultaneously. Therefore the oxidation alone and the reduction alone are each called a half-reaction, because two half-reactions always occur together to form a whole reaction [84].

Each half-reaction has a standard reduction potential (E^0), which is equal to the potential difference at equilibrium under standard conditions of an electrochemical cell in which the cathode reaction is the half-reaction considered, and the anode is a standard hydrogen electrode (SHE). For a redox reaction, the potential of the cell is defined by: $E^0_{cell} = E^0_{cathode} - E^0_{anode}$. If the potential of a redox reaction (E^0_{cell}) is positive, this reaction will spontaneous [84].

For example, consider the redox reaction of Ag^{2+} with Ag:

$$Ag^{2+} + Ag \longrightarrow Ag^{+}.$$

We can write two half-reactions for this reaction:

(1) $Ag^{2+} + e \longrightarrow Ag^{+}$,

(2) $Ag \longrightarrow Ag^{+} + e$.

The E^0 of the first reaction $(E^0_{cathode})$ is 1.98 V (vs. SHE) and the E^0 of the second reaction (E^0_{anode}) is 0.799 V (vs. SHE) [101]. Therefore, in this case, the E^0_{cell} ($E^0_{cathode} - E^0_{anode} = 1.181$) is positive and the above redox reaction between Ag^{2+} and Ag is spontaneous.

Silver (Ag) is a transition metal and has a large number of applications in jewelry, electrical contacts and conductors, catalysis of chemical reactions, disinfectants and microbiocides. Silver plays no known natural biological role in humans and itself is not toxic, but most silver salts are toxic, and some may be carcinogenic. Ag can be in three oxidation state: Ag (0), Ag (I) and Ag (II). Among Ag (I) and Ag (II), Ag (I) is very well characterized and many simple ionic compounds are known containing Ag^{+}. However, AgF_2 is known which Ag has oxidation state of II in it. AgF_2 is strongly oxidizing and a good fluorimating agent. But Ag (II) is more stable in complex forms. A number of Ag (II) complexes have been obtained by oxidation of Ag (I) salts is aqueous solution in the presences of the ligand. For example, $[Ag\,(pyridine)_4]^{2+}$ and $[Ag\,(bi\,pyridine)_2]^{2+}$

Table 7.7 Redox reactions Ag.

$\oplus$	Ag^{2+}	Ag^+	Ag
Ag^{2+}	Ag^{2+}	Ag^+, Ag^{2+}	Ag^+
Ag^+	Ag^+, Ag^{2+}	Ag^+	Ag, Ag^+
Ag	Ag^+	Ag^+, Ag	Ag

are quite stable. The $+1$ oxidation state is the best known oxidation state of silver. Ag^+ salts are generally insoluble in water with the exception of nitrate, fluoride and perchlorate. Most stable Ag (I) complexes have a linear structure [100].

As described above, Ag species with different oxidation state can react with themselves. All possible products for spontaneous reactions are presented in Table 7.7.

Table 7.7 is isomorphic to Table 7.3 of dismutation reactions. Therefore, $\oplus$ is weak associative. Also, we conclude that $(\{Ag^{2+},\ Ag^+\}, \oplus)$ and $(\{Ag^+,\ Ag\}, \oplus)$ are hypergroups.

Copper (Cu) is a ductile metal with very high thermal and electrical conductivity. It is used as a conductor of heat and electricity, a building material, and a constituent of various metal alloys. Cu can be in four oxidation state: Cu (0), Cu (I), Cu (II) and Cu (III). In nature, copper mainly is as $CuFeS_2$, with oxidation state of II for Cu. Also, Cu can be as Cu_2S or Cu_2O with the oxidation state of I. Pure copper is obtained by electrolytic refining using sheets of pure copper as cathode and impure copper as anode. In this process different ions of Cu, Cu (II) or Cu (I), reduced to Cu (0) at cathode. Cu (III) is generally uncommon, however some its complexes are known [100].

The standard reduction potential (E^0) for conversion of each oxidation state to another are: E^0 (Cu^{3+}/Cu^{2+}) = 2.4 V, E^0 (Cu^{2+}/Cu^+)= 0.153 V, E^0 (Cu^{2+}/Cu)= 0.342 V and E^0 (Cu^+/Cu)= 0.521 V, where potentials are versus SHE [101]. According to these standard potentials, and similar to example of Ag, the following reactions are spontaneous:

(1) $Cu^{3+} + Cu^+ \longrightarrow Cu^{2+}$,
(2) $Cu^{3+} + Cu \longrightarrow Cu^{2+} + Cu^+$.

Therefore, all possible products in reactions between oxidation states of Cu which can be produced spontaneously are listed in Table 7.8.

Table 7.8   Redox reactions Cu.

$\odot$	Cu	Cu^+	Cu^{2+}	Cu^{3+}
Cu	Cu	$Cu,\ Cu^+$	$Cu^{2+},\ Cu$	$Cu^{2+},\ Cu^+$
Cu^+	$Cu,\ Cu^+$	Cu^+	$Cu^{2+},\ Cu^+$	Cu^{2+}
Cu^{2+}	$Cu,\ Cu^{2+}$	$Cu^{2+},\ Cu^+$	Cu^{2+}	$Cu^{2+},\ Cu^{3+}$
Cu^{3+}	$Cu^+,\ Cu^{2+}$	Cu^{2+}	$Cu^{2+},\ Cu^{3+}$	Cu^{3+}

Table 7.9   Redox reactions Am.

$\otimes$	Am	Am^{2+}	Am^{3+}	Am^{4+}
Am	Am	$Am,\ Am^{2+}$	$Am,\ Am^{3+}$	$Am,\ Am^{4+}$
Am^{2+}	$Am,\ Am^{2+}$	Am^{2+}	$Am^{2+},\ Am^{3+}$	Am^{3+}
Am^{3+}	$Am,\ Am^{3+}$	$Am^{2+},\ Am^{3+}$	Am^{3+}	$Am^{3+},\ Am^{4+}$
Am^{4+}	$Am,\ Am^{4+}$	Am^{3+}	$Am^{3+},\ Am^{4+}$	Am^{4+}

In Table 7.8, the hyperoperation $\odot$ is weak associative. Hence, we have an H_v-semigroup. The hyperstructures $(\{Cu,\ Cu^+\}, \odot)$, $(\{Cu,\ Cu^{2+}\}, \odot)$, $(\{Cu^+,\ Cu^{2+}\}, \odot)$ and $(\{Cu^{2+},\ Cu^{3+}\}, \odot)$ are hypergroups. Let H be a set with three elements. On H, we define the following hyperoperation:

$$x \star y = \{x, y\}, \text{ for all } x, y \in H.$$

It is easy to see that $\star$ is associative and so $(H, \star)$ is a hypergroup. Now, we have

$$(\{Cu,\ Cu^+,\ Cu^{2+}\}, \odot) \cong (H, \star).$$

Note that $(\{Cu^+,\ Cu^{2+},\ Cu^{3+}\}, \odot)$ is not semihypergroup.

Americium (Am) is a transuranic radioactive chemical element in actinide series. It have four oxidation states of 0, 2, 3 and 4. The standard reduction potential (E^0) for conversion of each oxidation state to another are: $E^0\ (Am^{4+}/Am^{3+})= 2.6\ V$, $E^0\ (Am^{3+}/Am^{2+})= -2.3\ V$, $E^0\ (Am^{3+}/Am)= -2.048\ V$ and $E^0\ (Am^{2+}/Am)= -1.9\ V$, where potentials are versus SHE [101]. Therefore, the following reaction is spontaneous:

$$Am^{4+} + Am^{2+} \longrightarrow Am^{3+}.$$

Therefore, all possible combinations for different oxidation states of Am which can be produced without energy are presented in Table 7.9.

Regarding to Table 7.9, similar to Table 7.8, we have

$$(\{Am,\ Am^{2+},\ Am^{3+}\}, \otimes) \cong (H, \star).$$

Note that $(\{Am^{2+},\ Am^{3+},\ Am^{4+}\}, \otimes)$ is not semihypergroup.

Table 7.10   Redox reactions Au.

⊎	Au	Au^+	Au^{2+}	Au^{3+}
Au	Au	Au, Au^+	Au^+	Au, Au^{3+}
Au^+	Au, Au^+	Au^+	Au^+, Au^{2+}	Au^+, Au^{3+}
Au^{2+}	Au^+	Au^+, Au^{2+}	Au^{2+}	Au^{2+}, Au^{3+}
Au^{3+}	Au, Au^{3+}	Au^+, Au^{3+}	Au^{2+}, Au^{3+}	Au^{3+}

Gold (Au) is a dense, soft, shiny, malleable and ductile metal and can be in four oxidation states of Au (0), Au (I), Au (II) and Au (III). Au (III) is common for gold compounds and exist as: Au_2O_3, AuF_3, $AuCl_3$, $AuBr_3$ and Au $(OH)_3$. Au (I) is much less stable in solution and is stabilized in complexes [100].

The standard reduction potential (E^0) for conversion of each oxidation state to another are: E^0 (Au^{3+}/Au^+)= 1.401 V, E^0 (Au^{3+}/Au)= 1.498 V, E^0 (Au^{2+}/Au^+)= 1.8 V and E^0 (Au^+/Au)= 1.692 V, where potentials are versus SHE [101]. According to these standard potentials, the following reaction is spontaneous:

$$Au^{2+} + Au \longrightarrow Au^+.$$

Therefore, the major products in reactions between oxidation states of Au which can be produced spontaneously are listed in Table 7.10.

The H_v-semigroups defined in Table 7.9 and Table 7.10 are isomorphic.

7.1.4   Galvanic cell

Chemical reactions involving the transfer of electrons from one reactant to another are called oxidation-reduction reactions or redox reactions. In a redox reaction, two half-reactions occur; one reactant (with less electronegativity) gives up electrons (undergoes oxidation) and another reactant (with higher electronegativity) gains electrons (undergoes reduction). For example, a piece of zinc going into a solution as zinc ions, with each Zn atom giving up 2 electrons, is an example of an oxidation half-reaction.

$$Zn \longrightarrow Zn^{2+} + 2e^-.$$

In contrast, the reverse reaction, in which Zn^{2+} ions gain 2 electrons to become Zn atoms, is an example of a reduction half-reaction.

$$Zn^{2+} + 2e^- \longrightarrow Zn.$$

A redox reaction results when an oxidation and a reduction half-reactions are combined to complete a transfer of electrons as in the following example:

$$Zn + Cu^{2+} \longrightarrow Zn^{2+} + Cu.$$

The electrons are not shown in the above redox reaction because they are neither reactants nor products but have simply been transferred from one species to another (from Zn to Cu^{2+} in this case). In this redox reaction, the Zn is referred to as the reducing agent because it causes the Cu^{2+} to be reduced to Cu. The Cu^{2+} is called the oxidizing agent because it causes the Zn to be oxidized to Zn^{2+}.

A Galvanic cell or voltaic cell is a device in which a redox reaction spontaneously occurs and produces an electric current. In order for the transfer of electrons in a redox reaction to produce an electric current and be useful, the electrons are made to pass through an external electrically conducting wire instead of being directly transferred between the oxidizing and reducing agents. The design of a Galvanic cell allows this to occur. In a Galvanic cell, two solutions, one containing the ions of the oxidation half-reaction and the other containing the ions of the reduction half-reaction, are placed in separated compartments called half-cells. For each half-cell, the metal, which is called an electrode, is placed in the solution and connected to an external wire. The electrode at which oxidation occurs is called the anode (Zn in the above example) and the electrode at which reduction occurs is called the cathode (Cu in the above example). The two half-cells are connected by a salt-bridge that allows a current of ions from one half-cell to the other to complete the circuit of electron current in the external wires. When the two electrodes are connected to an electric load (such as a light bulb or voltmeter) the circuit is completed, the oxidation-reduction reaction occurs, and electrons move from the anode ($-$) to the cathode ($+$), producing an electric current.

Galvanic cell consists of two half-cells, such that the electrode of one half-cell is composed of metal A (with larger electronegativity) and the electrode of the other half-cell is composed of metal B (with smaller electronegativity). The redox reactions for the two separate half-cells are given as follows:

$$A^{n+} + ne^- \longrightarrow A,$$

$$B \longrightarrow B^{m+} + me^-.$$

The two metals A and B can react with each other according to the following balanced equation:

$$nB + mA^{n+} \longrightarrow mA + nB^{m+}.$$

Table 7.11   Galvanic cell.

$\oplus_1$	A	B	A^{n+}	B^{m+}
A	A	A, B	A, A^{n+}	A, B^{m+}
B	A, B	B	B^{m+}, A	B, B^{m+}
A^{n+}	A, A^{n+}	B^{m+}, A	A^{n+}	A^{n+}, B^{m+}
B^{m+}	A, B^{m+}	B^{m+}, B	A^{n+}, B^{m+}	B^{m+}

Table 7.12   Table for Theorem 7.1.

$\oplus_1$	a	b	c	d
a	a	a, b	a, c	a, d
b	a, b	b	a, d	b, d
c	a, c	a, d	c	c, d
d	a, d	b, d	c, d	d

Having the element Cu with greater electronegativity than that of Zn, we get that $Zn + Cu^{2+} \longrightarrow Zn^{2+} + Cu$ is an example of a redox reaction occurring in a Galvanic cell.

For more details about Galvanic cells, see [142].

Next, we present a commutative hyperstructure related to Galvanic cell and investigate its properties.

We consider the set $H = \{A, B, A^{n+}, B^{+m}\}$ and we define a hyperoperation $\oplus_1$ on H as follows:

$x \oplus_1 y$ is the result of a possible reaction between x and y in a Galvanic cell. If x and y do not react in a Galvanic cell then we set $x \oplus_1 y = \{x, y\}$.

All possible spontaneous redox reactions of $\{A, B, A^{n+}, B^{+m}\}$ in a Galvanic cell are summarized in Table 7.11.

In Table 7.11, if we change the names from A, B, A^{n+}, B^{m+} to a, b, c, d respectively, then the following theorem holds.

Theorem 7.1. *Let $H = \{a, b, c, d\}$, $\oplus_1$ be the hyperoperation on H and consider Table 7.12 corresponding to $(H, \oplus_1)$:*

Then $(H, \oplus_1)$ is a commutative H_v-semigroup.

Proof. It is clear from the above table that $(H, \oplus_1)$ is a commutative hypergroupoid. We need to show that $(H, \oplus_1)$ is a weak associative hypergroupoid, i.e, $x \oplus_1 (y \oplus_1 z) \cap (x \oplus_1 y) \oplus_1 z \neq \emptyset$ for all $(x, y, z) \in H^3$. We have three cases for x; $x = a$ or d, $x = b$ and $x = c$:

- Case $x = a$ or d. We have that $x \in x \oplus_1 (y \oplus_1 z) \cap (x \oplus_1 y) \oplus_1 z \neq \emptyset$.
- Case $x = b$. We have that $b \oplus_1 (c \oplus_1 c) = b \oplus_1 c = \{a, d\}$ and that $(b \oplus c) \oplus c = \{a, d\} \oplus c = \{a, c, d\}$. Thus, $b \oplus_1 (c \oplus_1 c) \cap (b \oplus_1 c) \oplus_1 c \neq \emptyset$. Moreover, one can easily check that $b \oplus_1 (c \oplus_1 z) \cap (b \oplus_1 c) \oplus_1 z \neq \emptyset$ and that $b \oplus_1 (y \oplus_1 c) \cap (b \oplus_1 y) \oplus_1 c \neq \emptyset$. If $y \neq c$ and $z \neq c$ then $b \in b \oplus_1 (y \oplus_1 z) \cap (b \oplus_1 y) \oplus_1 z$.
- Case $x = c$. This case is similar to that of Case $x = b$. ∎

Remark 7.1. Since $a \oplus_1 (b \oplus_1 c) = \{a, d\} \neq (a \oplus_1 b) \oplus_1 c = \{a, c, d\}$, it follows that $(H, \oplus_1)$ is not a semihypergroup.

Remark 7.2. $(H, \oplus_1)$ admits two identities; a and d. Moreover, a and d are strong identities.

7.1.5   *Electrolytic cell*

Voltaic cells are driven by a spontaneous chemical reaction that produces an electric current through an outside circuit. These cells are important because they are the basis for the batteries that fuel modern society. But they aren't the only kind of electrochemical cells. The reverse reaction in each case is non-spontaneous and requires electrical energy to occur. It is possible to construct a cell that does work on a chemical system by driving an electric current through the system. These cells are called electrolytic cells (or reverse Galvanic cells), and operate through electrolysis.

Electrolysis is used to drive an oxidation-reduction reaction in a direction in which it does not occur spontaneously by driving an electric current through the system while doing work on the chemical system itself, and therefore is non-spontaneous. Electrolytic cells, like Galvanic cells, are composed of two half-cells; one is a reduction half-cell, the other is an oxidation half-cell. The direction of electron flow in electrolytic cells, however, may be reversed from the direction of spontaneous electron flow in Galvanic cells, but the definition of both cathode and anode remain the same, where reduction takes place at the cathode and oxidation occurs at the anode. Because the directions of both half-reactions have been reversed, the sign, but not the magnitude, of the cell potential has been reversed.

Electrolytic cells consist of two half-cells, such that the electrode of one half-cell is composed of metal A (with larger electronegativity) and the electrode of the other half-cell is composed of metal B (with smaller

Table 7.13 Electrolytic cell.

$\oplus_2$	A	B	A^{n+}	B^{m+}
A	A	A, B	A, A^{n+}	A^{n+}, B
B	A, B	B	A^{n+}, B	B, B^{m+}
A^{n+}	A, A^{n+}	A^{n+}, B	A^{n+}	A^{n+}, B^{m+}
B^{m+}	A^{n+}, B	B^{m+}, B	A^{n+}, B^{m+}	B^{m+}

electronegativity). The redox reactions for the two separate half-cells are given as follows:

$$A \longrightarrow A^{n+} + ne^-,$$

$$B^{m+} + me^- \longrightarrow B.$$

The two metals A and B can react with each other according to the following balanced equation:

$$mA + nB^{m+} \longrightarrow nB + mA^{n+}.$$

An example of a reaction in an Electrolytic cell is:

$$Zn^{2+} + Cu \longrightarrow Zn + Cu^{2+}$$

which is the reverse of the reaction described in Section 3. For more details about Electrolytic cells, see [142].

Next we present a hyperstructure related to Electrolytic cells and investigate its properties.

We consider the set $H = \{A, B, A^{n+}, B^{+m}\}$ and we define a hyperoperation $\oplus_2$ on H as follows:

$x \oplus_2 y$ is the result of a possible reaction between x and y in an Electrolytic cell. If x and y do not react in an Electrolytic cell then we set $x \oplus_2 y = \{x, y\}$.

All possible non-spontaneous redox reactions of $\{A, B, A^{n+}, B^{+m}\}$ in an Electrolytic cell are summarized in Table 7.13.

In Table 7.13, if we change the names from A, B, A^{n+}, B^{m+} to a, b, c, d respectively, then the following theorem holds.

Theorem 7.2. *Let $H = \{a, b, c, d\}$, $\oplus_2$ be the hyperoperation on H and consider Table 7.14 corresponding to $(H, \oplus_2)$.*

Then $(H, \oplus_2)$ is a commutative H_v-semigroup.

Table 7.14   Table for Theorem 7.2.

$\oplus_2$	a	b	c	d
a	a	a,b	a,c	b,c
b	a,b	b	b,c	b,d
c	a,c	b,c	c	c,d
d	b,c	b,d	c,d	d

Proof. Let $f : (H, \oplus_1) \longrightarrow (H, \oplus_2)$ defined as follows:

$$f(a) = b, f(b) = a, f(c) = d \text{ and } f(d) = c.$$

It is easy to see that f is an isomorphism and thus, $(H, \oplus_1) \cong (H, \oplus_2)$. The latter and Theorem 7.1 imply that $(H, \oplus_2)$ is a commutative H_v-semigroup. ∎

Remark 7.3. $(H, \oplus_2)$ admits two identities; b and c. Moreover, b and c are strong identities.

7.1.6   Galvanic/Electrolytic cells

In this section, we present a commutative hyperstructure related to Galvanic/Electrolytic cells and investigate its properties.

We consider the set $H = \{A, B, A^{n+}, B^{+m}\}$ and we define a hyperoperation $\oplus$ on H as follows:

$x \oplus y$ is the result of a possible reaction between x and y in either a Galvanic cell or in an Electrolytic cell. If x and y neither react in a Galvanic cell nor in an Electrolytic cell then we set $x \oplus y = \{x, y\}$.

All possible spontaneous/non-spontaneous redox reactions of $\{A, B, A^{n+}, B^{+m}\}$ in a Galvanic/Electrolytic cell are summarized in Table 7.15.

Table 7.15   Galvanic/Electrolytic cell.

$\oplus$	A	B	A^{n+}	B^{m+}
A	A	A,B	A, A^{n+}	A^{n+}, B
B	A, B	B	A, B^{m+}	B, B^{m+}
A^{n+}	A, A^{n+}	A, B^{m+}	A^{n+}	A^{n+}, B^{m+}
B^{m+}	A^{n+}, B	B^{m+}, B	A^{n+}, B^{m+}	B^{m+}

Table 7.16   Table of Theorem 7.3.

$\oplus$	a	b	c	d
a	a	a,b	a,c	b,c
b	a,b	b	a,d	b,d
c	a,c	a,d	c	c,d
d	b,c	b,d	c,d	d

Remark 7.4. We can define $(H, \oplus)$ as follows:

$$x \oplus y = \begin{cases} x \oplus_1 y, & \text{if } x \oplus_2 y = \{x,y\}; \\ x \oplus_2 y, & \text{if } x \oplus_1 y = \{x,y\}; \\ \{x,y\}, & \text{if } x \oplus_1 y = x \oplus_2 y. \end{cases}$$

In Table 7.15, if we change the names from A, B, A^{n+}, B^{m+} to a, b, c, d respectively, then the following theorem and propositions hold.

Theorem 7.3. *Let* $H = \{a, b, c, d\}$, $\oplus$ *be the hyperoperation on* H *and consider Table 7.16 corresponding to* $(H, \oplus)$. *Then* $(H, \oplus)$ *is a commutative* H_v-*semigroup.*

Proof. It is clear from the above table that $(H, \oplus)$ is a commutative hypergroupoid. We need to show that $(H, \oplus)$ is a weak associative hypergroupoid. Let $(x, y, z) \in H^3$. We have four cases for x; $x = a$, $x = b$, $x = c$ and $x = d$:

- Case $x = a$. We have that $a \oplus (d \oplus d) = a \oplus d = \{b, c\}$ and that $(a \oplus d) \oplus d = \{b, c\} \oplus d = \{b, c, d\}$. Thus, $a \oplus (d \oplus d) \cap (a \oplus d) \oplus d \neq \emptyset$. Moreover, one can easily check that $a \oplus (d \oplus z) \cap (a \oplus d) \oplus z \neq \emptyset$ and that $a \oplus (y \oplus d) \cap (a \oplus y) \oplus d \neq \emptyset$. If $y \neq d$ and $z \neq d$ then $a \in a \oplus (y \oplus z) \cap (a \oplus y) \oplus z$.
- Case $x = b$. We have that $b \oplus (c \oplus c) = b \oplus c = \{a, d\}$ and that $(b \oplus c) \oplus c = \{a, d\} \oplus c = \{a, c, d\}$. Thus, $b \oplus (c \oplus c) \cap (b \oplus c) \oplus c \neq \emptyset$. Moreover, one can easily check that $b \oplus (c \oplus z) \cap (b \oplus c) \oplus z \neq \emptyset$ and that $b \oplus (y \oplus c) \cap (b \oplus y) \oplus c \neq \emptyset$. If $y \neq c$ and $z \neq c$ then $b \in b \oplus (y \oplus z) \cap (b \oplus y) \oplus z$.
- Case $x = c$. We have that $c \oplus (b \oplus b) = c \oplus b = \{a, d\}$ and that $(c \oplus b) \oplus c = \{a, d\} \oplus c = \{b, c, d\}$. Thus, $c \oplus (b \oplus b) \cap (c \oplus b) \oplus b \neq \emptyset$. Moreover, one can easily check that $c \oplus (b \oplus z) \cap (c \oplus b) \oplus z \neq \emptyset$ and that $c \oplus (y \oplus b) \cap (c \oplus y) \oplus b \neq \emptyset$. If $y \neq b$ and $z \neq b$ then $c \in c \oplus (y \oplus z) \cap (c \oplus y) \oplus z$.

- Case $x = d$. We have that $d \oplus (a \oplus a) = a \oplus d = \{b, c\}$ and that $(d \oplus a) \oplus a = \{b, c\} \oplus a = \{a, b, c\}$. Thus, $d \oplus (a \oplus a) \cap (d \oplus a) \oplus a \neq \emptyset$. Moreover, one can easily check that $d \oplus (a \oplus z) \cap (d \oplus a) \oplus z \neq \emptyset$ and that $d \oplus (y \oplus a) \cap (d \oplus y) \oplus a \neq \emptyset$.
  If $y \neq a$ and $z \neq a$ then $d \in d \oplus (y \oplus z) \cap (d \oplus y) \oplus z$. ∎

Remark 7.5. Every element in $(H, \oplus)$ is idempotent. This is trivial from chemical point of view as no reaction exists in an electrochemical cell between two identical elements, so, the element is unchanged.

Proposition 7.1. $(H, \oplus)$ *is not a quasi-hypergroup nor a semihyperegroup.*

Proof.   Since d is not an element in $a \oplus H$, it follows that $(H, \oplus)$ is not a quasi-hypergroup.

Having $a \oplus (d \oplus d) = \{b, c\} \neq (a \oplus d) \oplus d = \{b, c\} \oplus d = \{b, c, d\}$ implies that $(H, \oplus)$ is not a semihypergroup. ∎

Proposition 7.2. $(H, \oplus)$ *does not admit an identity element.*

Proof.   Since d, c, b, a are not elements of $a \oplus d, b \oplus c, c \oplus b, d \oplus a$, it follows that none of our elements is an identity. ∎

Remark 7.6. Proposition 7.2 implies that there exist no element $x \in H$ (in a Galvanic/Electrolytic cell) such that the following reaction occurs for all $y \in H$ and some $x \in H$:

$$x + y \longrightarrow y + z.$$

Remark 7.7. Remark 7.2, Theorem 7.2 and Proposition 7.2 imply that $(H, \oplus)$ is not isomorphic neither to $(H, \oplus_1)$ nor to $(H, \oplus_2)$.

Proposition 7.3. *There are only two H_v-subsemigroups of $(H, \oplus)$ up to isomorphism.*

Proof.   It is easy to see that $(\{a\}, \oplus)$ and $(\{a, b\}, \oplus)$ are the only two H_v-subsemigroups of $(H, \oplus)$ up to isomorphism. Moreover, $(\{a\}, \oplus)$ and $(\{a, b\}, \oplus)$ are hypergroups. ∎

Definition 7.1. Let $(H, \circ)$ be an H_v-semigroup and A be a non-empty subset of H. A is a complete part of H if for any natural number n and for all hyperproducts $P \in H_H(n)$, the following implication holds:

$$A \cap P \neq \emptyset \Longrightarrow P \subseteq A.$$

Proposition 7.4. $(H, \oplus)$ *has no proper complete parts.*

Proof. Let $A \neq \emptyset$ be a complete part of $(H, \oplus)$. We consider the following cases for A:

- Case $a \in A$. Having $a \in a \oplus x$, $x \in a \oplus x$ for all $x \in \{a, b, c\}$ imply that $x \in a \oplus x \subseteq A$. We get now that $b \in A$. Since $b \in b \oplus d$ and $d \in b \oplus d$, it follows that $d \in b \oplus d \subseteq A$. Thus, $A = H$.
- Case $b \in A$. Having $b \in b \oplus a$ implies that $a \in b \oplus a \subseteq A$. The latter implies that $a \in A$ and thus $A = H$ by the first case.
- Case $c \in A$. Having $c \in c \oplus a$ implies that $a \in c \oplus a \subseteq A$. The latter implies that $a \in A$ and thus $A = H$ by the first case.
- Case $d \in A$. Having $d \in c \oplus d$ implies that $c \in c \oplus d \subseteq A$. The latter implies that $c \in A$ and thus $A = H$ by the previous case.

Therefore, $(H, \oplus)$ has no proper complete parts. ∎

Proposition 7.5. $(H, \oplus)$ *has a trivial fundamental group.*

Proof. Since $\{a, b\} \subseteq a \oplus b$, it follows that $a\beta_2 b$. Similarly, we obtain $a\beta_2 c$, $b\beta_2 d$, $c\beta_2 d$. Having β^* the transitive closure of β, one can easily see that $x\beta^* y$ for all $(x, y) \in H^2$. Thus, $|H/\beta^*| = 1$. ∎

7.2 Biological examples

7.2.1 Inheritance examples

Scientific studies of inheritance began in 1866 with the experiments of Gregor Mendel. Mendel worked out the mathematical rules for the inheritance of characteristics in the garden pea. The significance of his work did not become widely appreciated until 1900. He discovered the principles of heredity by crossing different varieties of pea plants and analyzing the transmission pattern of traits in subsequent generations. Mendel began by studying monohybrid crosses, those between parents that differed in a single characteristic. Mendel's approach to the study of heredity was effective for several reasons. The foremost was his choice of an experimental subject, the pea plant, *Pisum sativum*, which offered obvious advantages for genetic investigations. It is easy to cultivate, and Mendel had a monastery garden and a greenhouse at his disposal. Peas grow relatively rapidly, completing an entire generation in a single growing season

[51, 108]. Mendel started with 34 varieties of peas and spent 2 years selecting those varieties that he would use in his experiments.

In [37], Davvaz et al. considered specific examples of simple and incomplete inheritance and relate them to hyperstructures by studying only the monohybrid and dihybrid cases. In [1], Al Tahan and Davvaz discussed simple and incomplete inheritances for the n-hybrid case with $n \geq 1$. So that the examples presented in [37] can be considered as special cases of our work. The aim of our paper is to provide examples about different types of autosomal inheritance (Mendelian and Non-Mendelian inheritance) and relate them to hyperstructures and to generalize the work done in [37]. Also, in [59], Ghadiri et al. analyzed the second generation genotypes of monohybrid and a dihybrid with a mathematical structure. They used the concept of H_v-semigroup structure in the F_2-genotypes with cross operation and proved that this is an H_v-semigroup. They determined the kinds of number of the H_v-subsemigroups of F_2-genotypes. Here are some examples of mono- and dihybrid crosses for Mendelian and Neomendelian inheritance. In the following examples, "parents" is denoted by P, "filial generation" by F and "mating" by $\otimes$. The main references for this section are [1, 37, 59].

Example 7.1. The Monohybrid Cross in the Pea Plant

Mendel began by studding monohybrid crosses those between parents that differed in a single characteristic. In one experiment, Mendel crossed a pea plant homozygous for round seeds with one that was homozygous for wrinkled seeds [41, 108]. The first generation of the cross was the P (parental) generation. After crossing the two varieties in the P generation, Mendel observed the offspring that resulted from the cross. The results of this experiment can be summarized in the following way:

$$P : \text{Round } (RR \text{ genotype}) \otimes \text{Wrinkled } (rr \text{ genotype})$$
$$\downarrow$$
$$F_1 : \qquad \text{All Round } (Rr \text{ genotype})$$

and

$$F_1 \otimes F_1 : \text{Round } (Rr \text{ genotype}) \otimes \text{Round } (Rr \text{ genotype})$$
$$\downarrow$$
$$F_2 : \qquad \text{Round } (RR \text{ genotype}), \text{ Round } (Rr \text{ genotype}),$$
$$\text{Wrinkled } (rr \text{ genotype}).$$

Round is denoted by R, and Wrinkled by W. The process can be describe in Table 7.17.

Table 7.17

⊗	R	W
R	R, W	R
W	R	W

Table 7.18

⊗	T	D
T	T, D	T
D	T	D

Considering $H = \{R, W\}$, it is easy to see that $(H, \otimes)$ is a hypergroup.

In other experiment, Mendel crossed a pea plant homozygous for Tall with one that was homozygous for Short [41, 6]. The results are similar to those of the previous example:

$$P : \text{Tall } (TT \text{ genotype}) \otimes \text{Short } (tt \text{ genotype})$$
$$\downarrow$$
$$F_1 : \qquad \text{All Tall } (Tt \text{ genotype})$$

and

$$F_1 \otimes F_1 : \qquad \text{Tall } (Tt \text{ genotype}) \otimes \text{Tall } (Tt \text{ genotype})$$
$$\downarrow$$
$$F_2 : \quad \text{Tall } (TT \text{ genotype}), \text{Tall } (Tt \text{ genotype}), \text{Short } (tt \text{ genotype}).$$

Tall is denoted by T, and Dwarf by D. Hence, we have Table 7.18. Considering $H = \{T, D\}$, it is easy to see that $(H, \otimes)$ is a hypergroup.

Example 7.2. The Dihybrid Cross in the Pea Plant

In addition to his work on monohybrid crosses, Mendel crossed varieties of peas that differed in two characteristics (dihybrid crosses) [41, 108]. For example, he had a homozygous variety of peas that produced round seeds and tall plants. Another homozygous variety produced wrinkled seeds and short plants. When he crossed the two plants, all the $F1$ progenies had round seeds and tall plants. For example:

$$P : \text{Round, Tall } (RRTT \text{ genotype}) \otimes \text{Wrinkled, Short } (rrtt \text{ genotype})$$
$$\downarrow$$
$$F_1 : \qquad \text{All Round, Tall } (RrTt \text{ genotype})$$

Table 7.19

$\otimes$	A	B	C	D
A	A B C D	A B C D	A B C D	A B C D
B	A B C D	B D	A B C D	B D
C	A B C D	A B C D	C D	C D
D	A B C D	B D	C D	D

and

$F_1 \otimes F_1$: Round, Tall ($RrTt$ genotype) $\otimes$ Round, Tall ($RrTt$ genotype)
$$\downarrow$$
F_2 :   Round, Tall ($RRTT$ genotype)
Round, Short ($RRtt$ and $Rrtt$ genotypes)
Wrinkled, Tall ($rrTT$ and $rrTt$ genotypes)
Wrinkled, Short ($rrtt$ genotype).

Tall and Round are denoted by A, Tall and Wrinkled by B, Dwarf and Round by C, finally Dwarf and Wrinkled by D. Hence, we have Table 7.19. Considering $H = \{A, B, C, D\}$, $(H, \otimes)$ is a hypergroup. Obviously, $H_0 = \{C, D\}$ is a subhypergroup of H.

Example 7.3. Flower Color (*Mirabilis Jalapa*) **Inheritance in the Four-o'clock Plant**

Mirabilis jalapa (The four o'clock flower) which is the most commonly grown ornamental species of Mirabilis is available in a range of colors. The plant produces fragrant flowers in a range of colors from white to red over a course of a few months. It is a multi-branched perennial plant in southern and warm western regions, and an annual plant in cooler northern regions of its native tropical South America. It has been naturalized in many parts of the world [101].

Table 7.20

$\otimes$	R	P	W
R	R	$R,\ P$	P
P	$R,\ P$	$R,\ P,\ W$	$R,\ P$
W	P	$R,\ P$	W

Using four-o'clock plants, a plant with red flower petals can be crossed with another that has white flower petals; the offspring will have pink flower petals. If these pink-flowered F_1 plants are crossed, F_2 plants appear in a ratio of 1:2:1, with red, pink, or white flower petals respectively. The pink-flowered plants are heterozygotes that have a petal color between the red and the white colors of the homozygotes. In this case, one allele (R_1) specifies a red pigment color, and another allele specifies no color (R_2; the flower petals have a white background [51, 108]. If Red and White-flowered plants cross, then the following results obtain:

$$P:\ \text{Red }(R_1R_1\text{ genotype}) \otimes \text{White }(R_2R_2\text{ genotype})$$
$$\downarrow$$
$$F_1:\qquad \text{All Pink }(R_1R_2\text{ genotype})$$

and

$$F_1 \otimes F_1:\text{Pink }(R_1R_2\text{ genotype}) \otimes \text{Pink }(R_1R_2\text{ genotype})$$
$$\downarrow$$
$$F_2:\qquad \text{Red }(R_1R_1\text{ genotype}),\ \text{Pink }(R_1R_2\text{ genotype}),$$
$$\text{White }(R_2R_2\text{ genotype}).$$

Red is denoted by R, Pink by P, and White by W. Thus, we have Table 7.20.

Considering $H = \{R, P, W\}$, $(H, \otimes)$ is an H_v-semigroup. For example:

$$R \otimes (R \otimes W) = R \otimes P = \{R, P\}$$
$$(R \otimes R) \otimes W = R \otimes W = P.$$

Example 7.4. Coat Color of Shorthorn Cattle Inheritance

The *Shorthorn* breed of cattle originated in the northeast of England in the late 18th century. The breed was developed as a dual purpose, suitable for both dairy and beef production. All Shorthorn cattle are red, white or roan although roan cattle are preferred by some. Also, completely white animals are not common [111]. When homozygous red-haired cattle are crossed with the homozygous white-haired type, the F_1 has reddish grey

Table 7.21

$\otimes$	R	G	W
R	R	$R,\ G$	G
G	$R,\ G$	$R,\ G,\ W$	$G,\ W$
W	G	$G,\ W$	W

hair and is designated as "roan". It must be noted that there is no mixture of red and grey pigments in a roan. But some hair is all red, some other all white, so that the final result is a reddish grey coat color [108]. Thus, let two shorthorn cattle (Red and White hairs) cross. Then, we have:

$$P:\ \text{Red}\ (r_1 r_1\ \text{genotype})\ \otimes\ \text{White}\ (r_2 r_2\ \text{genotype})$$
$$\downarrow$$
$$F_1:\qquad \text{All Roan}\ (r_1 r_2\ \text{genotype})$$

and

$$F_1 \otimes F_1:\ \text{Roan}\ (r_1 r_2\ \text{genotype})\ \otimes\ \text{Roan}\ (r_1 r_2\ \text{genotype})$$
$$\downarrow$$
$$F_2:\qquad \text{Red}\ (r_1 r_1\ \text{genotype}),\ \text{Roan}\ (r_1 r_2\ \text{genotype}),$$
$$\text{White}\ (r_2 r_2\ \text{genotype}).$$

Red-haired is denoted by W, Reddish grey-haired (Roan) by G, and White-haired by W.

Considering $H = \{R,\ G,\ W\}$, $(H,\ \otimes)$ is an H_v-semigroup (see Table 7.21).

Example 7.5. *ABO* Blood Group Inheritance

In 1900, the Austrian physician Karl Landsteiner realized that human blood was of different types, and that only certain combinations were compatible [57]. The International Society of Blood Transfusion (ISBT) recognizes 285 blood group antigens of which 245 are classified as one of 29 blood group systems (http://blood.co.uk/ibgrl/).

Each blood group system represents either a single gene or a cluster of two or three closely linked genes of related sequence with little or no recognized recombination occurring between them. Consequently, each blood group system is a genetically discrete entity.

Blood groups are inherited from both parents. The *ABO* blood type is controlled by a single gene (the *ABO* gene) with three alleles: I^A, I^B and i. The gene encodes glycosyltransferase that is an enzyme that modifies the carbohydrate content of the red blood cell antigens. The gene is located on the long arm of the ninth chromosome (9q34) [41, 108].

Table 7.22

$\otimes$	O	A	B	AB
O	O	O A	O B	A B
A	O A	O A	AB A B O	AB A B
B	O B	AB A B O	O B	AB A B
AB	A B	AB A B	AB A B	AB A B

People with blood type A have antigen A on the surfaces of their blood cells, and may be of genotype $I^A I^A$ or $I^A i$. People with blood type B have antigen B on their red blood cell surfaces, and may be of genotype $I^B I^B$ or $I^B i$. People with the rare blood type AB have both antigens A and B on their cell surfaces, and are genotype $I^A I^B$. People with blood type O have neither antigen, and are genotype ii. A type A and a type B couple can also have a type O child if they are both heterozygous ($I^A i$ and $I^B i$, respectively).

Considering $H = \{O,\ A,\ B,\ AB\}$, $(H, \otimes)$ is an H_v-semigroup (see Table 7.22). For example:

$$(O \otimes B) \otimes AB = \{O, B\} \otimes AB = \{A, B, AB\}$$
$$O \times (B \otimes AB) = O \otimes \{A, B, AB\} = \{O, A, B\}.$$

If $H_0 = \{O, A\}$ and $H_1 = \{O, B\}$, then $(H_0, \otimes)$ and $(H_1, \otimes)$ are hypergroups.

ABO blood type is often further differentiated by a $+$ or $-$, which refers to another blood group antigen called the Rh factor. In this system, the Rh^+ phenotype (D allele) is dominant on the Rh^- phenotype (d allele). The antigen was originally identified in rhesus monkeys, hence the name [108]. For example:

$$P: Rh^+ \text{ (DD genotype)} \otimes Rh^- \text{ (dd } genotype)$$
$$\downarrow$$
$$F_1: \qquad\qquad Rh^+ \text{ (Dd genotype)}$$

Table 7.23

$\otimes$	Rh^+	Rh^-
Rh^+	$Rh^+,\ Rh^-$	$Rh^+,\ Rh^-$
Rh^-	$Rh^+,\ Rh^-$	Rh^-

and

$F_1 \otimes F_1$: Rh^+ (Dd genotype) $\otimes$ Rh^+ (Dd genotype)

$\downarrow$

F_2 : Rh^+ (Dd genotype), Rh^+ (Dd genotype), Rh^- (dd genotype).

The different crosses of Rhesus system are indicated in Table 7.23.

Considering $H = \{Rh^+, Rh^-\}$, $(H, \otimes)$ is a hypergroup.

Table 7.24 indicates different crosses of two blood antigenic phenotypes together (ABO and Rhesus systems). For example:

P : O^+ ($iiDd$ genotype) $\otimes$ O^+ ($iiDd$ genotype)

$\downarrow$

$F_1 : O^+$ ($iiDD$ genotype), O^+ ($iiDd$ genotype), O^-($iidd$ genotype).

Considering $H = \{O^+, O^-, A^+, A^-, B^+, B^-, AB^+, AB^-\}$, $(H, \otimes)$ is an H_v-semigroup. If $H_1 = \{O^+, O^-\}$, $H_2 = \{O^-, A^-\}$, $H_3 = \{O^+, O^-, A^+, A^-\}$ and $H_4 = \{O^+, O^-, B^+, B^-\}$, then $(H_1, \otimes)$, $(H_2, \otimes)$, $(H_3, \otimes)$ and $(H_4, \otimes)$ are commutative H_v-subgroups.

Example 7.6. MN Blood Group Inheritance

MN blood is distinct from the better-known ABO blood groups, but the principle is the same. Blood is typed according to what type(s) of antigen (a cellular product that induces antibody formation in a foreign host) is/are found on the surface of the red blood cells. Within the MN blood groups, there are two antigens, M and N, whose production is determined by a gene with two alleles, L^M and L^N. L^M confers the ability to produce the M antigen, while L^N confers the ability to produce the N antigen. Individuals who have genotype $L^M L^M$ will only have the M antigen on their red cells, and will be of type M. Individuals with genotype $L^N L^N$ will only have the N antigen on their red cells, and will be of type N. Heterozygotes ($L^M L^N$) produce both antigens, and are of type MN [41, 108]. The following example is suggestive:

P : M ($L^M L^M$ genotype) $\otimes$ N ($L^N L^N$ genotype)

$\downarrow$

F_1 :   All M ($L^M L^N$ genotype)

Table 7.24

$\otimes$	O^+	O^-	A^+	A^-	B^+	B^-	AB^+	AB^-
O^+	O^+ O^-	O^+ O^-	A^+ A^- O^+ O^-	A^+ A^- O^+ O^-	B^+ B^- O^+ O^-	B^+ B^- O^+ O^-	A^+ A^- B^+ B^-	A^+ A^- B^+ B^-
O^-	O^+ O^-	O^-	A^+ A^- O^+ O^-	A^- O^-	B^+ B^- O^+ O^-	B^- O^-	A^+ A^- B^+ B^-	A^- B^-
A^+	A^+ A^- O^+ O^-	A^+ A^- O^+ O^-	A^+ A^- O^+ O^-	A^+ A^- O^+ O^-	AB^+ AB^- A^+ A^- B^+ B^- O^+ O^-	AB^+ AB^- A^+ A^- B^+ B^- O^+ O^-	AB^+ AB^- A^+ A^- B^+ B^-	AB^+ AB^- A^+ A^- B^+ B^-
A^-	A^+ A^- O^+ O^-	A^- O^-	A^+ A^- O^+ O^-	A^- O^-	AB^+ AB^- A^+ A^- B^+ B^- O^+ O^-	AB^- A^- B^- O^-	AB^+ AB^- A^+ A^- B^+ B^-	AB^- A^- B^-
B^+	B^+ B^- O^+ O^-	B^+ B^- O^+ O^-	AB^+ AB^- A^+ A^- B^+ B^- O^+ O^-	AB^+ AB^- A^+ A^- B^+ B^- O^+ O^-	B^+ B^- O^+ O^-	B^+ B^- O^+ O^-	AB^+ AB^- A^+ A^- B^+ B^-	AB^+ AB^- A^+ A^- B^+ B^-
B^-	B^+ B^- O^+ O^-	B^- O^-	AB^+ AB^- A^+ A^- B^+ B^- O^+ O^-	AB^- A^- B^- O^-	B^+ B^- O^+ O^-	B^- O^-	AB^+ AB^- A^+ A^- B^+ B^-	AB^- A^- B^-
AB^+	A^+ A^- B^+ B^-	A^+ A^- B^+ B^-	AB^+ AB^- A^+ A^- B^+ B^-	AB^+ AB^- A^+ A^- B^+ B^-	AB^+ AB^- A^+ A^- B^+ B^-	AB^+ AB^- A^+ A^- B^+ B^-	AB^+ AB^- A^+ A^- B^+ B^-	AB^+ AB^- A^+ A^- B^+ B^-
AB^-	A^+ A^- B^+ B^-	A^- B^-	AB^+ AB^- A^+ A^- B^+ B^-	AB^- A^- B^-	AB^+ AB^- A^+ A^- B^+ B^-	AB^- A^- B^-	AB^+ AB^- A^+ A^- B^+ B^-	AB^- A^- B^-

Table 7.25

$\otimes$	M	MN	N
M	M	$M,\ MN$	MN
MN	$M,\ MN$	$M,\ MN,\ N$	$N,\ MN$
N	MN	$N,\ MN$	N

Table 7.26

$*$	$x = O$	$x = A$	$x = B$	$x = AB$
$\mu(O,O)(x)$	1	0	0	0
$\mu(O,A)(x)$	1/4	3/4	0	0
$\mu(O,B)(x)$	1/4	0	3/4	0
$\mu(O,AB)(x)$	0	1/2	1/2	0
$\mu(A,O)(x)$	1/4	3/4	0	0
$\mu(A,A)(x)$	2/16	14/16	0	0
$\mu(A,B)(x)$	1/16	3/16	3/16	9/16
$\mu(A,AB)(x)$	0	4/8	1/8	3/8
$\mu(B,O)(x)$	1/4	0	3/4	0
$\mu(B,A)(x)$	1/16	3/16	3/16	9/16
$\mu(B,B)(x)$	2/16	0	14/16	0
$\mu(B,AB)(x)$	0	1/8	4/8	3/8
$\mu(AB,O)(x)$	0	1/2	1/2	0
$\mu(AB,A)(x)$	0	4/8	1/8	3/8
$\mu(AB,B)(x)$	0	1/8	4/8	3/8
$\mu(AB,AB)(x)$	0	1/4	1/4	1/2

and

$$F_1 \otimes F_1 : M\ (L^M L^N\ \text{genotype})\ \otimes\ M\ (L^M L^N\ \text{genotype})$$
$$\downarrow$$
$$F_2 : \quad M\ (L^M L^M\ \text{genotype}),\ MN\ (L^M L^N\ \text{genotype}),$$
$$N\ (L^N L^N\ \text{genotype}).$$

The different crosses of MN system are indicated in Table 7.25. Considering $H = \{M, MN, N\}$, $(H, \otimes)$ is an H_v-semigroup.

Now, ABO Blood Group Inheritance is considered as a probabilistic hypergroupoid. The probabilities are indicated in Table 7.26. This table shows the calculated proportions of phenotypic classes expected from ABO system mating. For example, the A phenotype can arise from either of two genotypes, $I^A I^A$ or $I^A i$, and the frequency of each genotype is 1/2. Thus, if a person with blood type A mates with a person with blood type O, they can have type A or type O children. The occurrence probability of each of these events is:

$$P : 1/2 I^A I^A \otimes ii$$
$$\downarrow$$
$$F_1 : 1/2 I^A i \ (A \text{ phenotype})$$

or

$$P : \quad 1/2 I^A i \otimes ii$$
$$\downarrow$$
$$F_1 : 1/2 \times 1/2 = 1/4 I^A i \ (A \text{ phenotype}, \quad 1/2 \times 1/2 = 1/4 ii \ (O \text{ phenotype}).$$

Thus, the probability of A phenotype children is $1/2 + 1/4 = 3/4$, and the probability of O phenotype children is $1/4$.

7.2.2 Examples of different types of non-Mendelian inheritance

In [37], some examples of simple inheritance and incomplete inheritance were discussed. In this section, we study some examples of five different types of non-Mendelian inheritance (Epistasis, Supplementary gene, Inhibitory gene, Complementary gene, Supplementary and complementary gene) and relate them to hypergroup theory. In the Mendelian inheritance, the presence of the dominant allele A over the recessive allele a, corresponding to the phenotypes $\widehat{A}$ and $\widehat{a}$ respectively, in the genotype of an organism implies that its phenotype is $\widehat{A}$ (and its absence in the genotype implies that its phenotype is $\widehat{a}$).

The examples presented in [1] are not equivalent to any of those presented by the authors in [37]. The main reference for this paragraph and the next paragraphs is [1].

Example 7.7. *Epistasis of dominant gene in the coat color of dogs.* There are two allelomorphic pairs which may be named Aa and Bb, A and B are dominant over a and b respectively. They interact as follows: $AxBy$ and $Axbb$ have phenotype white, $aaBy$ has phenotype black and $aabb$ has phenotype brown. Here $x = A$ or a and $y = B$ or b. The results of this experiment can be summarized in the following way:

$$P: AABB \otimes aabb$$
$$F_1 : AaBb$$
and
$$F_1 \otimes F_1 : AaBb \otimes AaBb$$
$$F_2 : \text{White, Black, Brown.}$$

Table 7.27

⊗	A_1	A_2	A_3
A_1	H	H	H
A_2	H	A_2, A_3	A_2, A_3
A_3	H	A_2, A_3	A_3

White is denoted by A_1, Black by A_2 and Brown by A_3.

Considering $H = \{A_1, A_2, A_3\}$. We present $(H, \otimes)$ in Table 7.27. Since no new elements are presented in the above table and H is found in each row and column of the table then $(H, \otimes)$ is a hypergroupoid that satisfies the reproduction axiom. Moreover, having our table symmetric implies that $(H, \otimes)$ is commutative. It is easy to see that $(H, \otimes)$ is associative. Thus, $(H, \otimes)$ is a commutative hypergroup. Moreover, it is obvious from the table that A_3 is a unique identity of $(H, \otimes)$, A_3 is the only idempotent element of $(H, \otimes)$ and that $H = A_1^2$ is a single power cyclic hypergroup with period 2.

Remark 7.8. Example 7.7 is not an example of the codominance inheritance which is discussed in ([37], p.183). Our example is not equivalent to that of [37]; the set of phenotypes was proved to be an H_v-semigroup in [37] whereas in our example it is a cyclic hypergroup.

Example 7.8. *Supplementary gene, The anthocyanin pigmentation of flowers.* The red-type anthocyanin color of many flowers is caused by two alleles which may be termed as Aa and Bb. In the snapdragon (Antirrhinum) flower:

$AxBy$ is the genotype of magenta flower, $Axbb$ is the genotype of ivory flower and $aaBy$, $aabb$ are the genotypes of white flower where $x = A$ or a and $y = B$ or b. The results of this experiment can be summarized in the following way:

$$P: AABB \otimes aabb$$
$$F_1 : AaBb$$
$$\text{and}$$
$$F_1 \otimes F_1 : AaBb \otimes AaBb$$
$$F_2 : \text{Magneta, Ivory, White.}$$

Magneta is denoted by B_1, Ivory by B_2 and White by B_3.

Considering $K = \{B_1, B_2, B_3\}$. We present $(K, \otimes)$ in Table 7.28. Since no new elements are presented in Table 7.28 and K is found in each row and column of the table then $(K, \otimes)$ is a hypergroupoid that satisfies the

Table 7.28

⊗	B_1	B_2	B_3
B_1	K	K	K
B_2	K	B_2, B_3	K
B_3	K	K	B_3

reproduction axiom. Moreover, having our table symmetric implies that $(K, \otimes)$ is commutative. It is easy to see that $(K, \otimes)$ is associative. Thus, $(K, \otimes)$ is a commutative hypergroup.

It is obvious from the table that B_3 is the unique identity and unique idempotent of $(K, \otimes)$ and that $K = B_1^2$ is a single power cyclic hypergroup with period 2.

Remark 7.9. The two hypergroups $(H, \otimes)$ and $(K, \otimes)$ defined in Examples 7.7 and 7.8 are non isomorphic hypergroups. If $(H, \otimes)$ and $(K, \otimes)$ are isomorphic then there exists a bijective function $f : H \longrightarrow K$ satisfying $f(x \otimes y) = f(x) \otimes f(y)$ for all $x, y \in H$. It is easy to see that $f(A_3) = B_3$ and thus $f(A_2) = B_2, f(A_1) = B_1$ or $f(A_2) = B_1, f(A_1) = B_2$. If $f(A_2) = B_2, f(A_1) = B_1$ then $f(A_2 \otimes A_3) = \{f(A_2), f(A_3)\} = \{B_2, B_3\}$ and $f(A_2) \otimes f(A_3) = B_2 \otimes B_3 = K$. The latter two non equal expressions imply that f is not an isomorphism. If $f(A_2) = B_1, f(A_1) = B_2$ then $f(A_1 \otimes A_1) = f(H) = K$ and $f(A_1) \otimes f(A_1) = B_2 \otimes B_2 = \{B_2, B_3\}$. The latter two non equal expressions imply that f is not an isomorphism.

Example 7.9. *Inhibitory gene, Rice leaf.* In some rice variety the presence of the gene P causes its leaves to be colored deep purple. But if a gene I is present then the purple color is inhibited and the leaf becomes normal green. The I gene may be considered as epistatic over P. They interact as follows:

The genotypes $IxPy, Ixpp, iipp$ correspond to green and the genotype $iiPy$ corresponds to purple where $x = I$ or i and $y = P$ or p. The results of this experiment can be summarized in the following way:

$$\text{P: } IIPP \otimes iipp$$
$$F_1 : IiPp$$
$$\text{and}$$
$$F_1 \otimes F_1 : IiPp \otimes IiPp$$
$$F_2 : \text{Green, Purple.}$$

Green is denoted by A_1 and Purple by A_2.

Table 7.29

⊗	A_1	A_2
A_1	L	L
A_2	L	L

Table 7.30

⊗	A_1	A_2
A_1	M	M
A_2	M	M

Considering $L = \{A_1, A_2\}$. We present $(L, \otimes)$ in Table 7.29.

It is obvious from Table 7.29 that $(L, \otimes)$ is a cyclic hypergroup with finite period equal two.

Example 7.10. *Complementary gene, Some rice variety.* In some rice varieties, the following interaction of two pairs of genes has been noted:

$AxBy$ = red grain and $aaBy, Axbb, aabb$ = grey grain.

$$\text{P: } AABB \otimes aabb$$
$$F_1 : AaBb$$
$$\text{and}$$
$$F_1 \otimes F_1 : AaBb \otimes AaBb$$
$$F_2 : \text{Red, Grey.}$$

Red is denoted by A_1 and Grey by A_2.

Considering $M = \{A_1, A_2\}$. We present $(M, \otimes)$ in Table 7.30.

It is obvious from Table 7.30 that $(M, \otimes)$ is a cyclic hypergroup with finite period equal two and it is isomorphic to $(L, \otimes)$ defined in Example 7.9.

Example 7.11. *Supplementary and complementary gene, Seed-coat color.* We consider an interesting case of gene interaction involving three pairs of genes (A, B and C) instead of two. It has been noted that seed-coat color in maize is controlled by three pairs of genes. C and B are two complementary genes which have no action independently but together cause a brownish green colouration of the stem while the grain remains colorless. If the gene A is added to C and B some plant parts as well as the grain-coat become purple by the development of anthocyanin. We may suppose that the anthocyanin pigment is developed in two stages: C and B are complementary in forming the brown pigment and when A is supplemented the

Table 7.31

$\otimes$	A_1	A_2
A_1	N	N
A_2	N	N

full anthocyanin pigment is developed. A is, therefore, a supplementary gene.

$AxByCz$ = purple seed and

$\{AxBycc, AxbbCz, aaByCz, aabbCz, aaBycc, Axbbcc, aabbcc\}$ = colorless seed.

$$P:\ AABBCC \otimes aabbcc$$
$$F_1 : AaBbCc$$
$$\text{and}$$
$$F_1 \otimes F_1 : AaBbCc \otimes AaBbCc$$
$$F_2 : \text{Purple, Colorless.}$$

Purple is denoted by A_1 and Colorless by A_2.

Considering $N = \{A_1, A_2\}$. We present $(N, \otimes)$ in Table 7.31.

It is obvious from Table 7.31 that $(N, \otimes)$ is a cyclic hypergroup with finite period equal two and it is isomorphic to $(L, \otimes)$ and $(M, \otimes)$ defined in Examples 7.9 and 7.10.

Remark 7.10. Examples 7.9, 7.10 and 7.11 are not equivalent to the Mendelian examples discussed in [37]. In our examples, we got total hypergroups which is not the case in [37].

7.2.3 *Hyperstructures in second generation genotype*

In this section we generalize the results in [37] regarding the simple and incomplete dominance (Blending inheritance) by doing hypothetical crosses of homozygous with independent number of alleles. First, we present results for the hypothetical cross of simple dominance of n different traits given by

$$A_1 A_2 \ldots A_n \otimes a_1 a_2 \ldots a_n.$$

Next, we present results for the hypothetical cross of incomplete dominance of n different traits given by

$$B_1 B_2 \ldots B_n \otimes \overline{B_1}\ \overline{B_2} \ldots \overline{B_n}.$$

Finally, we present results for the hypothetical cross of simple and incomplete dominance combined together of $m + n$ different traits given by

Table 7.32

$\otimes$	$\widehat{A_1}$	$\widehat{A_2}$
$\widehat{A_1}$	H	H
$\widehat{A_2}$	H	$\{\widehat{A_2}\}$

$$B_1 B_2 \ldots B_n A_1 A_2 \ldots A_m \otimes \overline{B_1}\ \overline{B_2} \ldots \overline{B_n} a_1 a_2 \ldots a_m.$$

Throughout this section, for $i = 1, 2, \ldots, n$, we denote by A_i the dominant allele over the recessive allele a_i and by B_i and $\overline{B_i}$ the codominance alleles (the cross of 2 homozygous with different phenotypes leads to the presence of a new phenotype in the offspring).

7.2.4   The hypothetical cross of n different traits, case of simple dominance

The case of simple dominance can be given by:

$$A_1 A_2 \ldots A_n \otimes a_1 a_2 \ldots a_n \text{ where } n \geq 1.$$

We consider first results for the Monohybrid cross $(n = 1)$ that differs in a single trait; a homozygous dominant parent $(A_1 A_1) \otimes$ a homozygous recessive parent $(a_1 a_1)$. The results of this hypothetical experiment can be summarized in the following way:

$$\text{P: } A_1 A_1 \otimes a_1 a_1$$
$$F_1 : A_1 a_1$$
$$\text{and}$$
$$F_1 \otimes F_1 : A_1 a_1 \otimes A_1 a_1$$
$$F_2 : \widehat{A_1} \text{ (of genotype } A_1 x_1), \widehat{A_2} \text{ (of genotype } a_1 a_1).$$

Where $x_1 = A_1$ or a_1.

Proposition 7.6. *Let $H = \{\widehat{A_1}, \widehat{A_2}\}$. Then $(H, \otimes)$ is a regular and single power cyclic hypergroup with unique identity.*

Proof.   We present $(H, \otimes)$ in Table 7.32.

Since no new elements are presented in Table 7.32 and H is found in each row and column of the table then $(H, \otimes)$ is a hypergroupoid that satisfies the reproduction axiom. Moreover, having our table symmetric implies that $(H, \otimes)$ is commutative. It is easy to see that $(H, \otimes)$ is associative. Thus, $(H, \otimes)$ is a commutative hypergroup.

It is obvious from the table that $\widehat{A_2}$ is a unique identity of $(H, \otimes)$ and that $H = \widehat{A_1}^2$ is a single power cyclic hypergroup with period 2. Since $(H, \otimes)$ admits an identity and each element in it admits at least one inverse $(\widehat{A_1}^{-1} = \{\widehat{A_1}, \widehat{A_2}\}$ and $\widehat{A_2}^{-1} = \widehat{A_2})$, it follows that $(H, \otimes)$ is a regular hypergroup. ∎

Proposition 7.7. $(H, \otimes)$ *has no proper linear subsets.*

Proof. Let M be a proper linear subset of $(H, \otimes)$; i.e $M \neq H$. Then $x \otimes y$ and x/y are subsets of M for all $x, y \in M$. We have that $\widehat{A_1}$ is not an element of M else $H = \widehat{A_1} \otimes \widehat{A_1} \subset M$. Also, $\widehat{A_2}$ is not an element of M else $H = \widehat{A_2}/\widehat{A_2} \subset M$. ∎

Proposition 7.8. $(H, \otimes)$ *has one proper subhypergroup.*

Proof. It is easy to see that the only subhypergroups of $(H, \otimes)$ are $(H, \otimes)$ and $(\{\widehat{A_2}\}, \otimes)$. ∎

We consider now results for the Dihybrid cross $(n = 2)$ that differs in two traits; a homozygous dominant parent $(A_1 A_1 A_2 A_2) \otimes$ a homozygous recessive parent $(a_1 a_1 a_2 a_2)$. The results of this hypothetical experiment can be summarized in the following way:

$$\text{P: } A_1 A_1 A_2 A_2 \otimes a_1 a_1 a_2 a_2$$
$$F_1 : A_1 a_1 A_2 a_2$$
$$\text{and}$$
$$F_1 \otimes F_1 : A_1 a_1 A_2 a_2 \otimes A_1 a_1 A_2 a_2$$
$$F_2 : \widehat{A_1} \text{ (of genotype } A_1 x_1 A_2 x_2), \widehat{A_2} \text{ (of genotype } A_1 x_1 a_2 a_2), \widehat{A_3} \text{ (of genotype } a_1 a_1 A_2 x_2), \widehat{A_4} \text{ (of genotype } a_1 a_1 a_2 a_2).$$

Where $x_1 = A_1$ or a_1 and $x_2 = A_2$ or a_2.

Proposition 7.9. *Let* $H = \{\widehat{A_1}, \widehat{A_2}, \widehat{A_3}, \widehat{A_4}\}$. *Then* $(H, \otimes)$ *is a regular and single power cyclic hypergroup with unique identity.*

Proof. We present $(H, \otimes)$ in Table 7.33. Since no new elements are presented in Table 7.33 and H is found in each row and column of the table then $(H, \otimes)$ is a hypergroupoid that satisfies the reproduction axiom. Moreover, having our table symmetric implies that $(H, \otimes)$ is commutative. It is easy to see that $(H, \otimes)$ is associative. Thus, $(H, \otimes)$ is a commutative hypergroup.

Table 7.33

$\otimes$	$\widehat{A_1}$	$\widehat{A_2}$	$\widehat{A_3}$	$\widehat{A_4}$
$\widehat{A_1}$	H	H	H	H
$\widehat{A_2}$	H	$\widehat{A_2},\widehat{A_4}$	H	$\widehat{A_2},\widehat{A_4}$
$\widehat{A_3}$	H	H	$\widehat{A_3},\widehat{A_4}$	$\widehat{A_3},\widehat{A_4}$
$\widehat{A_4}$	H	$\widehat{A_2},\widehat{A_4}$	$\widehat{A_3},\widehat{A_4}$	$\widehat{A_4}$

It is obvious from the table that $\widehat{A_4}$ is a unique identity of $(H,\otimes)$ and that $H = \widehat{A_1}^2$ is a single power cyclic hypergroup with period 2. Since $(H,\otimes)$ admits an identity and each element in it admits at least one inverse $(\widehat{A_4} \in (\widehat{A_1} \otimes \widehat{A_4}) \cap (\widehat{A_2} \otimes \widehat{A_4}) \cap (\widehat{A_3} \otimes \widehat{A_4}) \cap (\widehat{A_4} \otimes \widehat{A_4}))$, it follows that $(H,\otimes)$ is a regular hypergroup. ∎

It is easy to see that the $\widehat{A_4}$ is the only idempotent element in $(H,\otimes)$.

Proposition 7.10. $(H,\otimes)$ *has no proper linear subsets.*

Proof. Let M be a proper linear subset of $(H,\otimes)$; i.e $M \neq H$. Then $x \otimes y$ and x/y are subsets of M for all $x,y \in M$. We have that $H = x \otimes \widehat{A_1}$ for all $x \in H$. We get now that $\widehat{A_1} \in M$ as $\widehat{A_1} \in x/y$ for all $x,y \in M$. Since $\widehat{A_1} \in M$, it follows that $H = \widehat{A_1} \otimes \widehat{A_1} \subset M$. ∎

Proposition 7.11. $(H,\otimes)$ *has only 2 proper subhypergroups up to isomorphism.*

Proof. It is easy to see that the only subhypergroups of $(H,\otimes)$ are H, $\{\widehat{A_2}\}$, $\{\widehat{A_2},\widehat{A_4}\}$ and $\{\widehat{A_3},\widehat{A_4}\}$. Having that $\{\widehat{A_2},\widehat{A_4}\}$ and $\{\widehat{A_3},\widehat{A_4}\}$ isomorphic implies that we have two proper subhypergroups. ∎

We consider now the n−hybrid cross $(n \geq 3)$ that differs in n traits; a homozygous dominant parent $(A_1A_1A_2A_2 \ldots A_nA_n) \otimes$ a homozygous recessive parent $(a_1a_1a_2a_2 \ldots a_na_n)$. The results of this hypothetical experiment can be summarized in the following way:

$$P: A_1A_1A_2A_2 \ldots A_{n-1}A_{n-1}A_nA_n \otimes a_1a_1a_2a_2 \ldots a_na_n$$
$$F_1 : A_1a_1A_2a_2 \ldots A_na_n$$
and
$$F_1 \otimes F_1 : A_1a_1A_2a_2 \ldots A_na_n \otimes A_1a_1A_2a_2 \ldots A_na_n$$
$$F_2 : \widehat{A_1} \text{ (of genotype } A_1x_1A_2x_2 \ldots A_nx_n), \widehat{A_2} \text{ (of genotype}$$
$$A_1x_1A_2x_2 \ldots A_{n-1}x_{n-1}a_na_n), \ldots, \widehat{A_k} \text{ (of genotype } a_1a_1a_2a_2 \ldots a_na_n).$$

Where $k = 2^n$ is the number of different phenotypes, $x_i = A_i$ or a_i.

Theorem 7.4. *Let* $H = \{\widehat{A_1}, \widehat{A_2}, \ldots, \widehat{A_k}\}$. *Then* $(H, \otimes)$ *is a regular and single power cyclic hypergroup with unique identity.*

Proof. The proof is similar to that of Propositions 7.6 and 7.9. It is easy to see that $H = \widehat{A_1}^2$ and $\widehat{A_k}$ is a unique identity of $(H, \otimes)$. ■

Proposition 7.12. $(H, \otimes)$ *has no proper linear subsets.*

Proof. The proof is the same as that of Proposition 7.10. ■

Proposition 7.13. *The only idempotent element in* $(H, \otimes)$ *is the identity element.*

Proof. It is easy to see that the identity $\widehat{A_k}$ is an idempotent in $(H, \otimes)$. Let $\widehat{A_r}$ be an idempotent in $(H, \otimes)$ with corresponding genotype $x_1 y_1 \ldots x_n y_n$. Having $\widehat{A_r}^2 = \widehat{A_r}$ implies that $x_i = y_i$ for all $i = 1, \ldots, n$. The latter implies that $x_i = y_i = A_i$ or a_i. If there exists $i \in [1, n]$ such that $x_i = y_i = A_i$ then the genotype $x_1 y_1 \ldots A_i a_i \ldots x_n y_n$ corresponds to the phenotype A_r. It is easy to see that $x_1 y_1 \ldots a_i a_i \ldots x_n y_n \in \widehat{A_r}^2$ which contradicts our assumption that $\widehat{A_r}^2 = \widehat{A_r}$. ■

Example 7.12. *Trihybrid case,* $n = 3$. Let $H = \{\widehat{A_1}, \widehat{A_2}, \widehat{A_3}, \widehat{A_4}, \widehat{A_5}, \widehat{A_6}, \widehat{A_7}, \widehat{A_8}\}$ be the set of different phenotypes in F_2 generation of the trihybrid case and $\widehat{A_1}$ corresponds to the genotype $A_1 x_1 A_2 x_2 A_3 x_3$, $\widehat{A_2}$ corresponds to the genotype $A_1 x_1 A_2 x_2 a_3 a_3$, $\widehat{A_3}$ corresponds to the genotype $A_1 x_1 a_2 a_2 A_3 x_3$, $\widehat{A_4}$ corresponds to the genotype $A_1 x_1 a_2 a_2 a_3 a_3$, $\widehat{A_5}$ corresponds to the genotype $a_1 a_1 A_2 x_2 A_3 x_3$, $\widehat{A_6}$ corresponds to the genotype $a_1 a_1 A_2 x_2 a_3 a_3$, $\widehat{A_7}$ corresponds to the genotype $a_1 a_1 a_2 a_2 A_3 x_3$ and $\widehat{A_8}$ corresponds to the genotype $a_1 a_1 a_2 a_2 a_3 a_3$ and $x_i = A_i$ or a_i for $i = 1, 2, 3$. Since $(H, \otimes)$ is commutative then we can represent it by Table 7.34.

Here, $K = \{\widehat{A_2}, \widehat{A_4}, \widehat{A_6}, \widehat{A_8}\}$, $L = \{\widehat{A_3}, \widehat{A_4}, \widehat{A_7}, \widehat{A_8}\}$ and $M = \{\widehat{A_5}, \widehat{A_6}, \widehat{A_7}, \widehat{A_8}\}$.

It is easy to see that $\{\widehat{A_8}\}, \{\widehat{A_4}, \widehat{A_8}\}, \{\widehat{A_6}, \widehat{A_8}\}, \{\widehat{A_7}, \widehat{A_8}\}$, K, L and M are the only proper subhypergroups of $(H, \otimes)$. Having $\{\widehat{A_4}, \widehat{A_8}\}, \{\widehat{A_6}, \widehat{A_8}\}, \{\widehat{A_7}, \widehat{A_8}\}$ isomorphic hypergroups and K, L and M isomorphic hypergroups implies that $(H, \otimes)$ has only 3 proper subhypergroups (up to isomorphism).

Table 7.34

$\otimes$	$\widehat{A_1}$	$\widehat{A_2}$	$\widehat{A_3}$	$\widehat{A_4}$	$\widehat{A_5}$	$\widehat{A_6}$	$\widehat{A_7}$	$\widehat{A_8}$
$\widehat{A_1}$	H	H	H	H	H	H	H	H
$\widehat{A_2}$		K	H	K	H	K	H	K
$\widehat{A_3}$			L	L	H	H	L	L
$\widehat{A_4}$				$\widehat{A_4},\widehat{A_8}$	H	K	L	$\widehat{A_4},\widehat{A_8}$
$\widehat{A_5}$					M	M	M	M
$\widehat{A_6}$						$\widehat{A_6},\widehat{A_8}$	M	$\widehat{A_6},\widehat{A_8}$
$\widehat{A_7}$							$\widehat{A_7},\widehat{A_8}$	$\widehat{A_7},\widehat{A_8}$
$\widehat{A_8}$								$\widehat{A_8}$

7.2.5　The hypothetical cross of n different traits, case of incomplete dominance

The case of incomplete dominance can be given by:

$$B_1 B_2 \ldots B_n \otimes \overline{B_1}\ \overline{B_2} \ldots \overline{B_n} \text{ with } n \geq 1.$$

We consider first results for the Monohybrid cross ($n = 1$) that differs in a single trait; a homozygous parent ($B_1 B_1$) $\otimes$ a homozygous parent ($\overline{B_1}\ \overline{B_1}$). The results of this hypothetical experiment can be summarized in the following way:

$$\text{P: } B_1 B_1 \otimes \overline{B_1} \overline{B_1}$$
$$F_1 : B_1 \overline{B_1}$$
$$\text{and}$$
$$F_1 \otimes F_1 : B_1 \overline{B_1} \otimes B_1 \overline{B_1}$$
$$F_2 : \widehat{A_1} \text{ (of genotype } B_1 B_1), \ \widehat{A_2} \text{ (of genotype } B_1 \overline{B_1}),$$
$$\widehat{A_3} \text{ (of genotype } \overline{B_1}\ \overline{B_1}).$$

Proposition 7.14. *Let $H = \{\widehat{A_1}, \widehat{A_2}, \widehat{A_3}\}$. Then $(H, \otimes)$ is a cyclic H_v-semigroup with identity and two idempotent elements.*

Proof. We present $(H, \otimes)$ in Table 7.35. Since no new elements are present in Table 7.35 and the table is symmetric then $(H, \otimes)$ is a commutative hypergroupoid. Having $H = \widehat{A_2}^2$ implies that H is cyclic of period two. It is easy to see that $\widehat{A_2}$ is the unique identity of $(H, \otimes)$ and that $\widehat{A_1}$ and $\widehat{A_3}$ are idempotents. ∎

Remark 7.11. $(H, \otimes)$ is not a hypergroup since H does not appear in the first row of Table 7.35.

Table 7.35

$\otimes$	$\widehat{A_1}$	$\widehat{A_2}$	$\widehat{A_3}$
$\widehat{A_1}$	$\widehat{A_1}$	$\widehat{A_1}, \widehat{A_2}$	$\widehat{A_2}$
$\widehat{A_2}$	$\widehat{A_1}, \widehat{A_2}$	H	$\widehat{A_2}, \widehat{A_3}$
$\widehat{A_3}$	$\widehat{A_2}$	$\widehat{A_2}, \widehat{A_3}$	$\widehat{A_3}$

Remark 7.12. An example of the monohybrid case is studied in ([37], p. 183).

Proposition 7.15. *There is only one hypergroup contained in* $(H, \otimes)$ *(up to isomorphism).*

Proof. It is clear that $\{\widehat{A_1}\}$ and $\{\widehat{A_3}\}$ are the only hypergroups contained in H and they are isomorphic. ∎

We consider now results for the dihybrid cross $(n = 2)$ that differs in 2 traits; a homozygous parent $(B_1 B_1 B_2 B_2) \otimes$ a homozygous parent $(\overline{B_1}\ \overline{B_1}\ \overline{B_2}\ \overline{B_2})$. The results of this hypothetical experiment can be summarized in the following way:

$$P: B_1 B_1 B_2 B_2 \otimes \overline{B_1}\ \overline{B_1}\ \overline{B_2}\ \overline{B_2}$$
$$F_1 : B_1 \overline{B_1} B_2 \overline{B_2}$$
and
$$F_1 \otimes F_1 : B_1 \overline{B_1} B_2 \overline{B_2} \otimes B_1 \overline{B_1} B_2 \overline{B_2}$$

$F_2 : \widehat{A_1}$ (of genotype $B_1 B_1 B_2 B_2$), $\widehat{A_2}$ (of genotype $B_1 B_1 B_2 \overline{B_2}$), $\widehat{A_3}$ (of genotype $B_1 B_1 \overline{B_2}\ \overline{B_2}$), $\widehat{A_4}$ (of genotype $B_1 \overline{B_1} B_2 \overline{B_2}$), $\widehat{A_5}$ (of genotype $B_1 \overline{B_1} B_2 \overline{B_2}$), $\widehat{A_6}$ (of genotype $B_1 \overline{B_1}\ \overline{B_2}\ \overline{B_2}$), $\widehat{A_7}$ (of genotype $\overline{B_1}\ \overline{B_1} B_2 B_2$), $\widehat{A_8}$ (of genotype $\overline{B_1}\ \overline{B_1} B_2 \overline{B_2}$) and $\widehat{A_9}$ (of genotype $\overline{B_1}\ \overline{B_1}\ \overline{B_2}\ \overline{B_2}$).

Proposition 7.16. *Let* $H = \{\widehat{A_1}, \widehat{A_2}, \ldots, \widehat{A_9}\}$. *Then* $(H, \otimes)$ *is a cyclic H_v-group with identity and four idempotent elements.*

Proof. It is easy to prove that $(H, \otimes)$ is an H_v-semigroup. Having $H = \widehat{A_5}^2$ implies that H is cyclic of period two. Also, we can easily show that $\widehat{A_5}$ is the unique identity of $(H, \otimes)$ and that $\widehat{A_1}$, $\widehat{A_3}$, $\widehat{A_7}$ and $\widehat{A_9}$ are the only idempotents in $(H, \otimes)$. ∎

We consider now results for the $n-$ hybrid cross $(n \geq 3)$ that differs in n traits; a homozygous parent $(B_1 B_1 B_2 B_2 \ldots B_n B_n) \otimes$ a homozygous parent $(\overline{B_1}\ \overline{B_1}\ \overline{B_2}\ \overline{B_2} \ldots \overline{B_n}\ \overline{B_n})$. The results of this hypothetical experiment can be summarized in the following way:

$$\text{P: } B_1 B_1 B_2 B_2 \ldots B_n B_n \otimes \overline{B_1}\ \overline{B_2}\ \overline{B_2} \ldots \overline{B_n}\ \overline{B_n}$$
$$F_1 : B_1 \overline{B_1} B_2 \overline{B_2} \ldots B_n \overline{B_n}$$
$$\text{and}$$
$$F_1 \otimes F_1 : B_1 \overline{B_1} B_2 \overline{B_2} \ldots B_n \overline{B_n} \otimes B_1 \overline{B_1} B_2 \overline{B_2} \ldots B_n \overline{B_n}$$
$$F_2 : \widehat{A_1} \text{ (of genotype } B_1 B_1 B_2 B_2 \ldots B_n B_n), \ \widehat{A_2} \text{ (of genotype}$$
$$B_1 B_1 \ldots B_{n-1} B_{n-1} B_n \overline{B_n}), \ldots) \ \widehat{A_{k-1}} \text{ (of genotype } \overline{B_1}\ \overline{B_1} \ldots \overline{B_{n-1}}$$
$$\overline{B_{n-1}} B_n \overline{B_n}) \text{ and } \widehat{A_k} \text{ (of genotype } \overline{B_1}\ \overline{B_1}\ \overline{B_2}\ \overline{B_2} \ldots \overline{B_n}\ \overline{B_n}).$$

It is easy to see that we have $k = 3^n$ different phenotypes. Let $r = \frac{k+1}{2} \in [1, k]$ such that $\widehat{A_r}$ is of genotype $B_1 \overline{B_1} B_2 \overline{B_2} \ldots B_n \overline{B_n}$.

Proposition 7.17. *Let* $H = \{\widehat{A_1}, \widehat{A_2}, \ldots, \widehat{A_k}\}$ *and* $k = 3^n$. *Then* $(H, \otimes)$ *is a cyclic* H_v-*semigroup with identity and* 2^n *idempotent elements.*

Proof. Let X, Y, $Z \in H$ such that $x_1 x_1' \ldots x_n x_n'$, $y_1 y_1' \ldots y_n y_n'$, $z_1 z_1' \ldots z_n z_n'$ are their corresponding genotypes. Since the phenotype responsible for $x_1 z_1' \ldots x_n z_n' \in ((X \otimes Y) \otimes Z) \cap (X \otimes (Y \otimes Z))$ and no new phenotypes appear in $(H, \otimes)$ then $(H, \otimes)$ is an H_v-semigroup. Having $H = \widehat{A_r}^2$ implies that H is cyclic of period two. It is easy to see that $\widehat{A_r}$ is the unique identity of $(H, \otimes)$ and that $\widehat{A_i}$ of genotypes $x_1 y_1 x_2 y_2 \ldots x_n y_n$ with $x_i = y_i$ for all $i = 1, 2 \ldots, n$ are the only idempotents in $(H, \otimes)$. It can be easily shown that the number of such element is 2^n. ■

7.2.6 The hypothetical cross of $m + n$ different traits, case of simple and incomplete dominance combined together

The case of combination of simple and incomplete dominance can be given by:

$$B_1 B_2 \ldots B_n A_1 A_2 \ldots A_m \otimes \overline{B_1}\ \overline{B_2} \ldots \overline{B_n} a_1 a_2 \ldots a_m \text{ with } m, n \geq 1.$$

We consider first results for the cross $(m = n = 1)$ that differs in two traits; a homozygous parent $(B_1 B_1 A_1 A_1) \otimes$ a homozygous parent $(\overline{B_1}\ \overline{B_1} a_1 a_1)$. The results of this hypothetical experiment can be summarized in the following way:

$$\text{P: } B_1 B_1 A_1 A_1 \otimes \overline{B_1}\ \overline{B_1} a_1 a_1$$
$$F_1 : B_1 \overline{B_1} A_1 a_1$$
$$\text{and}$$
$$F_1 \otimes F_1 : B_1 \overline{B_1} A_1 a_1 \otimes B_1 \overline{B_1} A_1 a_1$$

$F_2 : \widehat{A_1}$ (of genotype $B_1 B_1 A_1 y_1$), $\widehat{A_2}$ (of genotype $B_1 B_1 a_1 a_1$), $\widehat{A_3}$ (of genotype $B_1 \overline{B_1} A_1 y_1$), $\widehat{A_4}$ (of genotype $B_1 \overline{B_1} a_1 a_1$),$\widehat{A_5}$ (of genotype $\overline{B_1}$ $\overline{B_1} A_1 y_1$) and $\widehat{A_6}$ (of genotype $\overline{B_1}$ $\overline{B_1} a_1 a_1$).

Where $y_1 = A_1$ or a_1.

Proposition 7.18. *Let* $H = \{\widehat{A_1}, \widehat{A_2}, \ldots, \widehat{A_6}\}$. *Then* $(H, \otimes)$ *is a cyclic* H_v-*semigroup with identity and two idempotent elements.*

Proof. It is easy to see that $H = \widehat{A_3}^2$ and that $\widehat{A_3}$ is an identity of $(H, \otimes)$. Moreover, $\widehat{A_2}$ and $\widehat{A_6}$ are the only idempotents in $(H, \otimes)$. ∎

$\{\widehat{A_1}, \widehat{A_2}\}$ and $\{\widehat{A_5}, \widehat{A_6}\}$ are two hypergroups contained in $(H, \otimes)$.

Example 7.13. *MN blood group inheritance with the Rh factor.* The blood group can be differentiated by a $+$ and $-$ which refers to a blood group antigen called the *Rh factor*. In this system, the Rh^+ (which can be considered as A_1) is dominant over the Rh^- which can be considered as a_1. Furthermore, in the MN system, there are two antigens M and N, whose production is determined by a gene with two alleles, L^M and L^N that produce the M and N antigens respectively. Individuals having the $L^M L^M$ genotype will have only the M antigen on their red cells and will be of type M, those having the $L^N L^N$ genotype will have only the N antigen on their red cells and will be of type N. Hetrozygous (with $L^M L^N$ genotype) will have both antigens on their red cells and will be of type MN.
After denoting M by B_1 and N by $\overline{B_1}$, we can apply Proposition 7.18 to this blood system.

We consider results for the cross that differs in $m + n$ traits; a homozygous parent $(B_1 B_1 \ldots B_n B_n A_1 A_1 \ldots A_m A_m) \otimes$ a homozygous parent $(\overline{B_1} \overline{B_1} \ldots \overline{B_n} \ \overline{B_n} a_1 a_1 \ldots a_m a_m)$. The results of this hypothetical experiment can be summarized in the following way:

$$\text{P: } B_1 B_1 \ldots B_n B_n A_1 A_1 \ldots A_m A_m \otimes \overline{B_1} \ \overline{B_1} \ldots \overline{B_n} \ \overline{B_n} a_1 a_1 \ldots a_m a_m$$
$$F_1 : B_1 \overline{B_1} \ldots B_n \overline{B_n} A_1 a_1 \ldots A_m a_m$$
$$\text{and}$$
$$F_1 \otimes F_1 : B_1 \overline{B_1} \ldots B_n \overline{B_n} A_1 a_1 \ldots A_m a_m \otimes B_1 \overline{B_1} \ldots B_n \overline{B_n} A_1 a_1 \ldots A_m a_m$$
$$F_2 : \widehat{A_1} \text{ (of genotype } B_1 B_1 \ldots B_n B_n A_1 y_1 \ldots A_m y_m), \widehat{A_2} \text{ (of genotype }$$
$$B_1 B_1 \ldots B_n B_n A_1 y_1 \ldots A_{m-1} y_{m-1} a_m a_m), \ldots \text{ and } \widehat{A_k} \text{ (of genotype } \overline{B_1}$$
$$\overline{B_1} \ldots \overline{B_n} \ \overline{B_n} a_1 a_1 \ldots a_m a_m).$$

Where $k = 2^m 3^n$ is the number of different phenotypes and $y_i = A_i$ or a_i for $i = 1, \ldots, m$.

Theorem 7.5. *Let* $H = \{\widehat{A_1}, \widehat{A_2}, \ldots, \widehat{A_k}\}$. *Then* $(H, \otimes)$ *is a cyclic* H_v-*semigroup with identity and* 2^n *idempotent elements.*

Proof. The proof of $(H, \otimes)$ is an H_v-semigroup is similar to that of Proposition 7.17.

Suppose that $r \in [1, k]$ such that $\widehat{A_r}$ is the phenotype corresponding to the genotype $B_1\overline{B_1} \ldots B_n\overline{B_n}A_1a_1 \ldots A_ma_m$. It is clear that $H = \widehat{A_r}^2$ and that $\widehat{A_r}$ is an identity of $(H, \otimes)$. Moreover, the only idempotents elements of $(H, \otimes)$ are those having genotypes of the form $x_1x_1 \ldots x_nx_na_1a_1 \ldots a_ma_m$ where $x_i = B_i$ or $\overline{B_i}$. It can be easily shown that the number of such elements is 2^n.   ∎

7.3   Physical examples

In the elementary particle physics, a fundamental particle is known as a particle which have no substructure and it is one of the basic building blocks of the universe from which all other particles are made. Nowadays, the Standard Model (SM) of elementary particles is known to be a well established theory to describe the elementary particles and the interacting forces between them [86]. In the SM, the Quarks, Leptons and Gauge Bosons are introduced as the elementary particles. This model contains six types of quarks, known as flavors: Up, Down, Charm, Strange, Bottom and Top plus their corresponding antiparticles which is known as antiquark. Since the quarks are never found in isolation, therefore quarks combine to form composite particles which are called Hadrons and they appear into two families: Baryons (made of three quarks) and Mesons (made of one quark and one antiquark). In the SM, gauge bosons consist of the photons(γ), gluons(g), $W^{\pm}$ and Z bosons act as carriers of the fundamental forces of nature [60]. The third group of the elementary particles are Leptons. In the SM, there are six flavors of leptons along with six antileptons. Leptons are an important part of the SM, especially the electrons which are one of the components of atoms. Since the leptons can be found freely in the universe and they are one of the important groups of the elementary particles.

A motivation for the study of hyperstructures comes from physical phenomenon as the nuclear fission. This motivation and the results were presented by S. Hošková, J. Chvalina and P. Račková (see [72], [73]). In [48], the authors provided, for the first time, a physical example of hyperstructures associated with the elementary particle physics, Leptons. They have considered this important group of the elementary particles and shown that

Table 7.36 Leptons classification to three generations. Q stands for charge in unit of the electron charge. L_e, L_μ and L_τ stand for the electronic, muonic and tauonic numbers, respectively.

Classify	First Generation				Second Generation				Third Generation			
Leptons	e	ν_e	e^+	$\overline{\nu}_e$	μ	ν_μ	μ^+	$\overline{\nu}_\mu$	τ	ν_τ	τ^+	$\overline{\nu}_\tau$
Q	-1	0	$+1$	0	-1	0	$+1$	0	-1	0	$+1$	0
L_e	1	1	-1	-1	0	0	0	0	0	0	0	0
L_μ	0	0	0	0	1	1	-1	-1	0	0	0	0
L_τ	0	0	0	0	0	0	0	0	1	1	-1	-1

this set along with the interactions between its members can be described by the algebraic hyperstructures. The main reference for this section is [48].

7.3.1 *Leptons*

In the SM, leptons form three generations. The first generation includes the electronic leptons which are electron (e), electron neutrino (ν_e), positron (e^+) and electron antineutrino ($\overline{\nu}_e$). The second generation contains the muonic leptons, i.e. muon (μ), muon neutrino (ν_μ), antimuon (μ^+) and muon antineutrino ($\overline{\nu}_\mu$). The third generation comprising the tauonic leptons which are tau (τ), tau neutrino (ν_τ), antitau (τ^+) and tau antineutrino ($\overline{\nu}_\tau$). In the leptons group the electron, muon and tau have the electric charge $Q = -1$ (the charge of a particle is expressed in unit of the electron charge). According to the definition of antiparticle, the electric charge of positron, antimuon and antitau is $Q = +1$ but the antineutrinos are neutral as well as neutrinos. The main differences between the neutrinos and antineutrinos are in the other quantum numbers such as leptonic numbers [60, 62]. In the SM, leptonic numbers are assigned to the members of every generation of leptons. Electron and electron neutrino have an electronic number of $L_e = 1$ while muon and muon neutrino have a muonic number of $L_\mu = 1$ and tau and tau neutrino have a tauonic number of $L_\tau = 1$. Antileptons have their respective generation's leptonic numbers of -1. These numbers are classified in Table 7.36.

In every interaction, the leptonic numbers should be conserved. Conservation of the leptonic numbers implies that leptons and antileptons must be created in pairs of a single generation. For example, the following processes are allowed under conservation of the electronic and munic numbers, respectively:

$$e + \nu_e \rightarrow e + \nu_e$$
$$\mu + \nu_\mu \rightarrow \mu + \nu_\mu. \tag{7.1}$$

In this work we use a new notation for the outgoing productions. For example, we write down: $e + \nu_e \rightarrow \{e, \ \nu_e\}$.

In other interactions, outgoing particles might be different, therefore all leptonic numbers must be checked. In the following interaction, the conservation of the electronic and muonic numbers implies to have two outgoing modes:

$$e + \nu_\mu \rightarrow e + \nu_\mu, \quad \mu + \nu_e. \tag{7.2}$$

According to the introduced notation we write down $e + \nu_\mu \rightarrow \{e, \ \nu_\mu, \ \mu, \ \nu_e\}$. The conservation of the electric charges and the leptonic numbers are required to occur a leptonic interaction. Considering these conservation rules, for the electron-positron interaction the interacting modes are:

$$e + e^+ \rightarrow e + e^+, \quad \mu + \mu^+, \quad \tau + \tau^+, \quad \nu_e + \overline{\nu}_e, \quad \nu_\mu + \overline{\nu}_\mu, \quad \nu_\tau + \overline{\nu}_\tau.$$

We write: $e + e^+ \rightarrow \{e, \ e^+, \ \mu, \ \mu^+, \ \tau, \ \tau^+, \ \nu_e, \ \overline{\nu}_e, \ \nu_\mu, \ \overline{\nu}_\mu, \ \nu_\tau, \ \overline{\nu}_\tau\} = L$. Other interactions between the members of the leptons group are shown in Table 7.37. To arrange this table we avoided writing the repeated symbols. For example, in the productions of the electron-electron interaction we only write e instead of $e + e$. All interactions shown in Table 7.37 are in the first order. It means in higher orders other particles can be produced that we do not consider them. For example in the electron-electron scattering (Muller scattering) one or several photons might be appeared in productions of the interaction, i.e. $e + e \rightarrow e + e + \gamma$ or in the electron-positron scattering (Bhabha scattering [14]) we can have: $e + e^+ \rightarrow e + e^+ + \gamma$. There also exist other processes that we do not consider them in this work. For example: $e + e^+ \rightarrow \gamma + \gamma$, $e + e^+ \rightarrow W^- + W^+$, $\tau + \tau^+ \rightarrow Z^0 + Z^0$ and so on.

7.3.2 The algebraic hyperstructure of Leptons

In this section, by considering the definitions presented in Section 2, we investigate that the Leptons along with the interactions arranged in Table 7.37 found a hyperstructure.

If we assume, L is the set of Leptons and hyperoperation $\otimes$ is the Leptonic interactions arranged in Table 7.37, then $(L, \otimes)$ is a commutative H_v-group.

In order to show the weak associativity of this hyperstructure, consider the following example:

$$\nu_\mu \otimes (\overline{\nu}_e \otimes e^+) = \nu_\mu \otimes \{\overline{\nu}_e, e^+\} = \{e^+, \mu, \overline{\nu}_e, \nu_\mu\},$$
$$(\nu_\mu \otimes \overline{\nu}_e) \otimes e^+ = \{e^+, \mu, \overline{\nu}_e, \nu_\mu\} \otimes e^+ = \{e^+, \mu, \overline{\nu}_e, \nu_\mu\}.$$

Table 7.37   Interaction between leptons are shown.

$\otimes$	e	ν_e	e^+	$\bar\nu_e$	μ	ν_μ	μ^+	$\bar\nu_\mu$	τ	ν_τ	τ^+	$\bar\nu_\tau$
e	e	e ν_e	L	e μ τ $\bar\nu_e$ $\bar\nu_\mu$ $\bar\nu_\tau$	e μ	e μ ν_e ν_μ	e μ^+ $\bar\nu_\mu$ ν_e	e $\bar\nu_\mu$	e τ	e τ ν_e ν_τ	e τ^+ $\bar\nu_\tau$ ν_e	e $\bar\nu_\tau$
ν_e	e ν_e	ν_e	e^+ μ^+ τ^+ ν_e ν_μ ν_τ	L	e μ ν_e ν_μ	ν_e ν_μ	μ^+ ν_e	e μ^+ $\bar\nu_\mu$ ν_e	e τ ν_e ν_τ	ν_e ν_τ	τ^+ ν_e	e τ^+ $\bar\nu_\tau$ ν_e
e^+	L	e^+ μ^+ τ^+ ν_e ν_μ ν_τ	e^+	e^+ $\bar\nu_e$	e^+ μ $\bar\nu_e$ ν_μ	e^+ ν_μ	e^+ μ^+	e^+ μ^+ $\bar\nu_e$ $\bar\nu_\mu$	e^+ τ $\bar\nu_e$ ν_τ	e^+ ν_τ	e^+ τ^+	e^+ τ^+ $\bar\nu_e$ $\bar\nu_\tau$
$\bar\nu_e$	e μ τ $\bar\nu_e$ $\bar\nu_\mu$ $\bar\nu_\tau$	L	e^+ $\bar\nu_e$	$\bar\nu_e$	μ $\bar\nu_e$	e^+ μ $\bar\nu_e$ ν_μ	e^+ μ^+ $\bar\nu_e$ $\bar\nu_\mu$	$\bar\nu_e$ $\bar\nu_\mu$	τ $\bar\nu_e$	e^+ τ $\bar\nu_e$ ν_τ	e^+ τ^+ $\bar\nu_e$ $\bar\nu_\tau$	$\bar\nu_e$ $\bar\nu_\tau$
μ	e μ	e μ ν_e ν_μ	e^+ μ $\bar\nu_e$ ν_μ	μ $\bar\nu_e$	μ	μ ν_μ	L	e μ τ $\bar\nu_e$ $\bar\nu_\mu$ $\bar\nu_\tau$	μ τ	μ τ ν_μ ν_τ	μ τ^+ $\bar\nu_\tau$ ν_μ	μ $\bar\nu_\tau$
ν_μ	e μ ν_e ν_μ	ν_e ν_μ	e^+ ν_μ	e^+ μ $\bar\nu_e$ ν_μ	μ ν_μ	ν_μ	e^+ μ^+ τ^+ ν_e ν_μ ν_τ	L	μ τ ν_μ ν_τ	ν_μ ν_τ	τ^+ ν_μ	μ τ^+ $\bar\nu_\tau$ ν_μ
μ^+	e μ^+ $\bar\nu_\mu$ ν_e	μ^+ ν_e	e^+ μ^+	e^+ μ^+ $\bar\nu_e$ $\bar\nu_\mu$	L	e^+ μ^+ τ^+ ν_e ν_μ ν_τ	μ^+	μ^+ $\bar\nu_\mu$	μ^+ τ $\bar\nu_\mu$ ν_τ	μ^+ ν_τ	μ^+ τ^+	μ^+ τ^+ $\bar\nu_\mu$ $\bar\nu_\tau$
$\bar\nu_\mu$	e $\bar\nu_\mu$	e μ^+ $\bar\nu_\mu$ ν_e	e^+ μ^+ $\bar\nu_e$ $\bar\nu_\mu$	$\bar\nu_e$ $\bar\nu_\mu$	e μ τ $\bar\nu_e$ $\bar\nu_\mu$ $\bar\nu_\tau$	L	μ^+ $\bar\nu_\mu$	$\bar\nu_\mu$	τ $\bar\nu_\mu$	μ^+ τ $\bar\nu_\mu$ ν_τ	μ^+ τ^+ $\bar\nu_\mu$ $\bar\nu_\tau$	$\bar\nu_\mu$ $\bar\nu_\tau$
τ	e τ	e τ ν_e ν_τ	e^+ τ $\bar\nu_e$ ν_τ	τ $\bar\nu_e$	μ τ	μ τ ν_μ ν_τ	μ^+ τ $\bar\nu_\mu$ ν_τ	τ $\bar\nu_\mu$	τ	τ ν_τ	L	e μ τ $\bar\nu_e$ $\bar\nu_\mu$ $\bar\nu_\tau$
ν_τ	e τ ν_e ν_τ	ν_e ν_τ	e^+ ν_τ	e^+ τ $\bar\nu_e$ ν_τ	μ τ ν_μ ν_τ	ν_μ ν_τ	μ^+ ν_τ	μ^+ τ $\bar\nu_\mu$ ν_τ	τ ν_τ	ν_τ	e^+ μ^+ τ^+ ν_e ν_μ ν_τ	L
τ^+	e τ^+ $\bar\nu_\tau$ ν_e	τ^+ ν_e	e^+ τ^+	e^+ τ^+ $\bar\nu_e$ $\bar\nu_\tau$	μ τ^+ $\bar\nu_\tau$ ν_μ	τ^+ ν_μ	μ^+ τ^+	μ^+ τ^+ $\bar\nu_\mu$ $\bar\nu_\tau$	L	e^+ μ^+ τ^+ ν_e ν_μ ν_τ	τ^+	τ^+ $\bar\nu_\tau$
$\bar\nu_\tau$	e $\bar\nu_\tau$	e τ^+ $\bar\nu_\tau$ ν_e	e^+ τ^+ $\bar\nu_e$ $\bar\nu_\tau$	$\bar\nu_e$ $\bar\nu_\tau$	μ $\bar\nu_\tau$	μ τ^+ $\bar\nu_\tau$ ν_μ	μ^+ τ^+ $\bar\nu_\mu$ $\bar\nu_\tau$	$\bar\nu_\mu$ $\bar\nu_\tau$	e μ τ $\bar\nu_e$ $\bar\nu_\mu$ $\bar\nu_\tau$	L	τ^+ $\bar\nu_\tau$	$\bar\nu_\tau$

As it is seen, $\nu_\mu \otimes (\bar{\nu}_e \otimes e^+) \cap (\nu_\mu \otimes \bar{\nu}_e) \otimes e^+ \neq \emptyset$. To investigate the condition of weak associativity of this hyperstructure we used MAPLE 14 software. This property is established for all members.

Since, there is an antiparticle for every element in the Leptons set which their interactions can produce all Leptons, then the reproduction axiom holds automatically. In other words: $\mu \otimes L = \tau \otimes L = \nu_\mu \otimes L = \cdots = L$. All the elements of $(L, \otimes)$ are idempotents.

In the next step, we summarize some results on this hyperstructure in the following.

If $(L, \otimes)$ is the above H_v-group, then the following statements hold:

(i) There is not any H_v-subgroups of orders 5, 7, 8, 9, 10 and 11 for $(L, \otimes)$.

(ii) All the H_v-subgroups of order 1 (L_i^1, $i = 1, \ldots, 12$) are:
$\{e\}, \{e^+\}, \{\mu\}, \{\mu^+\}, \{\tau\}, \{\tau^+\}, \{\bar{\nu}_e\}, \{\bar{\nu}_\mu\}, \{\bar{\nu}_\tau\}, \{\nu_e\}, \{\nu_\mu\}, \{\nu_\tau\},$
respectively.

(iii) All the H_v-subgroups of order 2 (L_i^2, $i = 1, \ldots, 30$) are:
$\{e, \mu\}, \ \{e, \tau\}, \ \{e, \bar{\nu}_\mu\}, \ \{e, \bar{\nu}_\tau\}, \ \{e, \nu_e\}, \ \{e^+, \mu^+\}, \ \{e^+, \tau^+\}, \ \{e^+, \nu_\mu\},$
$\{e^+, \nu_\tau\}, \ \{\mu, \tau\}, \ \{\mu, \bar{\nu}_e\}, \ \{\mu, \bar{\nu}_\tau\}, \ \{\mu, \nu_\mu\}, \ \{\mu^+, \tau^+\}, \ \{\mu^+, \bar{\nu}_\mu\},$
$\{\mu^+, \nu_e\}, \ \{\mu^+, \nu_\tau\}, \ \{\tau, \bar{\nu}_e\}, \ \{\tau, \bar{\nu}_\mu\}, \ \{\tau, \nu_\tau\}, \ \{\tau^+, \bar{\nu}_\tau\}, \ \{\tau^+, \nu_e\},$
$\{\tau^+, \nu_\mu\}, \ \{\bar{\nu}_e, \bar{\nu}_\mu\}, \ \{\bar{\nu}_e, \bar{\nu}_\tau\}, \ \{\bar{\nu}_\mu, \bar{\nu}_\tau\}, \ \{\nu_e, \nu_\mu\}, \ \{\nu_e, \nu_\tau\}, \ \{\nu_\mu, \nu_\tau\},$
$\{e^+, \bar{\nu}_e\}$, respectively.

(iv) All the H_v-subgroups of order 3 (L_i^3, $i = 1, \ldots, 16$) are:
$\{e, \ \mu, \ \tau\}, \ \{e, \ \mu, \ \bar{\nu}_\tau\}, \ \{e, \ \tau, \ \bar{\nu}_\mu\}, \ \{e^+, \ \mu^+, \ \tau^+\}, \ \{e^+, \ \mu^+, \ \nu_\tau\},$
$\{e^+, \ \tau^+, \ \nu_\mu\}, \ \{e^+, \ \nu_\mu, \ \nu_e\}, \ \{\mu, \ \tau, \ \bar{\nu}_e\}, \ \{\mu, \ \bar{\nu}_e, \ \bar{\nu}_\tau\}, \ \{\mu^+, \ \tau^+, \ \nu_e\},$
$\{\mu^+, \ \nu_e, \ \nu_\tau\}, \ \{\tau, \ \bar{\nu}_e, \ \bar{\nu}_\mu\}, \ \{e, \ \bar{\nu}_\mu, \ \bar{\nu}_\tau\}, \ \{\tau^+, \ \nu_e, \ \nu_\mu\}, \ \{\bar{\nu}_e, \ \bar{\nu}_\mu, \ \bar{\nu}_\tau\},$
$\{\nu_e, \ \nu_\mu, \ \nu_\tau\}$, respectively.

(v) All the H_v-subgroups of order 4 (L_i^4, $i = 1, \ldots, 9$) are:
$\{e, \ \mu, \ \nu_e, \ \nu_\mu\}, \ \{e, \ \mu^+, \ \bar{\nu}_\mu, \ \nu_e\}, \ \{e, \ \tau, \ \nu_e, \ \nu_\tau\}, \ \{e, \ \tau^+, \ \bar{\nu}_\tau, \ \nu_e\},$
$\{e^+, \ \mu^+, \ \bar{\nu}_e, \ \bar{\nu}_\mu\}, \ \{\mu, \ \tau, \ \nu_\mu, \ \nu_\tau\}, \ \{\mu, \ \tau^+, \ \bar{\nu}_\tau, \ \nu_\mu\}, \ \{\mu^+, \ \tau, \ \bar{\nu}_\mu, \ \nu_\tau\},$
$\{\mu^+, \ \tau^+, \ \bar{\nu}_\mu, \ \bar{\nu}_\tau\}$, respectively.

(vi) All the H_v-subgroups of order 6 (L_i^6, $i = 1, \ldots, 8$) are:

(1) $L_1^6 = \{e^+, \mu^+, \tau^+, \nu_e, \nu_\mu, \nu_\tau\}$ with the multiplicative Table 7.38.
(2) $L_2^6 = \{e, \mu, \tau, \nu_e, \nu_\mu, \nu_\tau\}$ with the multiplicative Table 7.39.
(3) $L_3^6 = \{e, \mu, \tau^+, \bar{\nu}_\tau, \nu_e, \nu_\mu\}$ with the multiplicative Table 7.40.
(4) $L_4^6 = \{e, \mu^+, \tau, \bar{\nu}_\mu, \nu_e, \nu_\tau\}$ with the multiplicative Table 7.41.
(5) $L_5^6 = \{e, \mu^+, \tau^+, \bar{\nu}_\mu, \bar{\nu}_\tau, \nu_e\}$ with the multiplicative Table 7.42
(6) $L_6^6 = \{e^+, \mu^+, \tau, \bar{\nu}_e, \bar{\nu}_\mu, \nu_\tau\}$ with the multiplicative Table 7.43.

(7) $L_7^6 = \{e^+, \mu^+, \tau^+, \bar\nu_e, \bar\nu_\mu, \bar\nu_\tau\}$ with the multiplicative Table 7.44.

(8) $L_8^6 = \{e, \mu, \tau, \bar\nu_e, \bar\nu_\mu, \bar\nu_\tau\}$ with the multiplicative Table 7.45.

There exist two classes of non-isomorphic H_v-subgroups of order 6, i.e.,

(1) $L_1^6 \cong L_8^6$,

(2) $L_2^6 \cong L_3^6 \cong L_4^6 \cong L_5^6 \cong L_6^6 \cong L_7^6$.

Table 7.38   Multiplicative table for $L_1^6 = \{e^+, \mu^+, \tau^+, \nu_e, \nu_\mu, \nu_\tau\}$.

	e^+	μ^+	τ^+	ν_e	ν_μ	ν_τ
e^+	e^+	e^+, μ^+	e^+, τ^+	$e^+, \mu^+, \tau^+, \nu_e, \nu_\mu, \nu_\tau$	e^+, ν_μ	e^+, ν_τ
μ^+	e^+, μ^+	μ^+	μ^+, τ^+	μ^+, ν_e	$e^+, \mu^+, \tau^+, \nu_e, \nu_\mu, \nu_\tau$	μ^+, ν_τ
τ^+	τ^+, e^+	μ^+, τ^+	τ^+	τ^+, ν_e	τ^+, ν_μ	$e^+, \mu^+, \tau^+, \nu_e, \nu_\mu, \nu_\tau$
ν_e	$e^+, \mu^+, \tau^+, \nu_e, \nu_\mu, \nu_\tau$	μ^+, ν_e	τ^+, ν_e	ν_e	ν_e, ν_μ	ν_e, ν_τ
ν_μ	e^+, ν_μ	$e^+, \mu^+, \tau^+, \nu_e, \nu_\mu, \nu_\tau$	ν_μ, τ^+	ν_μ, ν_e	ν_μ	ν_μ, ν_τ
ν_τ	ν_τ, e^+	μ^+, ν_τ	$e^+, \mu^+, \tau^+, \nu_e, \nu_\mu, \nu_\tau$	ν_τ, ν_e	ν_τ, ν_μ	ν_τ

Table 7.39   Multiplicative table for $L_2^6 = \{e, \mu, \tau, \nu_e, \nu_\mu, \nu_\tau\}$.

	e	μ	τ	ν_e	ν_μ	ν_τ
e	e	e, μ	e, τ	e, ν_e	e, μ, ν_e, ν_μ	e, τ, ν_e, ν_τ
μ	e, μ	μ	μ, τ	e, μ, ν_e, ν_μ	μ, ν_μ	$\mu, \tau, \nu_\mu, \nu_\tau$
τ	e, τ	τ, μ	τ	e, τ, ν_e, ν_τ	$\mu, \tau, \nu_\mu, \nu_\tau$	τ, ν_τ
ν_e	e, ν_e	e, μ, ν_e, ν_μ	e, τ, ν_e, ν_τ	ν_e	ν_e, ν_μ	ν_e, ν_τ
ν_μ	e, μ, ν_e, ν_μ	ν_μ, μ	$\mu, \tau, \nu_\mu, \nu_\tau$	ν_μ, ν_e	ν_μ	ν_μ, ν_τ
ν_τ	e, τ, ν_e, ν_τ	$\mu, \tau, \nu_\mu, \nu_e$	ν_τ, τ	ν_τ, ν_e	ν_τ, ν_μ	ν_τ

Table 7.40   Multiplicative table for $L_3^6 = \{e, \mu, \tau^+, \bar{\nu}_\tau, \nu_e, \nu_\mu\}$.

	e	μ	τ^+	$\bar{\nu}_\tau$	ν_e	ν_μ
e	e	e,μ	$e,\tau^+,$ $\bar{\nu}_\tau,\nu_e$	$e,\bar{\nu}_\tau$	e,ν_e	$e,\mu,$ ν_e,ν_μ
μ	e,μ	μ	$\mu,\tau^+,$ $\bar{\nu}_\tau,\nu_\mu$	$\mu,\bar{\nu}_\tau$	$e,\mu,$ ν_e,ν_μ	μ,ν_μ
τ^+	$e,\tau^+,$ $\bar{\nu}_\tau,\nu_e$	μ,τ^+ $\bar{\nu}_\tau,\nu_\mu$	τ^+	$\tau^+,\bar{\nu}_\tau$	τ^+,ν_e	τ^+,ν_μ
$\bar{\nu}_\tau$	$e,\bar{\nu}_\tau$	$\bar{\nu}_\tau,\mu$	$\bar{\nu}_\tau,\tau^+$	$\bar{\nu}_\tau$	$e,\tau^+,$ $\bar{\nu}_\tau,\nu_e$	$\mu,\tau^+,$ $\bar{\nu}_\tau,\nu_\mu$
ν_e	e,ν_e	$e,\mu,$ ν_e,ν_μ	τ^+,ν_e	$e,\tau^+,$ $\bar{\nu}_\tau,\nu_e$	ν_e	ν_e,ν_μ
ν_μ	$e,\mu,$ ν_e,ν_μ	ν_μ,μ	ν_μ,τ^+	$\mu,\tau^+,$ $\bar{\nu}_\tau,\nu_\mu$	ν_μ,ν_e	ν_μ

Table 7.41   Multiplicative table for $L_4^6 = \{e, \mu^+, \tau, \bar{\nu}_\mu, \nu_e, \nu_\tau\}$.

	e	μ^+	τ	$\bar{\nu}_\mu$	ν_e	ν_τ
e	e	$e,\mu^+,$ $\bar{\nu}_\mu,\nu_e$	e,τ	$e,\bar{\nu}_\mu$	e,ν_e	$e,\tau,$ ν_e,ν_τ
μ^+	$e,\mu^+,$ $\bar{\nu}_\mu,\nu_e$	μ^+	μ^+,τ $\bar{\nu}_\mu,\nu_\tau$	$\mu^+,\bar{\nu}_\mu$	μ^+,ν_e	μ^+,ν_τ
τ	e,τ	$\mu^+,\tau,$ $\bar{\nu}_\mu,\nu_\tau$	τ	$\tau,\bar{\nu}_\mu$	$e,\tau,$ ν_e,ν_τ	τ,ν_τ
$\bar{\nu}_\mu$	$e,\bar{\nu}_\mu$	$\bar{\nu}_\mu,\mu^+$	$\bar{\nu}_\mu,\tau$	$\bar{\nu}_\mu$	$e,\mu^+,$ $\bar{\nu}_\mu,\nu_e$	$\mu^+,\tau,$ $\bar{\nu}_\mu,\nu_\tau$
ν_e	ν_e,e	ν_e,μ^+	$e,\tau,$ ν_e,ν_τ	$e,\mu^+,$ $\bar{\nu}_\mu,\nu_e$	ν_e	ν_e,ν_τ
ν_τ	$e,\tau,$ ν_e,ν_τ	μ^+,ν_τ	ν_τ,τ	$\mu^+,\tau,$ $\bar{\nu}_\mu,\nu_\tau$	ν_τ,ν_e	ν_τ

Table 7.42   Multiplicative table for $L_5^6 = \{e, \mu^+, \tau^+, \bar{\nu}_\mu, \bar{\nu}_\tau, \nu_e\}$.

	e	μ^+	τ^+	$\bar{\nu}_\mu$	$\bar{\nu}_\tau$	ν_e
e	e	$e,\mu^+,$ $\bar{\nu}_\mu,\nu_e$	$e,\tau^+,$ $\bar{\nu}_\tau,\nu_e$	$e,\bar{\nu}_\mu$	$e,\bar{\nu}_\tau$	e,ν_e
μ^+	$e,\mu^+,$ $\bar{\nu}_\mu,\nu_e$	μ^+	μ^+,τ^+	$\mu^+,\bar{\nu}_\mu$	$\mu^+,\tau^+,$ $\bar{\nu}_\mu,\bar{\nu}_\tau$	μ^+,ν_e
τ^+	$e,\tau^+,$ $\bar{\nu}_\tau,\nu_e$	τ^+,μ^+	τ^+	$\mu^+,\tau^+,$ $\bar{\nu}_\mu,\bar{\nu}_\tau$	$\tau^+,\bar{\nu}_\tau$	τ^+,ν_e
$\bar{\nu}_\mu$	$\bar{\nu}_\mu,e$	$\bar{\nu}_\mu,\mu^+$	$\mu^+,\tau^+,$ $\bar{\nu}_\mu,\bar{\nu}_\tau$	$\bar{\nu}_\mu$	$\bar{\nu}_\mu,\bar{\nu}_\tau$	$e,\mu^+,$ $\bar{\nu}_\mu,\nu_e$
$\bar{\nu}_\tau$	$\bar{\nu}_\tau,e$	$\mu^+,\tau^+,$ $\bar{\nu}_\mu,\bar{\nu}_\tau$	$\bar{\nu}_\tau,\tau^+$	$\bar{\nu}_\mu,\bar{\nu}_\tau$	$\bar{\nu}_\tau$	$e,\tau^+,$ $\bar{\nu}_\tau,\nu_e$
ν_e	ν_e,e	ν_e,μ^+	ν_e,τ^+	$e,\mu^+,$ $\bar{\nu}_\mu,\nu_e$	$e,\tau^+,$ $\bar{\nu}_\tau,\nu_e$	ν_e

Table 7.43   Multiplicative table for $L_6^6 = \{e^+, \mu^+, \tau, \bar{\nu}_e, \bar{\nu}_\mu, \nu_\tau\}$.

	e^+	μ^+	τ	$\bar{\nu}_e$	$\bar{\nu}_\mu$	ν_τ
e^+	e^+	e^+, μ^+	$e^+, \tau,$ $\bar{\nu}_e, \nu_\tau$	$e^+, \bar{\nu}_e$	$e^+, \mu^+,$ $\bar{\nu}_e, \bar{\nu}_\mu$	e^+, ν_τ
μ^+	μ^+, e^+	μ^+	$\mu^+, \tau,$ $\bar{\nu}_\mu, \nu_\tau$	$e^+, \mu^+,$ $\bar{\nu}_e, \bar{\nu}_\mu$	$\mu^+, \bar{\nu}_\mu$	μ^+, ν_τ
τ	$e^+, \tau,$ $\bar{\nu}_e, \nu_\tau$	$\mu^+, \tau,$ $\bar{\nu}_\mu, \nu_\tau$	τ	$\tau, \bar{\nu}_e$	$\tau, \bar{\nu}_\mu$	τ, ν_τ
$\bar{\nu}_e$	$\bar{\nu}_e, e^+$	$e^+, \mu^+,$ $\bar{\nu}_e, \bar{\nu}_\mu$	$\bar{\nu}_e, \tau$	$\bar{\nu}_e$	$\bar{\nu}_e, \bar{\nu}_\mu$	$e^+, \tau,$ $\bar{\nu}_e, \nu_\tau$
$\bar{\nu}_\mu$	$e^+, \mu^+,$ $\bar{\nu}_e, \bar{\nu}_\mu$	$\bar{\nu}_\mu, \mu^+$	$\bar{\nu}_\mu, \tau$	$\bar{\nu}_\mu, \bar{\nu}_e$	$\bar{\nu}_\mu$	$\mu^+, \tau,$ $\bar{\nu}_\mu.\nu_\tau$
ν_τ	ν_τ, e^+	ν_τ, μ^+	ν_τ, τ	$e^+, \tau,$ $\bar{\nu}_e, \nu_\tau$	$\mu^+, \tau,$ $\bar{\nu}_\mu.\nu_\tau$	ν_τ

Table 7.44   Multiplicative table for $L_7^6 = \{e^+, \mu^+, \tau^+, \bar{\nu}_e, \bar{\nu}_\mu, \bar{\nu}_\tau\}$.

	e^+	μ^+	τ^+	$\bar{\nu}_e$	$\bar{\nu}_\mu$	$\bar{\nu}_\tau$
e^+	e^+	e^+, μ^+	e^+, τ^+	$e^+, \bar{\nu}_e$	$e^+, \mu^+,$ $\bar{\nu}_e, \bar{\nu}_\mu$	$e^+, \tau^+,$ $\bar{\nu}_e, \bar{\nu}_\tau$
μ^+	μ^+, e^+	μ^+	μ^+, τ^+	$e^+, \mu^+,$ $\bar{\nu}_e, \bar{\nu}_\mu$	$\mu^+, \bar{\nu}_\mu$	$\mu^+, \tau^+,$ $\bar{\nu}_\mu, \bar{\nu}_\tau$
τ^+	τ^+, e^+	τ^+, μ^+	τ^+	$e^+, \tau^+,$ $\bar{\nu}_e, \bar{\nu}_\tau$	$\mu^+, \tau^+,$ $\bar{\nu}_\mu, \bar{\nu}_\tau$	$\tau^+, \bar{\nu}_\tau$
$\bar{\nu}_e$	$\bar{\nu}_e, e^+$	$e^+, \mu^+,$ $\bar{\nu}_e, \bar{\nu}_\mu$	$e^+, \tau^+,$ $\bar{\nu}_e, \bar{\nu}_\tau$	$\bar{\nu}_e$	$\bar{\nu}_e, \bar{\nu}_\mu$	$\bar{\nu}_e, \bar{\nu}_\tau$
$\bar{\nu}_\mu$	$e^+, \mu^+,$ $\bar{\nu}_e, \bar{\nu}_\mu$	$\bar{\nu}_\mu, \mu^+$	$\mu^+, \tau^+,$ $\bar{\nu}_\mu, \bar{\nu}_\tau$	$\bar{\nu}_\mu, \bar{\nu}_e$	$\bar{\nu}_\mu$	$\bar{\nu}_\mu, \bar{\nu}_\tau$
$\bar{\nu}_\tau$	$e^+, \tau^+,$ $\bar{\nu}_e, \bar{\nu}_\tau$	$\mu^+, \tau^+,$ $\bar{\nu}_\mu, \bar{\nu}_\tau$	$\bar{\nu}_\tau, \tau^+$	$\bar{\nu}_\tau, \bar{\nu}_e$	$\bar{\nu}_\tau, \bar{\nu}_\mu$	$\bar{\nu}_\tau$

Table 7.45   Multiplicative table for $L_8^6 = \{e, \mu, \tau, \bar{\nu}_e, \bar{\nu}_\mu, \bar{\nu}_\tau\}$.

	e	μ	τ	$\bar{\nu}_e$	$\bar{\nu}_\mu$	$\bar{\nu}_\tau$
e	e	e, μ	e, τ	$e, \mu, \tau, \bar{\nu}_e, \bar{\nu}_\mu, \bar{\nu}_\tau$	$e, \bar{\nu}_\mu$	$e, \bar{\nu}_\tau$
μ	e, μ	μ	μ, τ	$\mu, \bar{\nu}_e$	$e, \mu, \tau, \bar{\nu}_e, \bar{\nu}_\mu, \bar{\nu}_\tau$	$\mu, \bar{\nu}_\tau$
τ	τ, e	μ, τ	τ	$\tau, \bar{\nu}_e$	$\tau, \bar{\nu}_\mu$	$e, \mu, \tau, \bar{\nu}_e, \bar{\nu}_\mu, \bar{\nu}_\tau$
$\bar{\nu}_e$	$e, \mu, \tau, \bar{\nu}_e, \bar{\nu}_\mu, \bar{\nu}_\tau$	$\mu, \bar{\nu}_e$	$\tau, \bar{\nu}_e$	$\bar{\nu}_e$	$\bar{\nu}_e, \bar{\nu}_\mu$	$\bar{\nu}_e, \bar{\nu}_\tau$
$\bar{\nu}_\mu$	$e, \bar{\nu}_\mu$	$e, \mu, \tau, \bar{\nu}_e, \bar{\nu}_\mu, \bar{\nu}_\tau$	$\bar{\nu}_\mu, \tau$	$\bar{\nu}_\mu, \bar{\nu}_e$	$\bar{\nu}_\mu$	$\bar{\nu}_\mu, \bar{\nu}_\tau$
$\bar{\nu}_\tau$	$\bar{\nu}_\tau, e$	$\mu, \bar{\nu}_\tau$	$e, \mu, \tau, \bar{\nu}_e, \bar{\nu}_\mu, \bar{\nu}_\tau$	$\bar{\nu}_\tau, \bar{\nu}_e$	$\bar{\nu}_\tau, \bar{\nu}_\mu$	$\bar{\nu}_\tau$

Bibliography

[1] M. Al Tahan and B. Davvaz, *Hyperstructures associated to Biological inheritance*, Mathematical Biosciences, 285 (2017), 112-118.

[2] M. Al Tahan and B. Davvaz, *Weak chemical hyperstructures associated to electrochemical cells*, Iranian Journal of Mathematical Chemistry, 9(1) (2018), 65-75

[3] A.A. Albert, *Power-associative rings*, Trans. Amer. Math. Soc., 64 (1948), 552-593.

[4] F.W. Anderson and K.R. Frank, *Rings and Categories of Modules*, Second edition. Graduate Texts in Mathematics, 13. Springer-Verlag, New York, 1992.

[5] R. Anderson, A. A. Bhalekar, B. Davvaz, P. S. Muktibodh, T. Vougiouklis, *An introduction to Santilli's isodual theory of antimatter and the open problem of detecting antimatter asteroids*, NUMTA Bulletin., 6 (2012), 1-33.

[6] N. Antampoufis, *Hypergroups and $\flat$-groups in complex numbers*, Proceedings of 9th AHA Congress, Journal of basic Science (JBSUM), 4(1), Babolsar, Iran, (2008), 17-25.

[7] N. Antampoufis and A. Dramalidis, *On a sequence of finite H_v-groups*, Int. J. of Algebraic Hyperstructures and Its Applications, 2(1) (2016), 151-159.

[8] S.M. Anvariyeh and B. Davvaz, *θ-Closure and θ-parts of hypermodules*, Algebra Colloquium, 18(4) (2011), 629-638.

[9] K.T. Atanassov, *Intuitionistic fuzzy sets*, Fuzzy Sets Syst. 20 (1986), 87-96.

[10] K.T. Atanassov, *New operations defined over the intuitionistic fuzzy sets*, Fuzzy Sets Syst., 61 (1994), 137-142.

[11] J. Barreiro, M. Snchez and M. Viladrich-Grau, *How much are people willing to pay for silence? A contingent valuation study*, Applied Economics, 37 (11) (2005), 1233-1246.

[12] R. Bayon and N. Lygeros, *Les hypergroupes et H_v-groupes d'ordre 3* (French) [Hypergroups and H_v-groups of order 3], Bull. Greek Math. Soc. 53 (2007), 15-22.

[13] R. Bayon and N. Lygeros, *The H_v-groups and Marty-Moufang hypergroups*, Proceedings of the 1st International Conference on Algebraic Informatics, 285-294, Aristotle Univ. Thessaloniki, Thessaloniki, 2005.

[14] H. Bhabha, *The scattering of positrons by electrons with exchange on Dirac's theory of the positron*, Proc. Roy. Soc., A154 (1936), 195-206.

[15] S.-C. Chung, K.M. Chun, N.J. Kim, S.Y. Jeong, H. Sim, J. Lee and H. Maeng, *Chemical hyperalgebras for three consecutive oxidation states of elements*, MATCH Communications in Mathematical and in Computer Chemistry, 72 (2014), 389-402.

[16] P. Corsini, *Prolegomena of Hypergroup Theory*, Aviani Editore, Italy, 1993.

[17] P. Corsini and V. Leoreanu, *Applications of Hyperstructure Theory*, Advances in Mathematics (Dordrecht), 5. Kluwer Academic Publishers, Dordrecht, 2003.

[18] P. Corsini and T. Vougiouklis, *From groupoids to groups through hypergroups*, Rendiconti Mat. S. VII, 9 (1989), 173-181.

[19] M.R. Darafsheh and B. Davvaz, H_v-*ring of fractions*, Italian J. Pure Appl. Math., 5 (1999), 25-34.

[20] C. Darwin, *The Variation of Animals and Plants Under Domestication* (1st American edition), New York: Orange Judd and Co., 1968.

[21] B. Davvaz, H_v-*module of fractions*, Proc. 8th Algebra Seminar of Iranian Math. Soc., University of Tehran, December 17-18, (1996) 37-46.

[22] B. Davvaz, *On H_v-rings and fuzzy H_v-ideals*, J. Fuzzy Math., 6(1) (1998), 33-42.

[23] B. Davvaz, *Fuzzy H_v-groups*, Fuzzy Sets and Systems, 101 (1999), 191-195.

[24] B. Davvaz, H_v-*near rings*, Math. Japonica, 52(3) (2000), 387-392.

[25] B. Davvaz, *Fuzzy H_v-submodules*, Fuzzy Sets Syst. 117 (2001), 477-484.

[26] B. Davvaz, *On H_v-groups and fuzzy homomorphisms*, J. Fuzzy Math., 9(2) (2001), 271-278.

[27] B. Davvaz, *A note on fuzzy H_v-submodules*, J. Appl. Math. Computing, 11 (2003), 265-271.

[28] B. Davvaz, *A study on the structure of H_v-near ring modules*, Indian J. Pure Appl. Math., 34(5) (2003), 693-700.

[29] B. Davvaz, *Polygroup Theory and Related Systems*, World Scientific Publishing Co. Pte. Ltd., Hackensack, NJ, 2013.

[30] B. Davvaz, *Semihypergroup Theory*, Elsevier/Academic Press, London, 2016.

[31] B. Davvaz, *Weak algebraic hyperstructures as a model for interpretation of chemical reactions*, Iranian Journal of Mathematical Chemistry, 7(2) (2016), 267-283.

[32] B. Davvaz, *Vougiouklis contributions in the field of algebraic hyperstructures*, Ratio Mathematica, 33 (2017), 77-88.

[33] B. Davvaz and A. Dehgan-nezad, *Chemical examples in hypergroups*, Ratio Matematica, 14 (2003), 71-74.

[34] B. Davvaz and A. Dehghan Nezhad, *Dismutation reactions as experimental verifications of ternary algebraic hyperstructures*, MATCH Communications in Mathematical and in Computer Chemistry, 68 (2012), 551-559.

[35] B. Davvaz, A. Dehghan Nezad and A. Benvidi, *Chain reactions as experimental examples of ternary algebraic hyperstructures*, MATCH Communications in Mathematical and in Computer Chemistry, 65(2) (2011), 491-499.

[36] B. Davvaz, A. Dehghan Nezhad and A. Benvidi, *Chemical hyperalgebra: Dismutation reactions*, MATCH Communications in Mathematical and in Computer Chemistry, 67 (2012), 55-63.

[37] B. Davvaz, A. Dehghan Nezad and M. M. Heidari, *Inheritance examples of algebraic hyperstructures*, Information Sciences, 224 (2013), 180-187.

[38] B. Davvaz, A. Dehghan Nezad and M. Mazloum-Ardakani, *Chemical hyperalgebra: Redox reactions*, MATCH Communications in Mathematical and in Computer Chemistry, 71 (2014), 323-331.

[39] B. Davvaz, A. Dehghan Nezad and M. Mazloum-Ardakani, *Describing the algebraic hyperstructure of all elements in radiolytic processes in cement medium*, MATCH Commun. Math. Comput. Chem., 72(2), (2014), 375-388.

[40] B. Davvaz, W.A. Dudek and Y.B. Jun, *Intuitionistic fuzzy H_v-submodules*, Information Sciences, 176 (2006), 285-300.

[41] B. Davvaz and M. Ghadiri, *Weak equality and exact sequences in H_v-modules*, Southeast Asian Bull. Math., 25 (2001), 403-411.

[42] B. Davvaz and V. Leoreanu-Fotea, *Hyperring theory and Applications*, International Academic Press, 115, Palm Harber, USA, 2007.

[43] B. Davvaz, R. M. Santilli and T. Vougiouklis, *Studies of multivalued hyperstructures for the characterization of matter-antimatter systems and their extension*, Algebras Groups and Geometries, 28 (2011), 105-116.

[44] B. Davvaz, R. M. Santilli and T. Vougiouklis, *Multi-valued hypermathematics for characterization of matter and antimatter systems*, Journal of Computational Methods in Sciences and Engineering (JCMSE), 13 (2013), 37-50.

[45] B. Davvaz, R.M. Santilli, T. Vougiouklis, *Algebra, Hyperalgebra and Lie-Santilli Theory*, J. Generalized Lie Theory Appl., 9(2) (2015), 1-5.

[46] B. Davvaz and T. Vougiouklis, *Commutative rings obtained from hyperrings (H_v-rings) with α^*-relations*, Communications in Algebra, 35 (2007), 3307-3320.

[47] B. Davvaz, S. Vougiouklis and T. Vougiouklis, *On the multiplicative H_v-rings derived from helix hyperoperations*, Utilitas Mathematica, 84 (2011), 53-63.

[48] A. Dehghan Nezhad, S.M. Moosavi Nejad, M. Nadjafikhah and B. Davvaz, *A physical example of algebraic hyperstructures: Leptons*, Indian Journal of Physics, 86(11) (2012), 1027-1032.

[49] S. Ostadhadi- Dehkordi and B. Davvaz, *On quotient hypermodules*, African Diaspora Journal of Mathematics, 18(1) (2015), 9097.

[50] M. De Salvo, *Feebly canonical hypergroups*, J. Combin. Inform. System Sci., 15 (1990), 133-150.

[51] A. Dramalidis, *Dual H_v-rings*, Rivista di Mathematica Pura ed Applicata, 17 (1996), 55-62.

[52] A. Dramalidis, *Geometrical H_v-structures*, Proc. of the "Structures Elements of Hyperstructures", Alexandroupolis, Greece, Spanidis Press, (2005), 41-51.

[53] A. Dramalidis and T. Vougiouklis, *H_v-semigroups as noise pollution models*, Ratio Matematica, 23 (2012), 39-50.

[54] J.M. Field, *Effect of personal and situational variables upon noise annoyance in residential areas*, Journal of the Acoustical Society of America, 93 (1993), 2753-2763.

[55] D. Freni, *Une note sur le cour d'un hypergroupe et sur la clôture transitive β^* de β. (French) [A note on the core of a hypergroup and the transitive closure β^* of β]*, Riv. Mat. Pura Appl., 8 (1991), 153-156.

[56] D. Freni, *Strongly transitive geometric spaces: applications to hypergroups and semigroups theory*, Comm. Algebra, 32(3) (2004), 969-988.

[57] D. Geoff, B. Imelda, *Essential Guide to Blood Groups*, Blackwell, 2007.

[58] M. Ghadiri and B. Davvaz, *Direct system and direct limit of H_v-modules*, Iran. J. Sci. Technol. Trans. A Sci., 28 (2004), 267-275.

[59] M. Ghadiri, B. Davvaz and R. Nekouian, *H_v-Semigroup structure on F_2-offspring of a gene pool*, International Journal of Biomathematics, 5(4) (2012), 1250011 (13 pages).

[60] D. Griffiths, *Introduction to Elementary Particles*, John Wiley & Sons, 1987.

[61] A.J.F. Griffith et al., it *An Introduction to Genetic Analysis*, 7th edn. WH Freeman, New York, 2000.

[62] F. Halzen and A. Martin, *Quarks & Leptons: An Introductory Course in Modern Particle Physics*, John Wiley & Sons, 1984.

[63] D.L. Hartl, E.W. Jones, *Genetics, Principles and Analysis*, Jones and Bartlett Publishers, 1998.

[64] M. Koskas, *Groupoids, demi-hypergroupes et hypergroupes*, J. Math. Pures Appl., 49(9) (1970), 155-192.

[65] M. Krasner, *A class of hyperrings and hyperfields*, International J. Math. and Math. Sci., 6 (1983), 307-312.

[66] M. Hamidi and A.R. Ashrafi, *Fundamental relation and automorphism group of very thin H_v-groups*, Comm. Algebra, 45(1) (2017), 130-140.

[67] C. Harding, D.A. Johnson, R. Janes, *Elements of the P Block*, Published by Royal Society of Chemistry, 2002.

[68] L.H. Hartwell, L. Hood, M.L. Goldberg, A.E. Reynolds, L.M. Silver and R.C. Veres, *Genetics: From Genes to Genomes*, 3rd edn., Mc Graw Hill, 2008.

[69] I.N. Herstein, *Topics in Algebra*, Second edition, Xerox College Publishing, Lexington, Mass.-Toronto, Ont., 1975.

[70] Š. Hošková, *Abelization of quasi-hypergroups, H_v-rings and transposition H_v-groups as a categorial reflection*, Glob. J. Pure Appl. Math., 3(2) (2007), 105-112.

[71] Š. Hošková and J. Chvalina, *Abelizations of proximal H_v-rings using graphs of good homomorphisms and diagonals of direct squares of hyperstructures*, Algebraic hyperstructures and applications (Alexandroupoli-Orestiada, 2002), 147-158, Spanidis, Xanthi, 2003.

[72] Š. Hošková, J. Chvalina and P. Račková, *Transposition hypergroups of Fredholm integral operators and related hyperstructures. Part I*, J. Basic Science, 4 (2008), 43-54.

[73] Š. Hošková, J. Chvalina and P. Račková, *Transposition hypergroups of*

Fredholm integral operators and related hyperstructures. Part II, J. Basic Science, 4 (2008), 55-60.

[74] Š. Hošková-Mayerova, *Construction of subquasi - ordered hypergroup*, International Colloguim on the management of Educational Process aimed at curent issues in science, education and creative thinking development, (2011), 159-165.

[75] T.W. Hungerford, *Algebra*, Reprint of the 1974 original, Graduate Texts in Mathematics, 73. Springer-Verlag, New York-Berlin, 1980.

[76] H.L. Ibsen, *Cattle inheritance*, Genetics, 18 (1999), 441482.

[77] O. Kazanci, S. Yamak and B. Davvaz, *The lower and upper approximations in a quotient hypermodule with respect to fuzzy sets*, Information Sciences, 178(10) (2008), 2349-2359.

[78] V. Leoreanu, *Direct limit and inverse limits of join soace associated to fuzzy set*, Pure Math. Appl., 11 (2000), 509-516.

[79] V. Leoreanu and R. Gheorghe, *Direct limit and inverse limit of join spaces associated with lattices*, Ital. J. Pure Appl. Math., 11 (2002), 121-130.

[80] W.J. Liu, *Fuzzy invariant subgroups and fuzzy ideals*, Fuzzy sets and systems, 8 (1982), 133-139.

[81] B.J. Longley, S.M. Greg, T. Lynda, G.D. Tie, M.A. Dirk, E.W. Douglas and H. Ruth, *Altered Metabolism of Mast-Cell Growth Factor (c-kit Ligand) in Cutaneous Mastocytosis*, New England Journal of Medicine, 328 (18) (1993), 13021307.

[82] S. Mirvakili and B. Davvaz, *Strongly transitive geometric spaces: Applications to hyperrings*, Revista de la Union Matematica Argentina, 53(1) (2012), 43-53.

[83] R.T. Morrison and R.N. Boyd, *Organic Chemistry*, Sixth Eddition, Prentice-Hall, Inc, 1992.

[84] C.E. Mortimer, *Chemistry*, 6th edit., 1986.

[85] C.V. Negoita and D.A. Ralescu, *Applications of Fuzzy Sets and Systems Analysis*, Birkhauser, Basel, 1975.

[86] T. Muta, *Foundations of Quantum Chromodynamics*, Third edition, World Sci. Lect. Notes Phys. 78, 2009.

[87] B.A. Pierce, *Genetics: A Conceptual Approach*, Third Edition, W.H. Freeman, 2007.

[88] G. Romeo, *Limite diretto di semi-ipergruppi di associativita*, Riv. Math. Univ. Parma, 8 (1982), 281-288.

[89] A. Rosenfeld, *Fuzzy groups*, J. Math. Anal. Appl., 35 (1971), 512-517.

[90] R. Rota, *Multiplicative hyperrings* (Italian) Rend. Mat. (7) 2 (1982), 711-724.

[91] J.J. Rotman, *An Introduction to Homological Algebra*. Academic Press Inc, 1979.

[92] R.M. Santilli, *An introduction to Lie-admissible algebras*, Nuovo Cimento Suppl. (1), 6 (1968), 1225-1249.

[93] R.M. Santilli, *Dissipativity and Lie-admissible algebras*, Meccanica 4(1) (1969) 3-11.

[94] R.M. Santilli, *Representation of antiparticles via isodual numbers, spaces and geometries*, Comm. Theor. Phys., 3 (1994), 153-181.

[95] R.M. Santilli, *Isotopic, genotopic and hyperstructural liftings of Lie's theory and their isoduals*, Algebras, Groups and Geometries, 15 (1998), 473-498.

[96] R.M. Santilli, *Hadronic Mathematics, Mechanics and Chemistry*, Volumes I, II, III, IV and V, International Academic Press, USA, 2007.

[97] R.M. Santilli and T. Vougiouklis, *Isotopies, genotopies, hyperstructures and their applications*, New frontiers in Hyperstructures, Hadronic, (1996), 1-48.

[98] R.M. Santilli and T. Vougiouklis, *Lie-admissible hyperalgebras*, Italian J. Pure Appl. Math., 31 (2013), 239-254.

[99] R.M. Santilli and T. Vougiouklis, *Hyperstructures in Lie-Santilli admissibility and iso-theories*, Ratio Matematica, 33 (2017), 151-165.

[100] M.S. Sethi and M. Satake, *Chemistry of Transition Elements*, South Asian publishers, New Delhi, 1988.

[101] D.A. Skoog, D.M. West and F.J. Hollers, *Fundamentals of Analytical Chemistry*, 5th edit. (1988), p 282.

[102] S. Spartalis, *On reversible H_v-groups*, Algebraic hyperstructures and applications (Iasi, 1993), 163-170, Hadronic Press, Palm Harbor, FL, (1994).

[103] S. Spartalis, *On the number of H_v-rings with P-hyperoperations*, Combinatorics (Acireale, 1992), Discrete Math. 155 (1996), no. 1-3, 225-231.

[104] S. Spartalis, *Homomorphisms on (H, R)-H_v-rings*, Sixth Int. Congress on AHA (1996), Prague. Check Rep., Democ. Univ. Press, 133-138.

[105] S. Spartalis, *Quotients of P-H_v-rings*, New frontiers in hyperstructures (Molise, 1995), 167-177, Ser. New Front. Adv. Math. Ist. Ric. Base, Hadronic Press, Palm Harbor, FL, 1996.

[106] S. Spartalis, A. Dramalides and T. Vougiouklis, *On H_v-group rings*, Algebra Group Geom., 15(1) (1998), 47-54.

[107] M.S. Tallini, *Hypervector spaces*, Proc. Fourth Int. Congress on Algebraic Hyperstructures and Applications (AHA 1990), World Scientific, (1991) 167-174.

[108] [108] R.H. Tamarin, *Principles of Genetics*, Seventh Edition, The McGraw-Hill Companies, 2001.

[109] Y. Vaziri, M. Ghadiri and B. Davvaz, *The M[-] and -[M] functors and five short lemma in H_v-modules*, Turk. J. Math., 40 (2016), 397-410.

[110] Y. Vaziri and M. Ghadiri, *Schanuel's lemma, the snake lemma, and product and direct sum in H_v-modules*, Turk. J. Math., 41 (5), 1121-1132.

[111] S. Vougioukli, *H_v-vector spaces from helix hyperoperations*, Int. J. Math. Anal. N. S., 1(2) (2009), 109-120.

[112] T. Vougiouklis, *Cyclicity in a special class of hypergroups*, Acta Univ. Carolin.Math. Phys., 22(1) (1981), 3-6.

[113] T. Vougiouklis, *Representations of hypergroups. Hypergroup algebra*, Hypergroups, other multivalued structures and their applications (Italian) (Udine, 1985), 59-73, Univ. Studi Udine, Udine, 1985.

[114] T. Vougiouklis, *Generalization of P-hypergroups*, Rend. Circ. Mat. Palermo (2), 36(1) (1987), 114-121.

[115] T. Vougiouklis, *Representations of hypergroups by hypermatrices*, Riv. Mat. Pura Appl., 2 (1987), 7-19.

[116] T. Vougiouklis, *Groups in hypergroups*, Annals Discrete Math., 37 (1988), 459-468.

[117] T. Vougiouklis, *The very thin hypergroups and the S-construction*, Combinatorics'88, Vol. 2 (Ravello, 1988), 471-477, Res. Lecture Notes Math., Mediterranean, Rende, 1991.

[118] T. Vougiouklis, *The fundamental relation in hyperrings. The general hyperfield*, Algebraic hyperstructures and applications (Xanthi, 1990), 203-211, World Sci. Publishing, Teaneck, NJ, 1991.

[119] T. Vougiouklis, H_v-*vector spaces*, Algebraic hyperstructures and applications (Iasi, 1993), 181-190, Hadronic Press, Palm Harbor, FL, 1994.

[120] T. Vougiouklis, *Hyperstructures and Their Representations*, Hadronic Press Monographs in Mathematics. Hadronic Press, Inc., Palm Harbor, FL, 1994.

[121] T. Vougiouklis, *A new class of hyperstructures*, J. Combin. Inform. System Sci., 20(1-4) (1995), 229-235.

[122] T. Vougiouklis, H_v-*groups defined on the same set*, Discrete Math., 155 (1996) 259-265.

[123] T. Vougiouklis, *On* H_v-*fields*, Proc. of 6th AHA Congress, Prague, 1996, Democritus Un. Press (1997), 151-159.

[124] T. Vougiouklis, *Enlarging* H_v-*structures*, Algebras and Combinatorics (Hong Kong, 1997), 455-463, Springer, Singapore, 1999.

[125] T. Vougiouklis, *On* H_v-*rings and* H_v-*representations*, Combinatorics (Assisi, 1996). Discrete Math., 208/209 (1999), 615-620.

[126] T. Vougiouklis, *On hyperstructures obtained by attaching elements*, Constantin Carathéodory in his ... origins (Vissa-Orestiada, 2000), 197-206, Hadronic Press, Palm Harbor, FL, 2001.

[127] T. Vougiouklis, *A hyperoperation defined on a groupoid equipped with a map*, Ratio Mathematica, 1 (2005), 25-36.

[128] T. Vougiouklis, *The* ∂ *hyperoperation*, Structure elements of hyperstructures, 53–64, Spanidis, Xanthi, 2005.

[129] T. Vougiouklis, *A hyperoperation defined on a groupoid equipped with a map*, Ration Matematica, 1 (2005), 25-36.

[130] T. Vougiouklis, *The hyper theta-algebras*, Advances in Algebra, 1 (2008), 67-78.

[131] T. Vougiouklis, H_v-*fields and* H_v-*vector spaces with* ∂-*operations*, 6th Pan. Conf. Algebra, Number Theory, Thessaloniki, Greece, (2006), 95-102.

[132] T. Vougiouklis, ∂-*operations and* H_v-*fields*, Acta Math. Sinica, 24(7) (2008), 1067-1078.

[133] T. Vougiouklis, *The relation of the theta-hyperoperation* (∂) *with the other classes of hyperstructures*, Journal of Basic Science, 4(1) (2008), 135-145.

[134] T. Vougiouklis, *Bar and Theta Hyperoperations*, Ratio Mathematica, 21 (2011), 27-42.

[135] T. Vougiouklis, *The e-hyperstructures*, Journal of Mahani Mathematical Research Center, 1 (2012), 13-28.

[136] T. Vougiouklis, *The Lie-hyperalgebras and their fundamental relations*, Southeast Asian Bull. Math., 37(4) (2013), 601-613.

[137] T. Vougiouklis, H_v-*fields, h/v-fields*, Ratio Matematica, 33 (2017), 181-201.

[138] T. Vougiouklis and P. Kambaki, *Algebraic models in applied research*, Jordan Journal of Mathematics and Statistics (JJMS), 1(1) (2008), 81-90.

[139] T. Vougiouklis and S. Vougiouklis, *The helix hyperoperations*, Ital. J. Pure Appl. Math., 18 (2005), 197-206.

[140] M. Yazer, M. Olsson and M. Palcic, *The cis-AB blood group phenotype: fundamental lessons in glycobiology*, Transfus Med. Rev., 20(3) (2006), 20717.

[141] Zadeh, L.A., *Fuzzy sets*, Inform. Control., 8 (1965), 338-353.

[142] S. Zumdahl, *Chemistry*, Seventh edition, Houghton Mifflin Compan, New York, 2007.

Index

four o'clock flower, 292
functor, 203
  natural, 203
fuzzy
  H_v-homomorphism, 103
  H_v-subgroup, 96
  H_v-submodule, 225
  left
    reproduction axiom, 96
  right
    reproduction axiom, 96
  anti
    H_v-subgroup, 99
  closed, 99
  grade, 94
  inclusion
    homomorphism, 103
  left
    closed, 99
  max, 186
  right
    closed, 99
  strong
    homomorphism, 103
  subgroup, 95
  submodule, 225
  subset, 94

Galois Field, 23
Galvanic cell, 281
general enlargement, 251
genofield on the left, 250
genotopies, 250
genotype
  second generation, 303
good representation, 263
group, 4
  abelian, 6
  automorphism, 16
  commutative, 6
  cyclic, 6
  fundamental, 54
  homomorphic, 13
  homomorphism, 13
    kernel, 13
  inner automorphism, 16

integers modulo n, 6
  isomorphic, 13
  isomorphism, 13
  order, 6
  product, 9
  quaternion, 6
  quotient, 11
group ring, 20
groupoid, 1

halogens, 273
heart, 194
height, 102
helix-addition, 133
helix-multiplication, 268
helix-projection, 132
helix-sum, 133
homomorphism, 37, 50, 67
  H_v-homomorphism, 70
  good, 60
  good homomorphism, 50
  inclusion, 122
  inclusion homomorphism, 50, 70
  natural, 29
  strong, 122
  strong homomorphism, 71
  weak, 122
  weak homomorphism, 70
hyperalgebra, 254
  Jordan, 256
hypergroup, 46
  P-hypergroup, 46
  canonical, 55
  heart, 54
  regular, 46
  total, 46
hypergroupoid, 43
  H_v-algebra, 173
hyperideal, 56, 59
  left, 56
  right, 56
hypermodule, 61
  generated, 63
  multiplicative, 63
  quotient, 63
hyperoperation